New Philosophical Perspectives on Scientific Progress

This collection of original essays offers a comprehensive examination of scientific progress, which has been a central topic in recent debates in philosophy of science.

Traditionally, debates over scientific progress have focused on different methodological approaches, notably the epistemic and semantic approaches. The chapters in Part I of the book examine these two traditional approaches, as well as the newly revived functional and newly developed noetic approaches. Part II features in-depth case studies of scientific progress from the history of science. The chapters cover individual sciences including physics, chemistry, evolutionary biology, seismology, psychology, sociology, economics, and medicine. Finally, Part III of the book explores important issues from contemporary philosophy of science. These chapters address the implications of scientific progress for the scientific realism/anti-realism debate, incommensurability, values in science, idealisation, scientific speculation, interdisciplinarity, and scientific perspectivalism.

New Philosophical Perspectives on Scientific Progress will be of interest to researchers and advanced students working on the history and philosophy of science.

Yafeng Shan is Research Associate in Philosophy at the University of Kent. He is the author of *Doing Integrated History and Philosophy of Science: A Case Study of the Origin of Genetics* (2020) and the editor of *Examining Philosophy Itself* (2022).

Routledge Studies in the Philosophy of Science

For more information about this series, please visit: www.routledge.com/
Routledge-Studies-in-the-Philosophy-of-Science/book-series/POS

New Philosophical Perspectives on Scientific Progress

Edited by Yafeng Shan

Routledge
Taylor & Francis Group

NEW YORK AND LONDON

First published 2023
by Routledge
605 Third Avenue, New York, NY 10158

and by Routledge
4 Park Square, Milton Park, Abingdon, Oxon, OX14 4RN

Routledge is an imprint of the Taylor & Francis Group, an informa business

ISBN: 978-0-367-76055-7 (hbk)
ISBN: 978-0-367-76184-4 (pbk)
ISBN: 978-1-003-16585-9 (ebk)

DOI: 10.4324/9781003165859

Typeset in Sabon
by Apex CoVantage, LLC

Contents

Acknowledgements

I would like to express my gratitude to all my colleagues in the Department of Philosophy and Centre for Reasoning, University of Kent. In particular, I wish to thank Jon Williamson for his support and advice. I am indebted to all the authors and referees for their hard work on this volume and to Andrew Weckenmann and Allie Simmons at Routledge for their help and expertise. I would also like to thank the support of the British Academy (SRG1920\101076) and the Leverhulme Trust (RPG-2019–059). In addition, I wish to thank Xianzhao He for his moral support. Finally, my thanks go to Zifei Li for her boundless support over the past few years, especially during the pandemic.

Contributors

Peter Achinstein, Department of Philosophy, Johns Hopkins University, USA

Hanne Andersen, Department of Science Education, University of Copenhagen, Denmark

Alexander Bird, Department of Philosophy, University of Cambridge, UK

Marcel Boumans, School of Economics, Utrecht University, Netherlands

Harold Cook, Department of History, Brown University, USA

Olivier Darrigol, Centre National de la Recherche Scientifique, Paris, France

Finnur Dellsén, School of Humanities, University of Iceland, Iceland

Uljana Feest, Institute of Philosophy, University of Hannover, Germany

David Harker, Department of Philosophy and Humanities, East Tennessee State University, USA

Robin Findlay Hendry, Department of Philosophy, Durham University, UK

Catherine Herfeld, Institute of Philosophy and Institute of Sociology, University of Zurich, Switzerland

Milena Ivanova, Department of History and Philosophy of Science, University of Cambridge, UK

Eva Jablonka, The Cohn Institute for the History and Philosophy of Science and Ideas, Tel Aviv University, Israel

Insa Lawler, Department of Philosophy, University of North Carolina at Greensboro, USA

Michela Massimi, School of Philosophy, Psychology and Language Sciences, University of Edinburgh, UK

Teru Miyake, School of Humanities, Nanyang Technological University, Singapore

Paul Needham, Department of Philosophy, Stockholm University, Sweden

Ilkka Niiniluoto, Department of Philosophy, History and Art Studies, University of Helsinki, Finland

Eric Oberheim, Institute for Philosophy, Humboldt University of Berlin, Germany

Yafeng Shan, Department of Philosophy, University of Kent, UK

Stephen Turner, Department of Philosophy, University of South Florida, USA

Introduction
Philosophical Analyses of Scientific Progress

Yafeng Shan

George Sarton, the first Professor of History of Science at Harvard University, once wrote: 'The history of science is the only history which can illustrate the progress of mankind. In fact, progress has no definite and unquestionable meaning in other fields than the field of science' (Sarton 1936, 5). This is a 'corollary' of his 'definition' and 'theorem' of science.

> *Definition.* Science is systematized positive knowledge, or what has been taken as such at different ages and in different places.
> *Theorem.* The acquisition and systematization of positive knowledge are the only human activities which are truly cumulative and progressive.
>
> (Sarton 1936, 5)

These big claims about the nature and development of science might seem controversial to many, but few would deny that there are good examples of progress in the history of the sciences. For example, it is widely accepted that the Copernican Revolution marks a progressive shift from Ptolemaic astronomy to Keplerian astronomy. By the end of the seventeenth century, our geocentric model of universe was abandoned and replaced by the Keplerian model in which the sun is located in one of the foci of an elliptical orbit of the earth. Other classical examples include the Chemical Revolution and the Einsteinian Revolution. As R. G. Collingwood (1965, 332) puts it, that scientific progress exists and is verifiable is 'the simplest and most obvious case'. However, there has been no consensus among historians and philosophers of science on what is the best way to characterise the nature and pattern of scientific progress.

In the current literature, there are four main approaches to the nature of scientific progress: the epistemic approach, the semantic approach, the functional approach, and the noetic approach. According to the epistemic approach (e.g. Bird 2007, 2022), science progresses if and only if scientific knowledge accumulates. According to the semantic approach (e.g. Niiniluoto 1980, 1984, 2014; Rowbottom 2008), science progresses if and only if our scientific theories are approximating the truth. According

DOI: 10.4324/9781003165859-1

to the functional approach (e.g. Laudan 1981; Shan 2019, 2020), science progresses if and only if the functions of science are better fulfilled. According to the noetic approach (e.g. Dellsén 2016, 2021), science progresses if and only if the scientific community has a better understanding of the phenomena in the world. In addition, the pattern of scientific progress has been widely debated (e.g. Bury 1920; Sarton 1936; Nagel 1961; Kuhn 1962; Lakatos and Musgrave 1970; Laudan 1977). A widespread view is that science progressed in a cumulative way: new theories, paradigms, or traditions always fully suppress old ones in some sense. For example, Heinz Post (1971, 229) maintains that 'as a matter of empirical historical fact', new theories always explained the whole of the well-confirmed part of their predecessors, while Kuhn (1970, 20) observes that in history a new paradigm typically solved 'all or almost all the quantitative, numerical puzzles that have been treated by its predecessor'. In contrast, Larry Laudan is highly sceptic of this view and argues for a non-cumulative account of scientific progress: 'the growth of [scientific] knowledge can be progressive even when we lose the capacity to solve certain problems' (Laudan 1977, 150).

What is worse, we lack a comprehensive philosophical examination of scientific progress. First, the recent debate pays too much attention to the epistemic approach and the semantic approach (see Rowbottom 2008, 2010; Bird 2008; Cevolani and Tambolo 2013; Niiniluoto 2014). Shan's new functional approach and Dellsén's noetic approach are still insufficiently assessed. Second, there is little in-depth analysis of the progress in the history of the sciences. It is unclear which of the main approaches best captures the historical development of a particular scientific discipline. Nor is it very clear whether different disciplines differ in the nature and pattern of their progress. It is also worth examining whether there is any progress in some disciplines. Third, many related philosophical issues are still to be explored: What are the implications of scientific progress for the scientific realism/antirealism debate? Is the incommensurability thesis a challenge to scientific progress? What role do aesthetic values play in scientific progress? Does idealisation impede scientific progress? How does scientific speculation contribute to scientific progress? Do interdisciplinary sciences progress in the same way as mono-disciplinary sciences? How does science progress through perspective-shifts?

This book fills this gap. It offers a new assessment of the four main approaches to scientific progress (Part I). It also features nine historical case studies to investigate the notion of progress in different disciplines: physics, chemistry, evolutionary biology, seismology, psychology, sociology, economics, and medicine respectively (Part II). It discusses some issues related to scientific progress: scientific realism, incommensurability, values in science, idealisation, scientific speculation, interdisciplinarity, and scientific perspectivalism (Part III).

1. Main Philosophical Approaches

In Chapter 1, Alexander Bird defends the epistemic approach. He maintains that scientific progress is just the accumulation of scientific knowledge. He further argues that many cases of scientific progress found in the history of biology and of astronomy cannot be accounted for by Dellsén's noetic and Shan's functional accounts. Bird claims that there are many instances of modest contributions to progress through the addition of new scientific knowledge that does not bring with it new understanding or new exemplary practices. He concludes that progress can be made also by knowing that some novel phenomenon, such as X-rays, exists, even when that knowledge does not include new understanding or new methods and practices.

In Chapter 2, Ilkka Niiniluoto argues for the semantic approach. He reviews the historical roots of the debate over scientific progress. On the basis of Popper's notion of verisimilitude, Niiniluoto develops an account of truthlikeness. He argues that an increase of estimated truthlikeness is a mark of the progressive development in science. He also argues that such a semantic account is more adequate and fundamental than its alternatives: the epistemic, functional, and noetic accounts.

In Chapter 3, Yafeng Shan defends his new functional approach. He argues that the functional approach should not be conflated with the Kuhn-Laudan functional approach. There are other versions of the functional approaches, such as the Popper-Lakatos approach and his new functional approach. According to his new functional approach, scientific progress is best characterised in terms of usefulness of exemplary practices. Shan shows that this new functional approach is immune to the main objections to old functional approaches. Moreover, he argues that the new functional approach is better than the epistemic, semantic, and noetic approaches by providing a fuller picture of scientific progress.

In Chapter 4, Finnur Dellsén argues for a version of the noetic approach according to which scientific progress on some phenomenon consists in making scientific information publicly available so as to enable relevant members of society to increase their understanding of that phenomenon. He compares this version of the noetic approach compared with four rival accounts of scientific progress, namely the truthlikeness approach, the problem-solving approach, the new functional approach, and the epistemic approach. In addition, Dellsén tries to precisify the question that a good account of scientific progress is (or should be) aiming to answer, namely 'what type of cognitive change with respect to a given topic or phenomenon X constitutes a (greater or lesser degree of) scientific improvement with respect to X?'

2. Historical Case Studies

In Chapter 5, Olivier Darrigol examines seven episodes in the history of physics, which are widely accepted as major progress. He shows that these cases involved conflicts between formal and empirical criteria of progress, with the latter kind, empirical adequacy, winning in the end. He argues that the physicists' ability to judge and compare the empirical adequacy of their theories crucially depends on a modular structure which provides a sound basis of comparison. Darrigol thus suggests that such a modular structure plays an essential role in physical theories. He argues that progress in physics should be best understood as an increase in the modular measure of empirical adequacy.

In Chapter 6, Robin Findlay Hendry defends a cumulative account of the progress in chemistry. He surveys seven key episodes dating from the 1790s to the 1980s: the chemical revolution, chemical atomism, the emergence of the periodic system, structure theory, the instrumental revolution, the application of quantum mechanics to chemistry, and the discovery of quasi-crystals. Hendry argues that chemistry has developed cumulatively in the sense that theoretical claims about the composition and structure of particular substances tend to be retained, enriched, and deepened by further research, rather than being radically revised or reinterpreted. The development of the general theoretical frameworks for understanding composition and structure has also been cumulative, with new frameworks tending to be conservative extensions of previous approaches. He concludes with a remark on the relationship of this cumulative development to the problem-solving and epistemic conceptions of scientific progress.

In Chapter 7, Paul Needham argues for an epistemic approach to progress in chemistry. Like Hendry, he contends that the pattern of progress in chemistry is cumulative. Unlike Hendry, he maintains that the epistemic account well captures the nature of progress in chemistry. By focussing on the development of theorising and conceptual clarification at the macroscopic and microscopic levels in the nineteenth and twentieth centuries, Needham argues that the accumulation of knowledge concerning the character and transformations of substances from ancient times constitutes the progress in chemistry. He concludes that the epistemic approach seems to give the best account of progress in chemistry amongst main philosophical approaches to scientific progress.

In Chapter 8, Eva Jablonka develops a developmental system approach to examining progress in the study of epigenetic inheritance. The approach combines Conrad Waddington's systems biology approach to investigating embryological development with Ludwik Fleck's sociological approach to analysing the development of scientific systems. She argues that the case study of epigenetic inheritance highlights the context-sensitive nature of scientific progress and suggests that scientific progress is relative to the delineation of the theoretical boundaries of the scientific system and the time scale that is chosen.

In Chapter 9, Teru Miyake offers a functional account of progress in seismology between 1889 and 1940. He argues that the main problem for seismologists in that period was that seismic wave recordings are extremely information-rich but extremely complex, and the progress in seismology during this period resulted from advances in methods for extracting information from complexly structured data. Miyake divides the rough half-century in question into three periods. In the first period, seismological research focused on the question of whether the waves that are recorded by seismographs are correctly theoretically characterised. In the second period, the research focused on accounting for anomalies in the seismic wave recordings by finding an interpretation for each significant anomaly. In the third period, the research focus was on making inferences from interpreted seismic wave recordings to features of the Earth's interior. In particular, he draws a contrast between British and German seismology and shows that the progress in British seismology was stifled by the lack of methods for properly interpreting seismic wave recordings.

In Chapter 10, Uljana Feest focusses on conceptual developments in psychology and explores the criteria by which such developments constitute progress. She distinguishes between the issue of what are units of psychological analysis and what are objects of psychological research, positing that the units of analysis are human (and animal) individuals and that the objects of research are (cognitive, behavioural, and experiential) capacities, which are often individuated by means of folk-psychological terms. While this suggests that conceptual progress occurs when concepts provide improved descriptions of the objects in their extension, Feest raises some doubts regarding the (seemingly intuitive) notion that are natural and/or ahistorical facts of the matter that settle what psychological concepts 'really' refer to. She argues that conceptual progress occurs when concepts track their (potentially changing) objects and such efforts rely on the availability of epistemic resources, which include both propositional and non-propositional knowledge. Moreover, Feest articulates a broad conception of progress in psychology as the accumulation of epistemic resources and argues that the history of psychology provides us with a trove of such resources.

In Chapter 11, Stephen Turner examines the questions of whether and how sociology progresses. He indicates that theories in sociology, which are not predictive systems that generate puzzles but second-order definitions and ideal types to abstract over intelligible world of the subjects. Accordingly, Turner suggests that progress in sociology can be understood in the way that theories in sociology provide new ways of framing in response to generically defined concerns, such as the stability of elites and novel social situations. Moreover, he highlights that progress in sociology is relative to the problems defined by local knowledge and arise for external reasons (such as changing normative and policy concerns).

In Chapter 12, Marcel Boumans and Catherine Herfeld focus on a specific kind of progress in economics, especially in the sub-branches that predominantly use mathematical models. They adopt Shan's functional account of progress to argue that economics progresses by means of the use of what they call 'common recipes' and the use of model templates to define and solve problems of relevance for economists. With a case study of twentieth-century business cycle research, Boumans and Herfeld show not only how model templates are re-applied to different phenomena but also how scientists come up with these model templates in the first place and how – once they are considered less useful – they are replaced with new ones. In a nutshell, Boumans and Herfeld argue that progress in economics occurs when a shared conceptual vision of the phenomenon allows a model template to define and solve problems, while progress is impeded when model templates are useless for certain problems.

In Chapter 13, Harold Cook examines the notion of progress in medicine. He proposes that impersonal methods of assessing weight, measure, and currency, which are fundamental for establishing commensurable norms in marketplaces, provide the foundation for considering how to apply material commensurability to knowledge of natural kinds. Once material sameness is defined and accepted, it disappears from the group of issues that are contested. The narrowly focused but powerful activities in biomedicine show how important material commensurability is for creating a scientific field that can include anyone, anywhere. Cook uses some historical examples to show how an understanding of medicinal substances (say, 'drugs') shifted from personal qualitative experience to impersonal materialistic experiment. He argues that the power to materially define and to extract or manufacture globally understood substances illuminates a widely understood version of progress of medicine. Its deep roots in market exchange also point to the limits of biomedicine to better the health of humans without other forms of governance.

3.   Related Issues

In Chapter 14, David Harker examines the relation of scientific progress and scientific realism. He argues that conceiving of scientific success in terms of scientific progress provides a distinct perspective, for purposes of defending a modest form of scientific realism. Harker suggests that scientific realists should be attending to patterns in the history of science for evidence that, over time, science does not just achieve more empirically adequate theories, but that it corrects the errors contained within previous scientific work, and thereby achieves greater truthlikeness. He concludes that to identify such patterns it is important that scientific realists pay more attention both to the varieties of progress and the means by which these are achieved.

In Chapter 15, Eric Oberheim addresses the following questions: What is incommensurability? What causes it? What are its consequences for intelligibility? What does it imply about theory comparison? And finally, what does incommensurability imply about truth, reality, and progress?

He argues that incommensurability implies there are at least two kinds of progress (commensurable and incommensurable), which correspond to two methods of writing history (hermeneutic historiography and present-centred historiography). According to the present-centred historiography, scientific progress is a series of better approximations to current theories assuming they are true (or at least approximately true and the closest to the truth available). The hermeneutic historiography, by contrast, characterises scientific progress as it happened.

In Chapter 16, Milena Ivanova explores the questions concerning the role of aesthetic values in science and how throughout scientific progress the questions we ask about the role of aesthetic values might change. Her examination starts with the traditional distinction between context of discovery and context of justification and shows that neither context is value-free. Then she illustrates how aesthetic values shape different levels of scientific activities, from designing experiments and reconstructing fossils, to evaluating data. Ivanova also explores how we could justify the epistemic import of aesthetic values and develop some concerns. Finally, she examines whether we should expect the questions surrounding aesthetic values in scientific practice to change with scientific progress, as we enter the era of post-empirical physics, big data science, and make more discoveries using machine learning and artificial intelligence (AI).

In Chapter 17, Insa Lawler focuses on a challenge to the philosophical accounts of scientific progress: idealisations with false assumptions are deliberately and ubiquitously used in science. Any account of scientific progress needs to account for this widely accepted scientific practice. She examines how the four main accounts – the functional account, the semantic account, the epistemic account, and the noetic account – can cope with the challenge from idealisation, with a focus on indispensable idealisations. Lawler concludes that, on all accounts, idealisations can promote progress, but only some accounts allow them to constitute progress.

In Chapter 18, Peter Achinstein discusses the role of scientific speculation in the development of science. He offers a broad definition of 'speculating', followed by an account of how scientific speculations are best evaluated, illustrated by the case of James Clerk Maxwell's kinetic-molecular theory of gases. He also examines the question of whether what will be called 'evidential progress', or the lack of it, in science generally can be appealed to in assessing the credibility of a speculative theory. By doing so, Achinstein offers a pragmatic solution to the problem of pessimistic meta-induction.

In Chapter 19, Hanne Andersen provides a philosophical qualification of the political discourse by examining how interdisciplinary progress can be characterised. She argues that in addition to the categories of incremental and transformative progress that are well known from mono-disciplinary science, interdisciplinary research can sometimes offer another category of progress that she calls quasi-transformative. In examining these three kinds of interdisciplinary progress, Andersen argues, first, that interdisciplinary progress does not necessarily require a specific type of integration between the involved disciplines or specialties, second, that social relations between scientists with different areas of expertise may play a crucial role in especially transformative progress, and third, that different disciplinary perspectives on what constitutes progress can draw wedges between scientists from different disciplines.

In Chapter 20, Michela Massimi develops a human rights approach to scientific progress. She starts with an analysis of the 'right to enjoy the benefits of scientific progress and its applications' (REBSP). Massimi offers a diagnosis for the patchy implementation to date, back to a number of assumptions about scientific knowledge and its progressive nature that are common to what she calls the 'manifest image' of progress and the 'philosophical image'. Thus, Massimi offers a different image of scientific knowledge and its growth, building on her work on perspectival realism. She urges replacing individuals with situated epistemic communities, the siloed picture with interlacing scientific perspectives and a view of progress sub specie aeternitatis or 'from here now' with one of progress 'from within'. She lays out the contours of a possible 'deontic framework' as a way of reinterpreting the core content of REBSP in light of perspectival realism. By doing so, Massimi suggests that the REBSP can be read as a 'cosmopolitan right' no longer trapped between the strictures of individual rights versus the rights of the communities to share in it. She concludes that this epistemic shift brings with it much-needed 'cosmopolitan obligations' when it comes to sharing in scientific knowledge and its advancements.

References

Bird, Alexander. 2007. "What Is Scientific Progress?" *Noûs* 41 (1): 64–89.
———. 2008. "Scientific Progress as Accumulation of Knowledge: A Reply to Rowbottom." *Studies in History and Philosophy of Science* 39: 279–281.
———. 2022. *Knowing Science*. Oxford: Oxford University Press.
Bury, J. B. 1920. *The Idea of Progress*. New York: Macmillan.
Cevolani, Gustavo, and Luca Tambolo. 2013. "Progress as Approximation to the Truth: A Defence of the Verisimilitudinarian Approach." *Erkenntnis* 78 (4): 921–935.
Collingwood, Robin George. 1965. *The Idea of History*. Oxford: Oxford University Press.
Dellsén, Finnur. 2016. "Scientific Progress: Knowledge versus Understanding." *Studies in History and Philosophy of Science* 56: 72–83.
———. 2021. "Understanding Scientific Progress: The Noetic Account." *Synthese* 199: 11249–11278.
Kuhn, Thomas Samuel. 1962. *The Structure of Scientific Revolutions*. 1st ed. Chicago, IL: The University of Chicago Press.

———. 1970. "Logic of Discovery or Psychology of Research?" In *Criticism and the Growth of Knowledge*, edited by Imre Lakatos and Alan Musgrave, 1–23. Cambridge: Cambridge University Press.

Lakatos, Imre, and Alan Musgrave, eds. 1970. *Criticism and the Growth of Knowledge*. Cambridge: Cambridge University Press.

Laudan, Larry. 1977. *Progress and Its Problems: Toward a Theory of Scientific Growth*. Berkeley and Los Angeles, CA: University of California Press.

———. 1981. "A Problem-Solving Approach to Scientific Progress." In *Scientific Revolutions*, edited by Ian Hacking, 144–155. Oxford: Oxford University Press.

Nagel, Ernest. 1961. *The Structure of Science: Problems in the Logic of Scientific Explanation*. New York: Harcourt, Brace & World.

Niiniluoto, Ilkka. 1980. "Scientific Progress." *Synthese* 45 (3): 427–462.

———. 1984. *Is Science Progressive?* Dordrecht: Springer.

———. 2014. "Scientific Progress as Increasing Verisimilitude." *Studies in History and Philosophy of Science Part A* 46: 73–77.

Post, Heinz R. 1971. "Correspondence, Invariance and Heuristics: In Praise of Conservative Induction." *Studies in History and Philosophy of Science* 2 (3): 213–255.

Rowbottom, Darrell P. 2008. "N-Rays and the Semantic View of Scientific Progress." *Studies in History and Philosophy of Science Part A* 39 (2): 277–278.

———. 2010. "What Scientific Progress Is Not: Against Bird's Epistemic View." *International Studies in the Philosophy of Science* 24 (3): 241–255.

Sarton, George. 1936. *The Study of the History of Science*. Cambridge, MA: Harvard University Press.

Shan, Yafeng. 2019. "A New Functional Approach to Scientific Progress." *Philosophy of Science* 86 (4): 739–758.

———. 2020. *Doing Integrated History and Philosophy of Science: A Case Study of the Origin of Genetics*. 1st ed. Boston Studies in the Philosophy and History of Science. Cham: Springer.

Part I

Main Philosophical Approaches

1 The Epistemic Approach

Scientific Progress as the Accumulation of Knowledge

Alexander Bird

1. Introduction: Approaches to Scientific Progress

One compelling approach to the nature of scientific progress focuses on truth. The semantic approach says that scientific progress is the accumulation of scientific truth or that it is the increasing proximity of scientific theory to the truth. Alternative accounts of progress can depart from this semantic approach in a number of ways. An anti-realist who nonetheless wants to be able to say that science does make progress will be less demanding than the semantic approach as regards truth. Science can progress without acquiring truth or getting closer to it. Without any truth-related requirement, an antirealist view will have to be more demanding than the semantic approach in some other respect. Typically, such a view will focus on some key feature of scientific practice that one might regard as the aim or function of science – this is the functional approach. This might be the development and solving of scientific problems. Science progresses when more problems are proposed and solved. This view is more demanding than the semantic approach in that the semantic approach does not require that contributions to scientific progress address scientific problems. The functional approach can be adopted without anti-realist intent. It remains the case that it is more demanding than the semantic approach in this respect.

Other approaches to scientific progress are also more demanding than the semantic approach, but in different respects. The epistemic and noetic approaches both say that scientific progress requires a cognitively more demanding state than belief that is true or truthlike. The epistemic view says that scientific progress requires an increase in scientific knowledge. It therefore requires that developments that contribute to progress have the epistemic justification characteristic of knowledge. The noetic approach says that scientific progress requires an increase in scientific understanding. Quite how the epistemic and noetic approaches relate to one another depends on how one sees the relationship between knowledge and understanding. If, as I do (Bird 2007: 84), one regards understanding as a special kind of knowledge, then the noetic view is

DOI: 10.4324/9781003165859-3

just an even more demanding approach than the epistemic approach. On the other hand, if like Dellsén (2016), Hills (2016), or Elgin (2007) one takes understanding to be a different sort of cognitive state, one that does not entail knowledge, then the noetic approach will be more demanding than the semantic approach, but not necessarily more demanding than the epistemic approach.

In my opinion, the fact that the epistemic approach is more demanding than the semantic approach by requiring progressive developments to be justified is an advantage, since adding a true proposition believed for bad reasons to the stock of science is not genuine progress. However, I think that the additional demands placed on progress by the functional and noetic approaches are misplaced. If you think that knowing is a simple, basic mental state (Williamson 1995) then the epistemic view is the simplest of the views on offer. In this chapter I argue that the additional complexity demanded by the functional and noetic approaches is mistaken, since discoveries, simply by coming to be known, can add to progress.

2.   The Noetic and the Functional Accounts of Dellsén and Shan

What I have called the noetic and the functional approaches are general approaches that encompass more specific accounts that characterise scientific progress in terms of understanding and in terms of the problem-solving function of science, respectively. I now turn to the leading specific accounts within each approach, the noetic account of Dellsén (2016 and this volume) and the functional account of Shan (2019 and this volume). These are articulated in detail elsewhere in this volume, and so I adumbrate their accounts as follows:

Dellsén's noetic account

Science progresses when there is an increase in scientific understanding. A scientist understands X if and only if they grasp a sufficiently accurate and comprehensive dependency model of X; their degree of understanding of X is proportional to the accuracy and comprehensiveness of their dependency model of X.

Shan's functional account

Science progresses when science proposes more useful exemplary practices. An exemplary practice is useful when it provides a way of defining and solving research problems that is repeatable, provides a reliable framework for looking for solutions to unsolved problems and generates more testable research problems across a wider range of fields.

(For simplicity, in this chapter, I will call these *the* noetic account and *the* functional account, while recognising that there are other accounts possible within the general approaches they exemplify.)

Without doubt the most important contributions to the progress of science conform to both of these accounts. The chemical revolution of the late eighteenth century led to a new and accurate understanding of the nature of matter and laid down the framework for setting and solving problems that dominated nineteenth century chemistry. Before Lavoisier, the theory of matter was comprised of the vestiges of the ancient four element theory, supplemented by the hypothesis of various principles, such as phlogiston, the principle of combustion. Lavoisier's contemporaries were debating, for example, whether water could be transmuted into earth, as suggested by the fact that water distilled several times would still leave a solid residue when evaporated.[1] Most famously Lavoisier disproved the hypothesis that combustion is a matter of a substance losing phlogiston. The new theory that combustion is a matter of a substance combining with a hitherto unknown substance, oxygen, led to a broader theory of matter: matter is composed of combinations of elements (or of an element in its pure form), where the number of such elements is not small. Clearly this constituted a new understanding of matter and of chemical reactions, and that understanding can be represented as grasping one or more dependency models. At the same time, the chemical revolution established a number of new and useful exemplary practices, most obviously the identification of new elements and showing of which elements particular substances are composed. When supplemented by the law of definite proportion, chemistry became centred on stoichiometry, the science of determining the proportions of reactants in a chemical reaction (and the laws of such reactions), and the related idea that a substance may be represented by a chemical formula.

The discovery of the structure of DNA in 1953 gave scientists not only an understanding of that structure but also a deeper understanding of how it was possible for genetic information to be duplicated in the process of mitosis and to be transferred from one generation to another. The discovery also opened a whole host of research questions in molecular biology, regarding, for example, the details of transcription and the relationship between RNA triples and amino acids, the nature of mutation, and the mechanisms for gene expression, among many others. Two decades after the discovery of DNA began the science and technology of genomics, with the first sequencing of a gene and then a whole organism, MS2, an RNA virus. This in particular initiated a range of exemplary practices in science for identifying and sequencing genes and then for modifying and synthesising genetic material.

Both the noetic account and the functional account can happily accommodate these undoubted instances of scientific progress. So too, for that matter, can the epistemic account. For the scientific claims of Lavoisier and of Crick and Watson were backed up by careful experiments, detailed analysis of the evidence, and rigorous argument that allowed them to know, for example, that water is a compound or that DNA has a double-helical structure.[2] The question for this chapter is whether the defining features of the noetic and functional accounts – understanding through grasp of dependency models and the initiation of new fields and techniques of research through exemplary practices – encompass *all* instances of scientific progress, including those that are not as dramatic as the revolutions in chemistry and in molecular biology. In the following, I argue that neither account can do this. In particular neither can account for the progress made by often smaller advances in the course of normal science. Nor can either account for the progress that is made by a significant but unexpected and baffling discovery. Whereas in these cases there is the accumulation of scientific knowledge, and so there is progress according to the epistemic account.

3.   Progress Through Knowledge Acquisition Alone

In this section I look at examples from the history of science that were contributions to scientific progress but which do not meet the descriptions required by one or other or both of the functional and noetic accounts described earlier.

3.1   *The Discovery of X-rays*

In 1895, in one of the best known episodes in the history of science, Wilhelm Roentgen found that with a high voltage he could use a Crookes cathode ray tube to produce a new kind of ray different in important respects from all those hitherto known. These X-rays (or Roentgen rays as they were also widely known) were an immediate sensation among both scientists and the general public. Roentgen himself published three papers on X-rays while scientists across the world published hundreds in the following months and years. There was a great deal of disagreement about the true nature of these rays. Many and diverse theories were proposed; as Alexi Assmus (1995) reports, 'Albert Michelson thought they might be vortices in the ether. Thomas Edison and Oliver Lodge suggested acoustical or gravitational waves'. Assmus goes on to say that opinion settled on the idea that X-rays were electromagnetic in nature. But in the physics of the day that left a lot of room for very different theories, especially as the aether theory of electromagnetism remained dominant. For example, some British scientists, including George Stokes and J. J. Thompson took X-rays to be impulses in the aether, not continuous waves, because X-rays seemed to be able to concentrate energy in a way that could not be explained by waves (as demonstrated in their ability to excite electrons). For that reason William Henry Bragg held X-rays to be particles. Roentgen's own view was that X-rays were longitudinal waves (like sound waves).

In 1901, Roentgen was awarded the first Nobel Prize in Physics for his discovery. This was before the correct theory of X-rays was known, following work by Johannes Stark, Max von Laue, and Charles Barkla among others. In particular, the development of the quantum theory was required before the concentration of energy effect could be reconciled with the idea that X-rays are also electromagnetic waves of the same kind as visible light except of much higher frequency.[3] The discovery of X-rays satisfies neither the functional nor the noetic account of progress, despite being a clear example of a contribution to progress.

The discovery of X-rays was not itself a useful exemplary practice. For sure, the discovery of X-rays led to exemplary practices, both within science and outside science. Very quickly X-rays were being used in medicine and also in engineering. It took longer before X-ray crystallography was developed by the Braggs as an exemplary practice in science *ne plus ultra*. But it was not these that made the discovery of X-rays a contribution to progress, it was the discovery itself. The 1901 Nobel Prize was given to Roentgen 'in recognition of the extraordinary services he has rendered by the discovery of the remarkable rays subsequently named after him'. (X-ray crystallography was developed after 1901 and merited its own Nobel Prize, in 1915.)

Nor did the discovery of X-rays involve a sufficiently accurate and comprehensive dependency model of X-rays. In 1901 several competing models of X-rays were on offer, none of them either accurate or comprehensive, since this date preceded Einstein's theory of the quantisation of light. What scientists knew in 1901 was that X-rays were important. But what they were remained largely a mystery. It is true that experiments had revealed a number of dependency relations. But for the reasons given, no one was able to accommodate them all in a satisfactory model. It is also true that all the physicists aimed to find that model and show it to be correct. But that shows only that understanding is a particularly valuable kind of progress sought by scientists, not that it is the only kind of progress.

In short, Roentgen's discovery of X-rays was a celebrated contribution to scientific progress because after 1895 scientists knew that these interesting and important rays existed, whereas before then they were ignorant of their existence. Progress here was made by the addition of knowledge and nothing more.

3.2 Astronomy, Star Catalogues, and the Discovery of Comets

Astronomy is one of the oldest sciences, with many societies attaching social and theological significance to the stars and planets as well as finding them intrinsically interesting. In particular the role of astronomy in constructing calendars, themselves important in many aspects of life from agriculture to religion, motivated study of the stars. Mayan astronomy, for example, was used in the organisation of religious rituals and in the planning of cities. The highly accurate data produced in their observatories enabled the Maya to distinguish the solar year from the tropical year and to predict eclipses. Indian astronomy also goes back thousands of years. It benefitted from contact with Greek astronomy following the invasion by Alexander the Great, and flourished especially in the mid first century with the work of Aryabhatta and Brahmagupta, which in turn influenced Islamic astronomy.

It is clear from looking at these texts that Mayan, Indian, Greek, and other astronomers had a great deal of knowledge. It is not so clear that they had a great deal of understanding. For example, both Eratosthenes and Aryabhatta gave accurate figures for the circumference of the Earth. It is true that they will have to have had mathematical understanding in order to calculate their figures, but that understanding alone does not account for the totality of their scientific achievement, for the latter involves them actually producing their numerical estimates from the data that they or others collected. Their understanding of the mathematics plus knowledge of the data enabled them to make the accurate calculations of the circumference, and the latter knowledge is *itself* a contribution to scientific progress. The noetic view, it would seem, has to deny that this knowledge is itself progress, and limit the latter to the mathematical understanding that preceded it. Ptolemy likewise was able to predict the observed positions of the planets with accuracy using his model of planetary motion. But this geocentric model was a highly *inaccurate* dependency model. The epistemic view takes Ptolemy's work to be progressive in virtue of the knowledge of the observable motions of the planets it provided, whereas the noetic view seems to deny that it constitutes any progress at all.

Hipparchus, the greatest of the ancient Greek observational astronomers, created one of the first and most important star catalogues. A star catalogue is a list of the fixed stars and their positions in the sky. It might also include a naming system and a record of the brightness of the star. Hipparchus's catalogue was adopted (perhaps fraudulently) by Ptolemy and thereby became the basis of subsequent star catalogues, such as the *Book of Fixed Stars* of the Persian astronomer Abd al-Rahman al-Sufi. Tycho Brahe, with his assistant, Johannes Kepler, drew up a new catalogue, the basis of the Rudolphine tables. The latter, which was both a star catalogue and the means of predicting planetary motions, incorporated 1,005 stars recorded by Brahe and Kepler as well as several hundred from Ptolemy (and so from Hipparchus). It also included stars observed by Johann Bayer, who himself produced the star catalogue and atlas, the *Uranometria Omnium Asterismorum*, which was the first of its kind to include stars observable only in the southern hemisphere, thus covering the whole celestial sphere. Bayer introduced a system of nomenclature for stars, using a Greek letter and the constellation in which it is found, hence Alpha Centauri, Gamma Cephei, and so on. The next great catalogue was drawn up over several decades by John Flamsteed, who introduced a new system of designation employing numbers instead of Greek letters, also still in use today. By using a telescope Flamsteed was able to increase the number of stars recorded, three times the number Tycho had listed, including many stars too faint to be seen with the naked eye. Following the tradition of over 2,000 years, subsequent astronomers corrected and added to Flamsteed's *Historia Coelestis Britannica*. Perhaps the most significant astronomer to do so was Caroline Herschel. Her corrections and additions to Flamsteed's catalogue were important to her and her brother William's project of searching for comets, nebulae, star clusters, and double stars. The Herschel's catalogue became the basis of the *New General Catalogue* (1888) of nebulae and star clusters, itself revised and updated to the present day. Caroline herself was able to identify numerous comets, but the most important discovery was a by-product of this activity, when she and William discovered the planet subsequently known as Uranus (which they initially took to be a comet).

The process of cataloguing stars is a contribution to scientific progress. It is so easy to see on the epistemic account (and on the semantic account). There is an accumulation of knowledge of the existence, position, and brightness of stars. This knowledge also increases as the catalogues become more accurate and when improvements in technology allow for greater precision in what is already known. Yet neither the noetic nor the functional views can account for this progress adequately. Clearly, adding more stars to the catalogue or recording the position and brightness of existing ones with greater precision and accuracy adds nothing in terms of understanding. There is no dependency model here, and so no improvement in a grasp of such a model. The only increase in understanding in the story told here came when Anders Lexell calculated the orbit of Uranus to confirm that it was indeed a planet. That too was a contribution to progress. But it was not the only one. Knowing that this was a new celestial object that was not a star was in itself a major advance in astronomy.

The story does include several instances of the addition of useful exemplary practices, such as the introduction of the telescope and of the naming systems of Bayer and Flamsteed. Numerous smaller advances were made that enabled astronomers to record the positions of stars more accurately or simply to identify stars more easily. One such innovation was made by Caroline Herschel, who reorganised Flamsteed's catalogue, so that stars are listed not by constellation but by North Polar distance (Hoskin 2011: 193). But for the most part the progress in astronomy described earlier was made not by the addition of new useful exemplary practices but by the application of existing practices laboriously and meticulously, again and again.

3.3   Natural History After Aristotle

The science formerly known as natural history was built on three basic practices: the careful observation of flora and fauna in their habitats, the collection and conservation or preservation of samples, and the naming and classification of species. These practices reinforced one another. For one scientist's observation of an animal could contribute to science only if other scientists know which animal the first is talking about. Samples might be conserved by breeding, such as the plants brought from all over the world and then bred at Kew Gardens. Or they might be preserved through taxidermy and other such techniques, giving us the glass cases full of carefully mounted and labelled insects and other animals that one can see at the Natural History Museum in Kensington. Such samples aided the shared recognition of species. Not everybody could have access to the physical samples, so printing became an important adjunct to collection. Kusukawa (1997, 2011) describes how in his *De Historia Plantarum* (1542), the Tübingen Professor of Medicine, Leonhart Fuchs compared the description of a plant given by the ancient Greek author Dioscorides in his *De Materia Medica* to a picture in his own book, to establish that they are the same plant, and was thereby able to add his own observations to those of Dioscorides in order to provide a more complete knowledge of this species.

The practices of observation, collection, and classification are all to be found in the work of Aristotle. Aristotle spent several years recording his observations of the fauna on the island of Lesbos, while his friend and successor as the scholarch of the peripatetic school at the Lyceum, Theophrastus, worked on its flora. Among his thousands of careful observations, Aristotle described the life cycle of the cicada, noted the changing colours of octopuses, and recorded the growth of different kinds of fish from their eggs. He removed sections from the shells of chickens' eggs to see how the parts of the embryo developed over time (although he drew the wrong inference from what he saw). Some of the samples with which he worked had been sent to him from Asia by Alexander the Great. Aristotle named and classified over 500 distinct animal types. Observing that dolphins gave birth to live young, he classified them with the quadrupedal mammals.

It is true that Aristotle was searching for causal relations among what he saw, although what he achieved in this respect was limited. In addition to the obstacles to establishing cause and effect in biology, Aristotle was limited by his own theories, such as his own theory of causation. He did not carry out experimental biology, in part because his distinction between natural and violent motions implied that what is seen under experimental conditions would not tell one about what things are like under normal conditions. So his ability to generate accurate dependency models was very limited. Although we know some of Aristotle's particular techniques, such as that with the chicken's eggs and dissections of animals, for the most part he did no more than observe carefully and systematically. So there is some but not a great deal in his work by way of novel exemplary practices. Furthermore, even if his biological practices were exemplary, they did not in fact serve as exemplars to any great extent, since his work and that of Theophrastus were not greatly added to in subsequent generations.[4] So it is difficult to see a great deal in Aristotle's *Historia Animalium* that represents progress according to either the noetic or the functional accounts. Aristotle himself distinguished between the *what* and the *why* in science. Only the latter counts as progress for the noetic view of progress. But it is instead our knowledge of the former that is greatly enhanced by Aristotle's work. That knowledge is an important contribution to science whether or not it contains solutions to scientific problems or exemplars of what a problem is and how it should be solved.

The next significant advances in zoology and botany (medicine aside) occurred in the sixteenth century and gathered pace in subsequent centuries. It was not until the work of Darwin and Mendel that biologists had anything like a grand explanatory theory that could be used to develop dependency models. And the scientific techniques involved were essentially the same as those used by Aristotle – fundamentally, keen observation and careful record keeping. Most naturalists were engaged in the study of morphology – the detailed description of the physical appearance of animals and plants – not least because morphology was the principal basis for naming and classification. Some, like Aristotle, were also interested in ethology, the study of animal behaviour. Edward Jenner was elected a Fellow of the Royal Society not for his work on vaccination but because he was the first to describe the behaviour of fledgling cuckoos in detail. Aristotle himself had noted that cuckoos lay their eggs in the nests of smaller birds. But before Jenner it was believed that it was the adult cuckoo that ejected the eggs of the foster-parent birds. Jenner's patient observations, recorded in a paper of 1788, demonstrated that it is the young cuckoo itself who pushes out the eggs and that the young cuckoo has a depression between its wings that enables it to do so. Perhaps the best known naturalist of Jenner's day, subsequently at least, was Gilbert White. White was the epitome of a certain type of naturalist – the country parson who studied and recorded his environment in detail and shared his findings with other scientifically minded colleagues, most famously in his letters to the leading British zoologist of the day, Thomas Pennant, and to the vice president of the Royal Society, Daines Barrington. The letters were published as *The Natural History of Selborne* (White 1789). The latter was hugely influential – the elderly Charles Darwin highlighted its significance in his own development as a young naturalist. These letters are full of novel facts, for example, the many facts deriving from the first bird census, one of White's innovations, including those observations that helped establish the fact of bird migration. And the fact that the bird known as the willow wren is in fact three species (the willow warbler, the chiffchaff, and the wood warbler), which he determined largely through differences in their songs (West 2021). And so on for hundreds of other valuable observations.

It is easy to find important elements in the work of White, Jenner, and other naturalists that satisfy the noetic and functional accounts of scientific progress. One can describe the nesting behaviour of the cuckoo and the migration of martins in terms of dependency relationships. As Britain's first ecologist, White noted many important connections of this kind. His famous lines on the earthworm, which particularly influenced Darwin (who shared his enthusiasm for this animal), establish the dependency of both vertebrates and plants upon it.[5] White's bird census is now a standard methodological tool for ornithologists. He also pioneered the use of pre-printed notebook for the systematic recording of observations, designed by Barrington. So his work certainly did employ new and useful exemplary practices.

Nonetheless, the contributions to understanding and to the development of exemplary practices made by these naturalists were small in comparison to the accumulation of knowledge that they added to science. It would be odd to think that their careful descriptive work added nothing to scientific progress in itself. One might also consider the voluminous and even more influential work of Alexander von Humboldt. Humboldt did make important methodological advances, but these made up only a small proportion of his work on the biogeography of Latin America and other regions of the world. Similarly, he was keen to understand the flora, fauna, landscape, climate, and people of these places and their interrelations. Yet, working before Darwin, his explanations and theories were only partially successful. Humboldt's significance for the progress of science does not rest on these alone. Rather the careful descriptions, including quantified analyses, of what he discovered were a major part of what made his reputation and what made his work important.

4. Mere Data or Observational Knowledge?

It should not be thought that these less glamorous additions to science are always mere data gathering, important only as fuel for scientific progress, not constituting progress itself. We can make a distinction between *data* and *observations*. Let us consider what happens when a scientist makes an observation. Often an observational practice will be subject to random error, so no single measurement can be taken to be an accurate representation of the true value. Precisely this problem arose in the determination of the orbits of the minor planets in the early nineteenth century, following the discovery of Ceres by Father Giuseppe Piazzi in 1801. And it was to solve this problem that Gauss developed the method of least squares and the theory of the normal distribution. The astronomer will record multiple measurements in her notebooks. These we may call her *data* (Bogen and Woodward 1988: 308). As noted, none are correct assertions about the quantity being measured and so none can be regarded as knowledge of such a quantity. So it is correct that the mere collection of data (thus understood) is no contribution to scientific progress.

Armed with sufficient data, the astronomer will then calculate the mean value of the quantity (or an orbit that satisfies the least squares condition, etc.). Now she has a value that can be regarded as likely to be the correct one, within a margin for error. Should she have carried out this process with sufficient rigour, then the resulting propositions concerning the value will be knowledge concerning the correct value of the quantity. These items of knowledge are her observations. And adding these to the stock of scientific knowledge is a contribution to the progress of science.

It might be replied that the data points might too be might generate knowledge, because they also will mostly be correct within a suitable (rather wider) margin for error. That margin for error, though, will typically not be known in advance – whereas the one for the finally recorded observation, the mean value of all the data points, can be calculated from the data itself. In any case, there are other examples where the data points clearly concern only the apparatus and process of observation and are not observations themselves. The LIGO detectors enabled an observation of a gravitational wave by measuring the strain in the arms of the detectors using an interferometer. That strain data would reflect both any gravitational wave plus environmental disturbances to the detectors (e.g. seismic activity). Those environmental disturbances are also measured independently. The strain data are taken to indicate a gravitational wave when a signal in the strain data is deemed to be a highly improbable occurrence given the environmental disturbances alone. So the strain data themselves cannot be thought to be measurements of anything outside the apparatus, not even within some margin for error. It is the inferred existence of the gravitational wave that constitutes the knowledge produced by the LIGO apparatus, and which in the scientists' parlance, constitutes the observation.

Many philosophers will balk at calling the latter item of knowledge an 'observation' since it is the product of inference, not of direct perception. Scientists, on the other hand, are entirely content to talk of the observation of imperceptible entities, such as quarks and magnetic fields, where that observation does involve inference. The present point, however, does not concern the correct analysis of 'observation'. Rather it is to point out that there is a principled way of distinguishing between 'mere' data, whose collection is not a contribution to scientific progress, and the result of processing that data to produce knowledge, which is a contribution to scientific progress. And the knowledge discussed in this chapter, from knowledge of the existence of X-rays to knowledge of the correct position of a star to knowledge of the differences between distinct species of warbler, is all of the latter type. These discoveries and piecemeal additions to scientific knowledge are not mere data.

As I have remarked, Aristotle himself made a distinction between the *what* and the *why*. For example, he writes:(Aristotle/Nussbaum 1985):

> Clearly there needs to be a study of all of these questions about animal locomotion and any others of the same kind; for that these things are thus is clear from our inquiries into nature; the reason why must now be investigated.

Science, according to this view, commences by establishing knowledge of the facts about what there is and what it is like (το 'οτι; to hoti), and then proceeds to ask why things are thus and so (το διοτι; to dioti).[6] While understanding is an important goal for science, it is not all of science. Gaining knowledge of what of things is its own scientific investigation and is itself a contribution to science's progress.

5.   Conclusion

Without doubt the most important contributions to the progress of science are those that add to our understanding and which expand our scope for setting and solving further scientific research problems.[7] Our initial examples of the discovery of DNA and the chemical revolution demonstrate this amply. Nonetheless, it would be fallacious to infer that what characterises the best examples of scientific progress must be present, albeit to a lesser degree, in all instances of scientific progress. Some worthwhile science is more prosaic, adding to the detail of our picture of the world, without deepening out understanding or opening new avenues for investigation. Knowing which species of flora and fauna there are or which stars, planets, and comets exist may be mundane by comparison with the greatest advances of science, but they are advances nonetheless. The naturalists and astronomers of the eighteenth and nineteenth centuries, like Aristotle and Hipparchus, in making careful and systematic observations of the world were not engaged in a fruitless or unscientific pursuit but were, by adding to the stock of scientific knowledge making small steps in the advancement of science. In his autobiography, Darwin asserted that Humboldt's *Personal Narrative* and John Herschel's *Introduction to the Study of Natural Philosophy* were the two books that had most influenced him 'stirring up in [him] a burning zeal to add even the most humble contribution to the noble structure of Natural Science' (Darwin 1958: 67–68). While the greatest scientific advances typically do add to our understanding and do furnish science with new exemplary practices, much of the work of science is not of this kind. To use Darwin's expression, even the most humble additions to scientific knowledge are also thereby contributions – even if only humble contributions – to scientific progress.

Notes

1. This Lavoisier disproved by showing that (after a hundred distillations) the weight of the residue is equal to the weight lost by the glass apparatus he used. By showing that the weight of the total apparatus and water remained the same throughout, he disproved Boyle's claim that fire particles are absorbed by it when heated.
2. See, for example, Footnote 1.
3. Stark was a committed Nazi who enthusiastically supported the removal of Jewish scientists from their posts. It is ironic therefore that his contribution to the history of X-rays was that he was among the first to accept Einstein's proposal regarding the quantisation of light and the first to apply it to X-rays.
4. Applications of zoology and botany to medicine are the exceptions to this, providing new knowledge of anatomy and of *materia medica* (plants used for medical purposes).
5. White (1789: 172) writes that

   > Earthworms, though in appearance a small and despicable link in the chain of nature, yet, if lost, would make a lamentable chasm. For, to say nothing of half the birds, and some quadrupeds, which are almost entirely supported by them, worms seem to be the great promoters of vegetation, which would proceed but lamely without them, by boring, perforating, and loosening the soil . . . and, most of all, by throwing up . . . worm-casts, which, being their excrement, is a fine manure for grain and grass.

6. Lennox (2009) argues that Aristotle's *History of Animals* is itself a work of το 'οτι investigation, while it is his *Parts of Animals* and *The Motion of Animals* are the subsequent το διοτι investigation. If so it is clear that Aristotle thought that the former, the investigation of the *what* of things is itself a worthwhile scientific goal.
7. This, I maintain, can be explained within the epistemic approach (Bird 2022z).

References

Aristotle/Nussbaum. 1985. *De Motu Animalium*. Princeton, NJ: Princeton University Press. Edited and translated by Martha Craven Nussbaum.

Assmus, A. 1995. Early history of X rays. *Beam Line* 25: 10–24.

Bird, A. 2007. What is scientific progress? *Noûs* 41: 64–89.

Bird, A. 2022. *Knowing Science*. Oxford: Oxford University Press, Forthcoming.

Bogen, J. and J. Woodward 1988. Saving the phenomena. *Philosophical Review* 97: 302–352.

Darwin, C. 1958. *Autobiography of Charles Darwin*. London: Collins. Edited by Nora Barlow.

Dellsén, F. 2016. Scientific progress: Knowledge versus understanding. *Studies in History and Philosophy of Science* 56: 72–83.

Elgin, C. Z. 2007. Understanding and the facts. *Philosophical Studies* 132: 33–42.

Hills, A. 2016. Understanding why. *Noûs* 50: 661–688.

Hoskin, M. 2011. *Discoverers of the Universe: William and Caroline Herschel*. Princeton, NJ: Princeton University Press.

Kusukawa, S. 1997. Leonhart Fuchs on the importance of pictures. *Journal of the History of Ideas* 58: 403–427.

Kusukawa, S. 2011. The role of images in the development of renaissance natural history. *Archives of Natural History* 38: 189–213.

Lennox, J. 2009. Aristotle's biology and Aristotle's philosophy. In M. L. Gill and P. Pellegrin (Eds.), *A Companion to Ancient Philosophy*, pp. 292–315. Chichester: Blackwell.

Shan, Y. 2019. A new functional approach to scientific progress. *Philosophy of Science* 86: 739–758.

West, S. 2021. Gilbert White: The modern naturalist. *Natural History Museum*. www.nhm.ac.uk/discover/gilbert-white.html, accessed 19 October 2021.

White, G. 1789. *The Natural History of Selborne*. Oxford: Oxford University Press. Edited by Anne Secord, published in 2013.

Williamson, T. 1995. Is knowing a state of mind? *Mind* 104: 533–565.

2 The Semantic Approach

Scientific Progress as Increased Truthlikeness

Ilkka Niiniluoto

This chapter outlines the so-called semantic account of scientific progress as increasing truthlikeness or verisimilitude.[1] Its philosophical background framework combines scientific realism (science seeks truth about mind-independent reality by means of explanatory theories) and fallibilism (the results of science are always corrigible and uncertain but may be probable or truthlike). The semantic view is defended by comparing it with rival accounts which define progress in terms of knowledge, problem-solving, or understanding.

1. The Philosophical Problem of Scientific Progress

Most historians and philosophers of science, as well as scientists themselves, are convinced that science is largely progressive. For example, George Sarton, the first professor of History of Science at Harvard University, claimed in 1936 that 'the acquisition and systematization of positive knowledge are the only human activities which are truly cumulative and progressive', and 'progress has no definite and unquestionable meaning in other fields than the field of science' (see Sarton 1957, 36). Karl Popper, one of the leading philosophers of science of the twentieth century, concurred that 'progress in science can be assessed rationally' and 'science seems to be the only field of human endeavour of which this can be said' (Popper 1975, 83). And when asked in interviews, scientists and even laypeople have firm opinions about the characteristics that make science progressive (see Mizrahi and Buckwalter 2014).

Yet, there is no consensus about the definition of scientific progress. This theme was lively debated in the 1960s and 1970s, when philosophers proposed various models of scientific change (see Lakatos and Musgrave 1970; Hacking 1981), and it has again become a hot topic among philosophers especially after Alexander Bird's paper in 2007. Bird distinguished three main approaches to scientific progress:

- epistemic: progress as increasing knowledge
- semantic: progress as increasing truthlikeness[2]
- functional: progress as increasing functional effectiveness.

DOI: 10.4324/9781003165859-4

A fourth approach has been thereafter added to the market:

- noetic: progress as increasing understanding.

These accounts agree that 'progress' is an axiological or a normative concept, which should be distinguished from neutral descriptive terms like 'change' and 'development': to say that step from stage A to stage B constitutes *progress* means that B is an improvement of A, or better than A relative to some standards, criteria, or aims (Niiniluoto 2019). But, as science is a many-splendored thing, its processes and products have many possible aspects X, which could be chosen as the defining feature of its progressive nature. Such aspect X should not only be prevailing (many historical episodes of scientific change have been progressive) but also non-trivial (the notion of progress is empty if it covers all changes in science). It should reflect the fact that science is in a broad sense a research-based cognitive activity which seeks reliable information about reality by observations and experiments, constructs and applies theories to explain and predict phenomena, and helps us to interact successfully with our natural and cultural environment. Therefore, X should express what *constitutes* scientific progress, instead of antecedent factors which merely *promote* or bring about progress (e.g. the education and skills of the scientists, the improved instruments and methods of scientific inquiry, the funding and organisation of research) (Dellsén 2016, 73, forthcoming). Likewise X should be distinguished from the short-term and long-term *consequences* of successful science (e.g. development of new technological innovations, economic wealth, science education in schools, social progress, quality of life, and justice in society). This still leaves room for many different choices of X, and the tasks of a philosopher include the explication of X (making X clear by analytic or logical means) and the argument why X is more important or fundamental than other rival accounts.

In the rival theories of truth, the realists insist on the distinction between the meaning of truth (e.g. correspondence) and the indicators of truth (e.g. empirical confirmation) (see Niiniluoto 1999). In the same way, one should distinguish the semantical or *conceptual* question of progress (what is meant by progress in science?) from the *epistemic* question (how can we recognise progress by some evidence?). Further, the *methodological* question (how can we promote progress in research and science policy?) should be distinguished from the *factual* question (is science progressive by our standards?) (see Niiniluoto 1979, 1980). All of these questions are significant, and could be addressed by methodologists, sociologists, and historians, but the conceptual issue is most fundamental. This chapter concentrates on the relative merits of the semantic, epistemic, functional, and noetic approaches as rival answers to the philosophical problem of defining scientific progress.

2. The Historical Roots of the Progress Debate

According to the traditional mainstream view, science is a collection of certified truths. For Aristotle, the premises of scientific syllogisms are necessary truths. The medieval conception of science was static: *scientia* is the possession and contemplation of knowledge, and its method is needed to organise the already established truths for the purpose of teaching. However, after the Scientific Revolution in the early seventeenth century, Francis Bacon and René Descartes advocated the dynamic view that the aim of science is to find *new* truths by inquiry. After this shift in the aims of science, it became relevant to raise the conceptual problem about scientific progress: what is meant by progress in science?

Both the empiricist Bacon and the rationalist Descartes adhered to the requirement that scientific knowledge is completely certain to be true – at least after the employment of right inductive or deductive methods – and hence they answered the conceptual problem by asserting that scientific progress is *the accumulation of knowledge*.

The historical roots of the debate about scientific progress go back to two important trends in epistemology and philosophy of science. First, in the ancient Greece, Plato defined knowledge (Gr. *episteme*) as true belief with justification, but the programme of pursuing such knowledge was challenged by the sceptics. From this controversy emerged the *fallibilist* school which argued in different ways that scientific knowledge is uncertain and at best approaches to the truth (see Niiniluoto 1984, 2000). Great names of this view include Nicholas of Cusa in the fifteenth century and Charles S. Peirce in the late nineteenth century. Peirce characterised truth as the limit of endless inquiry: 'the opinion which is fated to be ultimately agreed by all who investigate, is what we mean by truth' (*Collected Papers* 5.407), and science approaches to the truth with its 'self-corrective method' (*Collected Papers* 5.575). Even today this kind of fallibilism inspires the semantic account of scientific progress with the explication of Karl Popper's (1963) notion of truthlikeness.

Secondly, the nature of physical and astronomical theories was debated between the *realists* and the *instrumentalists*: the former required that such theories are true or false, the latter that they merely 'save the phenomena' (Duhem 1969). While the Aristotelians were realists, Ptolemy's geocentric theory was often interpreted in instrumentalist terms. This controversy was highlighted by the Copernican revolution. Pierre Duhem, who was on the side of the instrumentalists, denied in his 1906 work that theories as conceptual tools have truth values. Real progress (RP) occurs only slowly and constantly on the level of the increasing empirical content of theories (see Duhem 1954, 38–39). Thus, progress means the accumulation of observational statements covered by fluctuating theories. However, Duhem added to his instrumentalism the claim that physical theory makes progress by becoming 'more and more similar to a natural classification which is its ideal end' (*ibid.*, 298). One may wonder whether such a peculiar form of convergent realism makes sense without assuming that the classified laws refer to real theoretical entities. Henri Poincaré, who argued that theories are conventions without truth values, gave in 1902 a vivid picture of 'the man of the word', who sees theories to be 'abandoned one after another', 'ruins piled upon ruins', so that 'theories in fashion to-day will in short time succumb to their turn' (Poincaré 1952, 160). Yet, even Poincaré subscribed to a kind of structural realism, where Maxwell's equations express 'true relations between real objects' (*ibid.*, 161). His account of theories has inspired later discussions about 'the pessimistic meta-induction', where scientific theories up to our present and next ones constitute sequences of false hypotheses (Laudan 1984). The realist reply is that such sequences may still be progressive in the sense that they convergence to the truth.

While instrumentalists typically deny that there is any progress on the level of scientific theories, they usually assume the cumulative model on the level of observations. Progress as empirical accumulation is formulated in some empiricist accounts of reduction, where a new theory includes all true or verified empirical consequences of its predecessor (see e.g. Kemeny and Oppenheim 1956). Contemporary heirs of empiricist instrumentalism have not developed detailed definitions of scientific progress, but a similar account of progress could be a part of Bas van Fraassen's (1980) *constructive empiricism*, which demands that theories aim at empirical adequacy, that is all of their observational consequences are true. So the more such true empirical consequences, the merrier.[3] The same idea might be applied in Kyle Stanford's (2006) epistemic instrumentalism, as it claims that theoretical claims in science have unknown truth values which 'exceed our grasp'.

The model of cumulative growth was challenged in the 1960s also by two related theses: observations are theory-laden, and successive theories are incommensurable (Lakatos and Musgrave 1970). Thomas Kuhn (1970) argued against Popper that there is no theory-independent notion of truth so that convergence to ultimate truth does not make sense, even though paradigm-based normal science is cumulative on the level of 'puzzle-solving'. This inspired Larry Laudan (1977), who acknowledges that theories have truth values, to develop his account of progress for research traditions by the functional notion of problem-solving effectiveness. Popper (1975, 83) and Paul Feyerabend proclaimed that science should be in the state of 'permanent revolution', but paradoxically Feyerabend's anarchism led him to an extremely cumulative view of science: knowledge is 'an ever increasing ocean of mutually incompatible (and perhaps even incommensurable) alternatives', where 'nothing is ever settled, no view can ever be omitted from a comprehensive account' (Feyerabend 1975, 30). In his more serious mood, Feyerabend (1984) suggested that the progress of science has an interesting similarity with the changes of styles in the fine arts.

3. Scientific Progress as Increasing Truthlikeness

Karl Popper's *Logik der Forschung* in 1934 and its English translation *The Logic of Scientific Discovery* in 1959 attacked probabilistic versions of fallibilism by asserting that we can never prove that a theory is true or probable. Popper's insight was that science is interested in information content, that is bold rather than logically weak or probable hypotheses. In 1935 Popper learned about Alfred Tarski's semantic concept of truth, which in his view saves the classical correspondence theory of truth. Thus, his falsificationism is combined with the realist interpretation of theories: science grows when a scientist proposes bold conjectures and puts them in severe tests (Popper 1963). Popper sometimes seemed to claim that information content is *the* aim of science, but he qualified this with the requirement that an informative hypothesis has to be corroborated by severe tests: information content is a criterion of 'potential progressiveness', but the preferable theory should pass some crucial tests (see Popper 1963, 217). Also Imre Lakatos defined empirical progress in terms of the 'excess corroboration' of theories, that is the excess empirical content of the new theory is corroborated by experiments (Lakatos and Musgrave 1970, 118).

Popper's basic idea can be at least partly formulated within a probabilistic framework of cognitive decision theory. The basic idea, due to Carl G. Hempel (1965), is to treat the values of science – such as truth, information, explanatory power, predictive power, and simplicity – as *epistemic utilities* which scientists are trying to maximise under uncertainty so that the expected utility is calculated with respect to probabilities. A sophisticated formulation was given by Isaac Levi (1967), who showed that truth or information alone cannot be adequate epistemic utilities so that the aim of science should be taken to be *informative truth*. Levi defined epistemic utility as a weighted combination of truth value and information content, where the weight of the information factor as 'an index of boldness' indicates how much the scientist is willing to risk error, or 'gamble with truth', in an attempt to avoid agnosticism. It follows that the expected utility is then a weighted combination of posterior probability and information content. In this sense, the demand of high posterior probability is balanced with the interest in bold (informative, a priori improbable) hypotheses.

Popper is often mentioned as the leading advocate of the semantic account of scientific progress. But, in his Herbert Spencer Lecture in 1975, Popper defined 'striking progress' by two criteria: first, the new theory should contradict and overthrow its predecessor; secondly, it should be 'conservative' in the sense that it 'must be able to explain fully the success of its predecessor' (Popper 1975).[4] There is no mention of the new idea of comparative verisimilitude, introduced by Popper in 1960 (see Popper 1963). And, as we shall see, Popper's own explication of this idea does not quite match his two criteria of progress.

Popper's (1963) comparative and quantitative definitions of *truthlikeness* or *verisimilitude*, as 'the idea of approaching comprehensive truth' or 'better or worse correspondence with reality', belong to the non-probabilistic tradition of fallibilism. Treating theories as deductively closed sets of statements in an interpreted scientific language L, with the sets T and F of true and false statements, respectively, theory A is more truthlike than theory B if and only if $B \cap T \subseteq A \cap T$ (A has larger truth content than B) and $A \cap F \subseteq B \cap F$ (A has smaller falsity content than B), where at least one of these set-inclusions is strict. So, on the level of truth content, the conservative requirement states that the better theory should include all the successes of the other theory. But these conditions do not require that the two theories contradict each other: besides revisions, the definition allows expansions.

The important properties of Popper's definition include the following:

(Q1) T is at least as truthlike as any other theory A.
(Q2) $A \cap T$ is more truthlike than A, if A is false.
(Q3) A is more truthlike than B, if both A and B are true and A is logically stronger than B.

Popper's definition can be motived by the idea that fallibilism should be able to make sense of the idea of *scientific progress as increasing truthlikeness*. Principle (Q1) expresses the idea that complete truth is the ultimate aim of progress, and the principle (Q2) shows that the step from a false theory to its truth content is progressive. Principle (Q3) shows that this account includes as a special case progress by the accumulation of knowledge. The weakest true theory is a tautology, which represents ignorance as a cognitive state. It is not allowed that a false theory is more truthlike than a tautology (i.e. better than ignorance), but on the other hand a tautology cannot be more truthlike than a false theory. However, David Miller and Pavel Tichý proved in 1974 that Popper's definition fails dramatically: if A is more truthlike than B in Popper's sense, A must be true (see Oddie 2016). This means that it cannot be used for its intended use of comparing false theories.

After the failure of Popper's attempt, a number of alternative and still debated precise definitions of truthlikeness have been given. Some attempts to rescue Popper's approach in model-theoretical terms stumble on the implausible consequence that truthlikeness covaries with logical strength among false theories so that a false theory could be improved just by joining new falsities to it. A more promising alternative is to restrict the deductive content of theories to relevant consequences (see Schurz 2018).[5] But the most popular post-Popperian explication is *the similarity approach*, which starts from the observation that Popper's definition (just as Levi's epistemic utility) fails to recognise that some falsities are closer to the truth than others (see Oddie 1986; Niiniluoto 1987).[6] When similarity between states of affairs is included in the definition (besides logical entailment and truth values), then the following principles hold:

(Q4) Some false theories may be more truthlike than a tautology.
(Q5) Some false theories may be more truthlike than some other false theories.

Principle (Q4) indicates that science may progress from ignorance (i.e. trivial truth) to a false statement B, if B is sufficiently truthlike. For example, approximately true but false measurements by a reliable instrument are better than nothing. Principle (Q5), which was fatal to Popper's definition, shows that historical sequences of false theories may be progressive in their increasing truthlikeness (e.g. in mechanics, Aristotle – Buridan – Galileo – Newton – QM). Indeed, such sequences may converge to the truth in two different ways:

(c1) $F_1, F_2, \ldots, F_n, T_1, T_2, \ldots$
(c2) $F_1, F_2, \ldots, F_n, \ldots \to T$.

In (c1), truth is reached via falsities in a finite number of steps, and then improved by even better true theories, while in (c2) truth is the asymptotic limit of false theories (see Niiniluoto 1980).

In quantitative cases, similarity can be defined by the metric structure of the space of real numbers or real-valued functions. In qualitative cases, distance between statements can be defined by 'feature matching', that is by counting the agreement and disagreement between their claims. This allows us to define distances $\Delta(C_i, C_j)$ between complete answers in various kinds of cognitive problems, including singular statements about individual objects, existence claims, generalisations, and laws (for details, see Niiniluoto 1987).[7] Then the next stage is to extend this distance to partial answers so that the distance $\Delta(H, C_i)$ of a disjunction H from a given C_i is a function of the distances between the disjuncts of H and C_i. If C_i is chosen as the target C^*, which is the most informative truth in the given conceptual framework, then $\Delta(H, C^*)$ measures the closeness of H to the complete truth C^*. When this distance varies between 0 and 1, the *degree of truthlikeness* $Tr(H, C^*)$ of H relative to C^* is equal to $1 - \Delta(H, C^*)$. In agreement with Popper's (Q1), $Tr(H, C^*)$ is maximal if and only if H is equivalent to the complete truth C^*. This degree can be high even when H is false.

Graham Oddie (1986, 2016) defines the function $\Delta(H, C^*)$ as the average distance of the disjuncts of H from C^*. It follows that his explication fails to satisfy Popper's requirement (Q3): in some cases logically stronger true theories are farther from the truth than weaker truths. The recent generalisation of the feature matching account of Cevolani and Festa (2020) agrees with Oddie's average function, but as an account of scientific progress this is problematic. Instead, Niiniluoto's (1987) minsum measure satisfies (Q3). It defines $\Delta(H,C^*)$ as a weighted sum of the minimum distance $\Delta_{min}(H,C^*)$ of the disjuncts of H from C^* and the (normalised) sum $\Delta_{sum}(H,C^*)$ of all these distances. The minimum distance expresses our goal of being close to the target C^* (this is sometimes used as an explication of the notion of *approximate truth*), and the sum factor gives a penalty to all mistakes allowed by H with respect to C^*.

When the target C^* is unknown, degrees of truthlikeness are likewise unknown. The epistemic notion of truthlikeness attempts to tell how close we can estimate theory H to be to the target C^*, given our background knowledge and available evidence E. Niiniluoto's (1987) solution to the epistemic problem assumes that a rational probability measure P is defined for the language L so that the posterior epistemic probability $P(C_i/E)$ given evidence E is defined for each complete answer C_i in L. Then the unknown degree of truthlikeness $Tr(H,C^*)$ may be estimated by its *expected* value relative to the complete answers C_i and their posterior probabilities given evidence E:

$$ver(H/E) = \sum P(C_i/E)\, Tr(H,C_i).$$

It is important that $ver(H/E)$ may be high even when $P(H) = 0$ or $P(H/E) = 0$. Application of probabilities of Jaakko Hintikka's inductive logic shows that for successful increasing singular evidence E we may have asymptotically $ver(H/E) \to 1$ for a general theory H (Niiniluoto 2007).

With the functions Tr and ver, we may distinguish between real and estimated progress (EP):

(RP) Step from theory H to theory H' *is progressive* if and only if $Tr(H,C^*) < Tr(H',C^*)$.

(EP) Step from theory H to theory H' *seems progressive* on evidence E if and only if $ver(H/E) < ver(H'/E)$.

See Niiniluoto (1979). According to definition (RP), objective truthlikeness Tr gives an ahistorical standard for telling how close we really are from the target C^*, even when we don't know it, and likewise a standard of *RP* in science. According to definition (EP), estimated verisimilitude ver expresses our judgements about progress, sensitive to historically changing situations with variable evidence. In contrast, decrease of objective or estimated truthlikeness is a mark of *regressive* development in science.

For a fallibilist, the claim that the change from theory H to theory H' is progressive is always uncertain, but it can be confirmed by evidence (Niiniluoto 1984, 147). Support for this kind of comparative realism (Kuipers 2009) can be obtained from an abductive argument: increasing truthlikeness is the best explanation of the fact that H' is empirically more successful than H (Niiniluoto 2018b, 159). The history of science also provides an optimistic meta-induction for comparative realism: past false theories in Laudan's (1984) list (e.g. phlogiston, caloric, ether) have been replaced with more truthlike alternatives so that such progressive pairs of theories can be expected in the future (Niiniluoto 2017).

Definitions (RP) and (EP) presuppose that theories are assessed relative to the same target C^*. This is motivated by the idea that the compared theories have to be rival answers to the same cognitive problem – there is no point of comparing the truthlikeness of Darwin's theory of evolution and Einstein's theory of relativity. If H and H' are expressed in different languages L and L', (RP) has to be modified by translating these theories into a common conceptual framework. If the vocabulary of L' is an extension of L, this common framework can be the richer language L', and in other cases it might be the union of the vocabularies of L and L'. If the languages L and L' have different meaning postulates, then for the purposes of comparison, a theory H in L should include the specific meaning postulates of L and similarly for a theory H' in L'. In the richer framework, one may find continuity between theories H and H' so that it is possible to compare their degrees of truthlikeness and speak about convergence to the truth. Reference invariance in spite of meaning variance and incommensurability can be defended also on the basis of appropriate theories of *reference* (see Niiniluoto 1999, 128–132) so that it is meaningful to state that rival false theories refer to the same entity (e.g. electron) and one of them gives a more truthlike description of it.

The relativisation of truthlikeness to cognitive problems with specific targets is in harmony with the key idea of *conceptual pluralism*. All inquiry is relative to some conceptual framework which is used by the scientists for describing reality. Critical realists argue that such conceptual frameworks can be changed, revised, and enriched. If a language lacks expressive power, new terms can be added to its vocabulary. Our representations of reality are always piecemeal in the sense that we need not accept with Wilfrid Sellars (1968) the existence of an ideally adequate 'Peirceish' conceptual framework. If such an ideal language would exist, as metaphysical realists think, degrees of truthlikeness could be defined relative to its expressive vocabulary, but in practice the scientists have to work in languages which try to capture the most important aspects of the relevant fragment of reality. Contrary to Hilary Putnam's (1981) *internal realism*, which denies the idea of a 'ready-made world', conceptual pluralism does not imply that the notion of truth is epistemic (Niiniluoto 1980, 1999).

Progressive theory change in science often combines continuity and improvement: the new theory corrects the superseded theory in some respects but retains part of its content. This is illustrated by the principle that the old theory is obtained from the new one when some parameters approach a counterfactual value.[8] This principle is important when idealised theories are de-idealised or 'concretised'. For example, Boyle-Mariotte law pV = RT for ideal gas can be de-idealised by van der Waals law $(p + a/V^2)$ $(V - b) = RT$, which takes into account intermolecular attractive forces a and the finite size b of gas molecules.[9] This equation is a progressive improvement of the ideal gas law, but it has later been corrected in statistical thermodynamics (cf. Niiniluoto 1990; Garcia Lapeña forthcoming).

Darrell Rowbottom (2015) argues against the definitions (RP) and (EP) that scientific progress is possible in the absence of increasing verisimilitude. He asks us to imagine that the scientists in a specific area of physics have found the maximally verisimilar theory C*. Then, by the criterion (RP), no more progress is possible, but yet this general true theory could be used for further predictions and applications. Similarly, Dellsén (forthcoming) points out that sometimes an accepted theory is stable, but still it may lead deductively to new consequences. One reply to these arguments is that predictions from C* (or applications of a given theory to new directions) constitute new cognitive problems for the scientists – and the success in these problems is again measured by truthlikeness. These kinds of truth-seeking applications of basic theories are typical of applied research where the primary axiological aim is informative truth and secondary aims may include predictive power, simplicity, manageability, and social relevance (Niiniluoto 1999, 163). Moreover, on the basis of conceptual pluralism, in Rowbottom's thought experiment, it would still be possible for the physicists to achieve further progress by extending their conceptual framework in order find a still deeper complete truth about their research domain (Niiniluoto 2018a).

4.   Comparisons

4.1.   *The Epistemic Account*

Given the historical background about debates on scientific change, it may seem surprising that Bird (2007) was able to regain the cumulative model as a serious candidate in his *epistemic* account of scientific progress. According to Bird, progress means increasing knowledge. Even though he follows Timothy Williamson in thinking that truth cannot be explicitly defined, still knowledge presupposes truth and justification. Bird's position is realist, but it returns to the old model where progress means the same as accumulation of justified truths.

Among Bird's critics, Rowbottom (2008) contends against Bird that justification is instrumental rather than constitutive of progress, that is justification is a means for establishing a link between truth-seeking and truth-finding. Dellsén (2016) argues that progress as understanding does not presuppose knowledge. The definition of RP by increasing truthlikeness does not require knowledge, but Cevolani and Tambolo (2013) remind that the verisimilitudinarian approach handles issues about justification by means of the distinction between RP and EP.

Bird defends the epistemic view by arguing against the semantic view by the following thought experiment. Imagine that a scientific community has formed beliefs B by an irrational method M, such as astrology, and B happens to be true. M is then shown to be unreliable and the beliefs B are given up. He goes on to suggest that for the semantic view, the acquisition of accidentally true beliefs by an unreliable method is progress, and the rejection of unfounded but true beliefs is regressive, while the judgements of the epistemic view are opposite. Bird's argument gives attention to a hidden assumption that the primary application of the notion of scientific progress concerns successive theories, which have been seriously considered or accepted by the scientific community. So one response to Bird would be to point out that irrational beliefs and beliefs without any justification simply do not belong to the scope of *scientific* progress (Niiniluoto 2014).

Bird (2016) modifies his original thought experiment so that the true beliefs B are obtained by scientific but unreliable means, that is by derivation from an accepted theory which turns out to be false. In his example, the Renaissance doctors derived from the humoral theory H that infants and young children have more body water than adults. If a scientist reaches true or truthlike beliefs by such reasoning from mistaken premises, are we compelled to say that progress has been achieved? One option here is to say that the true conclusion B about body water was really progressive, but it was also justified by the available methods and accepted theories at the time, that is ver(B/H) was high, but the overthrow of the humoral theory showed that this fallible estimate was incorrect.

There are also important historical cases where the Tr-value is high but ver-value is initially low. In examples of anticipation a good theory is first suggested without sufficient justification and only much later is shown to be scientifically acceptable (e.g. Aristarchus on the heliocentric system, Wegener on continental drift). The initial evidence E for a such a hypothetical theory H may be weak, and the theory H is not yet accepted in science, but genuine progress is eventually achieved when new evidence E' increases its expected verisimilitude and thus gives reasons to claim that H is truthlike and leads to the acceptance of H.

Bird (2007) recognises the problem that scientific theories are often not in fact true, and sometimes are even known to be false. He mentions the historical transitions from Galileo to Newton to Einstein and from Ptolemy to Copernicus to Kepler (cf. sequences (c1) and (c2)), but argues that the cumulative epistemic view can handle them by the following trick: if H is approximately true, then the proposition 'approximately H' or A(H) is fully true. So replace the sequence of false theories $H_1, \ldots, H_k$ by the sequence $A(H_1), \ldots, A(H_k)$, which contains fully true theories adding to the truth provided by their predecessors. This attempt to save the cumulative model by transforming false theories into true ones raises a number of questions (Niiniluoto 2014). First, A(H) states that H is in an objective sense close to the truth, but do we really *know* such statements? Secondly, statements A(H) are not ones that we actually meet in the history of science. Scientists may formulate and even accept false theories without trying to specify such margins of error which would make them true or even probable so that we don't find A(NEWTON) in the *Principia*. Thirdly, the falsity of many past theories is not simply a matter of numerical approximation. What would be A(H) for a theory H with mistaken existence assumptions (like phlogiston theory) or implicit counterfactual idealisations (like the ideal gas law)? Fourthly, the most fatal problem with Bird's proposal is the inability of the cumulative epistemic view to distinguish progress and regress in science. Suppose that $H_1, \ldots, H_k$ is a regressive sequence of theories with increasing distances from the truth. By Bird's argument, even in this case the sequence $A(H_1), \ldots, A(H_k)$ would consist of true and known statements. But in spite of this cumulative knowledge, the original sequence is not progressive.

Bird (2016) modifies his argument by pointing out that scientists will believe many true consequences of accepted false theories: Dalton mistakenly assumed that water is HO, but correctly believed that water is a compound of hydrogen and oxygen. But in his example of the humoral theory, such nontrivial true consequences do not constitute knowledge, and their ability to save the epistemic account of progress does not appear promising.

In my view, the only fallibilist way of saving Bird's epistemic view would be to propose a notion of conjectural or *non-factive knowledge,* which does not presuppose truth (Niiniluoto 2018a, 195). This terminology is in fact a common practice in everyday life, where the currently accepted, so-far best theories are said to belong to 'scientific knowledge' in spite of their falsity. For centuries Newton's theory was believed to be true, even though it is at best approximately true. So let us say that the scientific community SC knows* that p if and only if p is truthlike and the estimated truthlikeness of p is larger than its rivals on available evidence (see Niiniluoto 1999, 84).[10] But even this weakened form of knowledge* would capture only some examples from the history of science. Furthermore, this account would differ from Bird's epistemic view of progress, since such Popperian sequences of known* but incompatible theories would not be cumulative in the sense that later theories entail the earlier ones.

4.2.   *The Functional Account*

Kuhn's puzzle-solving (1970) and Laudan's (1977) problem-solving are examples of the functional account of scientific progress. For Laudan, science is a problem-solving rather than a truth-seeking activity so that progress can be defined by the *problem-solving effectiveness* of a theory (the number of solved problems minus the anomalies and generated conceptual problems). As for him solving an empirical problem means that a 'statement of the problem' is deduced from a theory, such solutions include explanations (answering the question 'why q?' by deriving q from theory H) and predictions (deriving p or ~p from H). Thus, the number of solved problems for Laudan is virtually equivalent to what Hempel (1965) called the *systematic power* of a theory.

While Hempel demanded that an explanatory theory must be true, Laudan is an expressed anti-realist who operates only with the notion of potential explanation (i.e. explanation without the truth requirement). Truth for him is a 'utopian' aim of science (Laudan 1984). Bird (2016) objects that the derivation of false statements from a false theory does not count as progress. A scientific realist can argue against Laudan that the problem-solving ability is a fallible indicator of the truth of the theory. The Bayesian theory of confirmation shows that successful explanations and predictions of empirical phenomena confirm the theory by increasing the posterior probability that the theory is true (see Niiniluoto 2018b, 95). This treatment can be generalised to cases, where the empirical success is approximate and the abductive conclusion via the ver-function concerns the estimated truthlikeness of the theory (*ibid.*, 143–145; cf. Niiniluoto 1984, 179).

Another way of looking at problem-solving is to replace the syntactical formulation of a theory as one overall claim about the world by the semantic or structuralist notion of a theory as a network of applications to specific domains. With some simplification, a theory in this sense is a pair <C,J>, where the core C is claimed to be applicable to all members of J. Wolfgang Stegmüller (1976) hoped that this move helps to avoid the 'teleological metaphysics' of Popper's account of truthlikeness. But each such application z in J constitutes a cognitive problem in the realist sense, and the success of the theory <C,J> depends on the truthlikeness of its solutions of such problems. The local success of the theory for a particular application z in J is measured by the truthlikeness of C with respect to the target z (in terms of the Tr- or ver-function). An overall realist measure of the cognitive problem-solving ability of a theory can then be given by the weighted sum of its local successes (see Niiniluoto 1984, 98, 146).

Scientific theories are also *pragmatically successful* as guides of our actions. Laudan's treatment does not include problems of action, which have been extensively studied in Operations Research (e.g. decision theory, game theory, linear programming). But the success of research-based practical decisions and actions, which is different from truth as an aim of science, crucially depend on the truth or truthlikeness of the applied theoretical models: recommendations from false premises may be misleading and even disastrous (Niiniluoto 1984, 253). In this sense, the method of Operations Research is truth-using or quasi-factive. Similarly, predictive problems in applied science are truth-using and truth-seeking at the same time.

Some philosophers have proposed functional definitions of scientific progress in terms of successful action. Thus, Nicholas Rescher's (1977) 'methodological pragmatism' characterises scientific progress as 'the increased success of applications in problem solving and control'. A similar proposal by Heather Douglas (2014) defines scientific progress as 'the increased capacity to predict, control, manipulate, and intervene in various contexts'. Even though these pragmatic successes may be consequences of good science, a realist should resist such reduction of scientific progress to technological progress.

Yafeng Shan (2019) proposes a new functional account of scientific progress. He gives up the internalist requirement of Kuhn and Laudan that the scientific community is able to recognise whether it is making progress. His proposal also tries to avoid Laudan's need to introduce weights for the solved problems, but the issue with Kuhn-losses makes this difficult. According to Shan's main definition, 'science progresses if more useful research problems and their corresponding solutions are proposed', where usefulness requires repeatability in further investigation. The definition is well illustrated by the development of early theories of heredity from Darwin's pangenesis to Mendel, de Vries, and Bateson. Shan argues that the epistemic and semantic accounts of progress overlook the significance of practical aspects of science, while Douglas (2014) overlooks the significance of theoretical aspects to some extent. However, the scientific realists need not deny that there is lots of *know how* in the design of experiments and in the applications of theories, but they typically treat these kinds of aspects as belonging to the promotion and consequences of cognitive progress. While Shan gives emphasis to both problem-defining and problem-solving, the former belongs to factors promoting progress. Moreover, Shan admits that it is counterintuitive to separate progress from truth and knowledge, and links his approach to the notions of perspectival truth and know how, which blurs the sharp borderline between his new functional account and the semantic and epistemic accounts.

4.3.   The Noetic Account

Finnur Dellsén (2016) coined the term 'the noetic account' (from the Greek *nous*) for the view that scientific progress consists of increasing understanding.[11] He employs the objectual notion of understanding instead of understanding why, but proposes to analyse understanding as grasping 'how to correctly explain and/or predict some aspects of a target'. So according to the noetic account, 'science makes (cognitive) progress precisely when scientists grasp how to correctly explain or predict more aspects of the natural world than they did before' (*ibid.*, 75).

The combination of explanation and prediction as the main ingredients of understanding may seem problematic to some philosophers, who give exclusive stress to explanatory coherence (e.g. the hard-line realist

Sellars 1968) or to predictive success (e.g. the conventionalist Poincaré 1952). But already Hempel (1965) proposed that explanatory and predictive power can be combined in the notion of systematic power. An immediate concern for Dellsén is that (as we noted earlier) measures of systematic power are formally related to Laudan's functional notion of problem-solving. Probabilistic measures of systematic power do not give satisfactory rules for the acceptance of theories, but when used as truth-dependent epistemic utilities they give results which are equivalent to Levi's principle of combining truth (high posterior probability) and information content (low prior probability) (Niiniluoto 1999, 187). But in comparison to the semantic approach, these principles are limited as they do not recognise the fact that some falsities are closer to the truth than others.

Dellsén's main target has been Bird's epistemic account, since he thinks that progress can be achieved without knowledge by hypothetical explanatory arguments and by false idealised theories, and sometimes knowledge is accumulated without increase in understanding (Dellsén 2018). Park (2017) objects that scientists have to believe in the reality of their explananda and in the truth of their confirmed predictions. Dellsén (2016) also suggests that the simpler of two equivalent theories provides more understanding, but Park regards this only as an unwarranted psychological illusion.[12]

The noetic and semantic accounts agree against the epistemic view that justification is not necessary for (real) progress. By the no miracles argument, truthlikeness is the best explanation of the empirical success of a theory so that explanatory and predictive success confirms a theory (Niiniluoto 1984, 178–183, 1999, 185–198, 2018b, 156–163). It follows that noetic progress is an indicator of progress as increasing verisimilitude, while the latter explains noetic progress. In this sense, it may be argued that semantic progress is more fundamental than the noetic one.

A new turn of the noetic account is given by Dellsén's (forthcoming) revision of the notion of understanding in terms of 'dependency models'. Paradigm examples of dependence relations between variables are causation and grounding. According to the new noetic account, progress is 'a change due to scientific research in the publicly available information that enables relevant members of society to increase their understanding' of a topic or phenomenon Y, and 'an agent S understands Y if and only if S grasps a sufficiently accurate and comprehensive dependency model of Y'. Such understanding helps S to explain, manipulate, and predict relevant phenomena, but this is a consequence to the grasp of dependences. The revised noetic account is clearly an improvement of the earlier formulation, but it is even closer to the semantic account than the old one, as accuracy and comprehensiveness correspond to the demands of truth and information in the Popperian concept of truthlikeness. Dependencies include laws of nature – captured by the notion of legisimilitude – but there are subtle differences: besides mathematical or logical relations the grounding principles may include links of micro-reduction (e.g. water consists of H_2O). Dellsén goes on to argue that his new account is not too narrow by showing that classification systems (e.g. the periodic table of elements, biological taxonomy) and the discovery of new entities (e.g. quarks, Brownian motion, platypus) may directly contribute to progress or indirectly promote it. Both types of cases are covered by the verisimilitude account, where a typical target or complete truth C^* consists of existence and non-existence claims within a classification scheme. As a possible difference Dellsén mentions entirely spurious correlations which are fully true but do not advance understanding, but one may doubt that such artificial non-causal correlations would be an interesting topic of scientific research. Instead, massive systems of Big Data obtained by systematic reliable measurements (e.g. astronomy, high energy physics, climate, biological populations, medicine, and social statistics) do not merely promote progress: when the measurement values are construed as long conjunctions, such truthlike singular statements directly contribute to progress in the semantic sense.

Notes

1. We follow Popper (1963) in using 'truthlikeness' and 'verisimilitude' as synonyms. It is only a matter of convenience that the symbols Tr and ver are used for objective and estimated truthlikeness, respectively.
2. Bird (2007) gives also an alternative formulation of the semantic account as the accumulation of truths, but admits later that it is less popular and plausible than the verisimilitude account (Bird 2016, 550).
3. A more sophisticated empiricist account of progress is given by Hasok Chang (2004) in terms the 'epistemic iteration' of coherent systems of measurement values. According to Chang, this method is self-corrective, but he denies that progress means closer approach to the truth.
4. Kuhn (1970) challenged this condition by arguing that there are losses of meaningful problems in scientific revolutions.
5. Other realist approaches to scientific progress include approximate truth (Aronson, Harré, and Way 1994) and partial truth (da Costa and French 2003). Theo Kuipers (2019) defends his own version of 'nomic truth approximation'.
6. Similarity considerations are also included in the refined truth approximation of Kuipers (2000) and the quantitative version of the relevant consequence approach of Schurz and Weingartner (Schurz 2018).
7. Distance from the true (deterministic or probabilistic) law defines the notion of *legisimilitude*.
8. Following Niels Bohr's characterisation of the relation of classical mechanics and quantum mechanics, this requirement is often called 'the Principle of Correspondence', but it should not be confused with the correspondence theory of truth.
9. Note that van der Waals law approaches Boyle-Mariotte law, when a and b approach the value 0.
10. Adam Bricker (2018) defends a non-factive analysis of knowledge on the basis of cognitive science.

11. See also Bangu (2015), whose notion of understanding emphasises the role of unifying theories in scientific progress (cf. Niiniluoto 2018b).
12. Eino Kaila's notion of *relative simplicity* could be used as an explication of theoretical understanding. It is defined as the ratio between the explanatory power of a theory and its logical complexity. See Kaila (2014, 77–83) and Niiniluoto (1999, 167, 182).

References

Aronson, J. L., Harré, R. and Way, E. C. (1994) *Realism Rescued: How Scientific Progress Is Possible*, London: Duckworth.
Bangu, Sorin (2015) "Progress, Understanding, and Unification", in I. D. Toader, G. Sandu, and I. Parvu (eds.), *Romanian Studies in Philosophy of Science*, Cham: Springer, pp. 239–253.
Bird, Alexander (2007) "What Is Scientific Progress?", *Nous* 41: 92–117.
Bird, Alexander (2016) "Scientific Progress", in P. Humphreys (ed.), *The Oxford Handbook of Philosophy of Science*, Oxford: Oxford University Press, pp. 544–563.
Bricker, Adam (2018) *Visuomotor Noise and the Non-Factive Analysis of Knowledge*, Edinburgh: The University of Edinburgh.
Cevolani, Gustavo and Festa, Roberto (2020) "A Partial Consequence Account of Truthlikeness", *Synthese* 197: 1627–1646.
Cevolani, Gustavo and Tambolo, Luca (2013) "Progress as Approximation to the Truth: A Defence of the Verisimilitudinarian Approach", *Erkenntnis* 78: 921–935.
Chang, Hasok (2004) *Inventing Temperature: Measurement and Scientific Progress*, Oxford: Oxford University Press.
da Costa, Newton and French, Steven (2003) *Science and Partial Truth*, New York: Oxford University Press.
Dellsén, Finnur (2016) "Scientific Progress: Knowledge versus Understanding", *Studies in History and Philosophy of Science* 56: 72–83.
Dellsén, Finnur (2018) "Scientific Progress: Four Accounts", *Philosophy Compass* 13: e12525.
Dellsén, Finnur (forthcoming) "Understanding Scientific Progress", *Synthese*.
Douglas, Heather (2014) "Pure Science and the Problem of Progress", *Studies in History and Philosophy of Science* 46: 55–63.
Duhem, Pierre (1969) *To Save the Phenomena: The Idea of Physical Theory from Plato to Galileo*, Chicago: Chicago University Press.
Duhem, Pierre Maurice Marie (1954) *The Aim and Structure of Physical Theory*, Princeton, NJ: Princeton University Press. Translated by Philip P. Wiener.
Feyerabend, Paul (1975) *Against Method: Outline of an Anarchistic Theory of Knowledge*, London: New Left Books.
Feyerabend, Paul (1984) *Wissenschaft als Kunst*, Frankfurt am Main: Suhrkamp.
Garcia Lapeña, Alfonso (forthcoming) "Truthlikeness for Quantitative Deterministic Laws", *The British Journal for the Philosophy of Science*.
Hempel, Carl G. (1965) *Aspects of Scientific Explanation*, New York: The Free Press.
Kaila, Eino (2014) *Human Knowledge: A Classic Statement of Logical Empiricism*, Chicago: Open Court.

Kemeny, John and Oppenheim, Paul (1956) "On Reduction", *Philosophical Studies* 7: 6–19.

Kuhn, Thomas (1970) *The Structure of Scientific Revolutions*, 2nd ed. Chicago: The University of Chicago Press.

Kuipers, Theo (2000) *From Instrumentalism to Constructive Realism: On Some Relations between Confirmation, Empirical Progress, and Truth Approximation*, Dordrecht: Kluwer.

Kuipers, Theo (2009) "Comparative Realism as the Best Response to Antirealism", in Clark Glymour et al. (eds.), *Logic, Methodology and Philosophy of Science: Proceedings of the Thirteenth International Congress*, London: College Publications, pp. 211–240.

Kuipers, Theo (2019) *Nomic Truth Approximation Revisited*, Cham: Springer.

Lakatos, Imre and Musgrave, Alan (1970) *Criticism and the Growth of Knowledge*, Cambridge: Cambridge University Press.

Laudan, Larry (1977) *Progress and Its Problems: Toward a Theory of Scientific Growth*, London: Routledge and Kegan Paul.

Laudan, Larry (1984) *Science and Values: The Aims of Science and Their Role in Scientific Debate*, Berkeley: The University of California Press.

Levi, Isaac (1967) *Gambling with Truth: An Essay on Induction and the Aims of Science*, New York: Alfred A. Knopf.

Mizrahi, Moti and Buckwalter, Wesley (2014) "The Role of Justification in the Ordinary Concept of Scientific Progress", *Journal for General Philosophy of Science* 45: 151–166.

Niiniluoto, Ilkka (1979) "Verisimilitude, Theory-Change, and Scientific Progress", in Ilkka Niiniluoto and Raimo Tuomela (eds.), *The Logic and Epistemology of Scientific Change*, Acta Philosophica Fennica 30, Helsinki: The Philosophical Society of Finland, pp. 243–264.

Niiniluoto, Ilkka (1980) "Scientific Progress", *Synthese* 45: 427–464.

Niiniluoto, Ilkka (1984) *Is Science Progressive?* Dordrecht: D. Reidel.

Niiniluoto, Ilkka (1987) *Truthlikeness*, Dordrecht: D. Reidel.

Niiniluoto, Ilkka (1990) "Theories, Approximations, and Idealizations", in Jerzy Brzezinski et al. (eds.), *Idealization I: General Problems*, Amsterdam: Rodopi, pp. 9–57.

Niiniluoto, Ilkka (1999) *Critical Scientific Realism*, Oxford: Oxford University Press.

Niiniluoto, Ilkka (2000) "Scepticism, Fallibilism, and Verisimilitude", in Juha Sihvola (ed.), *Ancient Scepticism and the Sceptical Tradition*. Acta Philosophica Fennica 66. Helsinki: The Philosophical Society of Finland, pp. 145–169.

Niiniluoto, Ilkka (2007) "Evaluation of Theories", in Theo Kuipers (ed.), *Handbook of Philosophy of Science: General Philosophy of Science – Focal Issues*, Amsterdam: Elsevier, pp. 175–217.

Niiniluoto, Ilkka (2014) "Scientific Progress as Increasing Verisimilitude", *Studies in History and Philosophy of Science* 46: 73–77.

Niiniluoto, Ilkka (2017) "Optimistic Realism about Scientific Progress", *Synthese* 194: 3291–3309.

Niiniluoto, Ilkka (2018a) "Scientific Progress", in Juha Saatsi (ed.), *The Routledge Handbook of Scientific Realism*, Abington: Routledge, pp. 187–199.

Niiniluoto, Ilkka (2018b) *Truth-Seeking by Abduction*, Cham: Springer.

Niiniluoto, Ilkka (2019) "Scientific Progress", in E. Zalta (ed.), *The Stanford Encyclopedia of Philosophy*, Winter 2019 ed. https://plato.stanford.edu/entries/scientific-progress/

Oddie, Graham (1986) *Likeness to Truth*, Dordrecht: D. Reidel.

Oddie, Graham (2016) "Truthlikeness", in E. Zalta (ed.), *The Stanford Encyclopedia of Philosophy*, Winter 2016 ed. https://plato.stanford.edu/entries/truthlikeness/

Park, Seungbae (2017) "Does Scientific Progress Consist in Increasing Knowledge or Understanding?", *Journal for General Philosophy of Science* 48: 569–579.

Peirce, Charles S. (1931–35) *Collected Papers*, ed. by C. Hartshorne and P. Weiss, Cambridge, MA: Harvard University Press, pp. 1–6.

Poincaré, Henri (1952) *Science and Hypothesis*, New York: Dover.

Popper, Karl R. (1959). *The Logic of Scientific Discovery*, London: Hutchinson.

Popper, Karl R. (1963) *Conjectures and Refutations: The Growth of Scientific Knowledge*, London: Hutchinson.

Popper, Karl R. (1975) "The Rationality of Scientific Revolutions", in Rom Harré (ed.), *Problems of Scientific Revolution*, Oxford: Oxford University Press (Also in Hacking, 1981, 80–106).

Putnam, Hilary (1981) *Reason, Truth, and History*, Cambridge: Cambridge University Press.

Rescher, Nicholas (1977) *Methodological Pragmatism*, Oxford: Blackwell.

Rowbottom, Darrell (2008) "N-Rays and the Semantic View of Scientific Progress", *Studies in History and Philosophy of Science* 39: 277–278.

Rowbottom, Darrell (2015) "Scientific Progress without Increasing Verisimilitude: In Response to Niiniluoto", *Studies in History and Philosophy of Science Part A* 51: 100–104.

Sarton, George (1957) *The Study of the History of Science*, New York: Dover.

Schurz, Gerhard (2018) "Truthlikeness and Approximate Truth", in Juha Saatsi (ed.), *The Routledge Handbook of Scientific Realism*, Abington: Routledge, pp. 133–148.

Sellars, Wilfrid (1968) *Science and Metaphysics*, London: Routledge and Kegan Paul.

Shan, Yafeng (2019) "A New Functional Account of Scientific Progress", *Philosophy of Science* 86: 739–758.

Stanford, P. Kyle (2006) *Exceeding Our Grasp: Science, History, and the Problem of Unconceived Alternatives*, Oxford: Oxford University Press.

Stegmüller, Wolfgang (1976) *The Structure and Dynamics of Theories*, New York: Springer.

van Fraassen, Bas (1980) *The Scientific Image*, Oxford: Oxford University Press.

3 The Functional Approach
Scientific Progress as Increased Usefulness

Yafeng Shan

[E]very new discovery has led to new problems and new methods of solution, and opened up new fields for exploration.

(Bury 1920, 3)

1. Introduction

When talking of the functional approach, one tends to think of Thomas Kuhn's or Larry Laudan's account of scientific progress. Both emphasise the significance of problem-solving in science. For Kuhn (1970b) and Laudan (1977, 1981), science progresses if more problems are solved or problems are solved in a more effective and efficient way. This is probably why the functional approach is also sometimes called the problem-solving approach.

The principal proponents of the problem-solving approach to progress are Thomas Kuhn [...] and Larry Laudan [...].

(Bird 2016, 546)

There are other approaches to scientific progress in the literature. They are the semantic approach [...] and the problem-solving approach [...].

(Park 2017, 570)

Kuhn's idea is fleshed out by Larry Laudan in his problem-solving account of scientific progress.

(Dellsén 2018, 2)

Accordingly, the key feature of the functional approach has often been summarised as problem-solving.

The principal representatives of [the functional approach] are the puzzle- and problem-solving views of Kuhn and Laudan ... This view

DOI: 10.4324/9781003165859-5

is functional because it takes progress to be a matter of the success a scientific field has in fulfilling a function – that of solving problems.

(Bird 2007, 67)

Each account places its own distinctive type of cognitive achievement at the heart of scientific progress – truthlikeness, problem-solving, knowledge, or understanding.

(Dellsén 2018, 2)

Unfortunately, this is highly problematic to identify the functional approach with the problem-solving approach. It might be true that Kuhn's and Laudan's approaches are better known than other functional approaches, but it is incorrect to claim that the functional approach is just the problem-solving approach. Other representatives include the Popper-Lakatos functional approach (Popper 1963; Lakatos 1978) and my new functional approach (Shan 2019, 2020a). There is a danger of conflating the functional approach with the problem-solving approach: it seems to many that the functional approach is simply indefensible as both Kuhn's and Laudan's problem-solving approaches face serious challenges. In this chapter, I defend the functional approach to scientific progress. In Section 2, I critically examine two traditional versions of the functional approach. In Section 3, I elaborate my new functional approach. In Section 4, I argue that my new functional approach is better than the epistemic, semantic, and noetic approaches. In Section 5, I address two objections to my approach.

2. Traditional Functional Approaches

The most influential representative of the functional approach is first proposed by Kuhn (1962, 1970a) and further developed by Laudan (1977, 1981). Kuhn (1970b, 164) argues that the nature of scientific progress is the increase of 'both the effectiveness and the efficiency with the group as a whole solves new problems'. Laudan (1981, 145) is also explicit on the point that 'science progresses just in case successive theories solve more problems than their predecessors'. Kuhn and Laudan differ in the explication of problem-solving, though. For Kuhn (1970b, 189–191), a problem P is solved if its solution is sufficiently similar to a relevant paradigmatic problem solution. For Laudan (1977, 22–23), a problem P is solved by a theory T if T entails an approximate statement of P. Nevertheless, both Kuhn and Laudan maintain that scientific progress is nothing to do with truth or knowledge if truth or knowledge is construed in a classical way. More specifically, whether a problem is solved is independent of whether the paradigmatic solution assumes any paradigm-dependent truth (for Kuhn), or whether the background theory is true (for Laudan). Since the acceptance of a problem solution is determined independently of external factors like truth or knowledge, whether a progress is achieved can be judged by the scientific community itself. Thus, as I have summarised in some earlier work (Shan 2019, 2020a), there are four central tenets of the Kuhn-Laudan functional approach to scientific progress.[1]

T1. Scientific progress is solely determined by the problem-solving power.

T2. The problem-solving power is assessed by the amount and significance of the problems solved.

T3. The problem-solving power is independent of whether the solution is true or knowledge.

T4. Scientific progress is judged and known by the scientific community.

The Kuhn-Laudan approach has weathered a great deal of criticisms. As I have summarised in Shan (2019, 2020a), there are four main objections, which correspond to the four central tenets respectively. T1 faces the problem of sufficiency: an accumulation of problem solutions does not guarantee progress in science. As Alexander Bird (2007, 69–70) argues, it seems implausible for many to accept that there is an on-going progress in science, as the false solution statements (derived from the false theory) accumulate. T2 encounters the problem of quantitative weighing: it is difficult to find a proper quantitative way to identify and calculate the problems of different significance. T3 is challenged by the problem of counter-intuition: it seems counter-intuitive to many that scientific progress is conceptually independent of truth or knowledge. T4 is susceptible to the problem of internalism: the Kuhn-Laudan approach implies that a scientific community well recognises whether it is making progress or not by examining its problem-solving power, but this is difficult to hold from a historical point of view. It is not unusual for a scientific community to overlook the significance of some scientific work.[2] These problems are so serious that it seems to be a difficult task for one to defend the Kuhn-Laudan approach. To some extent, it is no wonder the Kuhn-Laudan approach has been taken for granted indefensible.

However, as I have emphasised, the Kuhn-Laudan approach should not be conflated with the functional approach. Therefore, the functional approach should not be simply rejected or neglected just because there are many problems of the Kuhn-Laudan approach.

Another representative of the functional approach is rooted in the work of Karl Popper (1959, 1963) and mainly developed by Imre Lakatos.[3] For Popper (1963, 217), the criterion of scientific progress is testability: a scientific theory is progressive if it 'contains the greater amount of empirical information or content', 'has greater explanatory and predictive power', and 'can therefore be more severely tested by comparing predicted facts with observations'. Popper's criterion consists of two requirements.

Logical requirement: a progressive theory should have more falsifiable content.

Empirical requirement: a progressive theory should pass new and severe tests.

Following Popper's idea, Lakatos (1978, 33–34) develops his functional account of scientific progress: a research programme is progressive if it generates novel and well corroborated predictions. Inspired by Popper's two requirements, Lakatos carefully distinguishes two types of progress in science: theoretical progress and empirical progress. If a research programme is merely generating uncorroborated novel predictions, it only counts as theoretical progress. Only when novel predications are corroborated by experiments, a research programme is making empirical progress. The real sense of scientific progress, for Lakatos, consists of both theoretical progress and empirical progress.

That is, I give for criteria of progress and stagnation within a programme and also rules for 'elimination' of whole research programmes. A research programme is said to be progressing as long as its theoretical growth anticipates its empirical growth, that is, as long as it keeps predicting novel facts with some success.

(Lakatos 1978, 112)

It is evident that there is a crucial difference between the Popper-Lakatos and Kuhn-Laudan approaches. They differ in the key function of science. For Kuhn (and Laudan to a less extent), science is basically about problem-solving. In contrast, Popper and Lakatos maintain that falsifiability is the key virtue of science and highlight the significance of novel predictability.[4] Such a difference makes the Popper-Lakatos approach less vulnerable to some challenges that the Kuhn-Laudan approach faces. It is clear that the problem of quantitative weighing is inapplicable, as the Popper-Lakatos approach does not define scientific progress in terms of problem-solving. Neither is there a problem of counter-intuition for the Popper-Lakatos approach. Novel predictability is arguably related to verisimilitude or truthlikeness. It is not very clear if the Popper-Lakatos approach suffers a problem of sufficiency. At least, it is not simply undermined by Bird's thought experiment (2007, 69–70).

That said, the Popper-Lakatos approach does face the problem of internalism, like the Kuhn-Laudan approach. Whether a research programme generates novel and corroborated predictions can be easily judged by a scientific community. In other words, the Popper-Lakatos approach is as internalist as the Kuhn-Laudan approach is.[5] But it would be too hasty for one to conclude that the functional approach is seriously challenged by the problem of internalism. In the next section, I shall introduce a non-internalist functional approach.

3.   The New Functional Approach

Recently, I have developed a new functional approach (Shan 2019, 2020a). In a nutshell, I define scientific progress as follows:

> *Science progresses if and only if more useful exemplary practices are proposed.*

This approach shares two basic assumptions behind the traditional functional approaches.

> A1. Scientific progress should be analysed and assessed in a holistic way.
> A2. Scientific progress is determined by the fulfilment of key functions of science.

Both the Kuhn-Laudan and Popper-Lakatos approaches assume that the nature of scientific progress is about how a scientific community fulfils the key functions of science. They both maintain that the unit of analysis for examining scientific progress should be a community-based consensus, though they differ in how to characterise it. The unit of analysis in the Kuhn-Laudan approach is a paradigm or a research tradition, while the unit of analysis in the Popper-Lakatos approach is a research programme. Thus, scientific progress cannot be applied to characterise some improvement of or advance in particular scientific activities. For example, it makes little sense to say that Galileo's improvement of telescope was progressive. In contrast, it only makes sense to argue that Galileo's improvement of telescope contributed to a progressive paradigm or research programme. I am sympathetic to this view that scientific progress should be analysed and assessed in a holistic rather than a piecemeal way.

However, my new functional approach differs from two traditional functional approaches in two main aspects: the unit of analysis and the key virtue of science. The unit of analysis in the new functional approach is an exemplary practice, which is defined as a particular way of problem-defining and problem-solving, typically by means of problem-proposing, problem-refining, problem-specification, conceptualisation, hypothesisation, experimentation, and reasoning (Shan 2020b). It should be highlighted that an exemplary practice is different from a paradigm, a research tradition, or a research programme. The former is an example of the unit of micro-scientific consensus, while the latter are cases of the unit of macro-scientific consensus.[6] A marco-scientific consensus is something general or universal, invariantly shared by the members of a scientific community, such as theories, laws, and models. For example, for Kuhn, the paradigm of Newtonian mechanics consists of universal generalisations like $F = ma$, while for Lakatos, the research programme of Newtonian mechanics includes Newton's three laws of motion and the law of gravitation as the hard core. In contrast, a unit of micro-scientific consensus is something local and context-dependent. Gregor Mendel's work on the development of pea hybrids is such a case. It was accepted by early Mendelians in the 1900s. It is worth noting that what was accepted is (at least some components of) Mendel's particular way of problem-defining and problem-solving rather than the generality of Mendel's laws of development. Such a difference suggests an advantage of the new functional approach: it better captures the actual cases in history. It is often difficult to identify the content of a macro-scientific consensus. For example, it is natural to identify the theory of evolution by natural selection as the hard core of the Darwinian research programme, but it is extremely difficult to articulate what the theory of evolution by natural selection is. In particular, it is not an easy task to characterise an account of that theory which was invariantly shared by the members of the research programme. This is a serious problem for the traditional functional approaches which analyse scientific progress in terms of macro-scientific consensus. But it is not a problem for the new functional approach. As I have shown in Mendel's case (Shan 2020a), it is not very difficult to identify the content of a micro-consensus among the members of a community. Early Mendelians did differ in the formulation of the Mendelian laws, but they all accepted Mendel's exemplary practice, which provides conceptual tools, experimental guidelines, problems, and patterns of reasoning for the study of heredity (Shan 2020a).

The other important difference is that I argue that the key virtue of science is not problem-solving success or predictive novelty, but usefulness. The definition of usefulness is as follows:

> *An exemplary practice is useful if and only if its way of defining and solving research problems is repeatable and provides a reliable framework for further investigation to solve problems and to generate novel research problems across different areas (or disciplines).*

The notion of usefulness encompasses four virtues: repeatability, problem-defining novelty, problem-solving promise, and interdisciplinarity. The repeatability of the way of defining and solving research problems is a prerequisite for the recognition of its usefulness. Consider the case of the Mendelian-Biometrician controversy in the history of genetics.[7] One main reason for W. F. R. Weldon (1902a, 1902b) to resist the Mendelian approach to the study of heredity was that he failed to repeat

Mendel's conceptualisation of dominance and recessiveness, that is his way of distinguishing the dominant and recessive characters. In contrast, Carl Correns' acceptance of (the usefulness of) the Mendelian approach (1900) was based on his successful repetition of Mendel's practice, including problem-defining, problem-refining, conceptualisation, hypothesisation, reasoning, and experimentation. Moreover, the mixed reception of the Mendelian approach in the first decade of the twentieth century was also due to the different results of the application of the Mendelian approach to studying the transmission of characters in different species. Hugo de Vries' acceptance of the Mendelian approach (1900a, 1900b) was because that it was successfully applied to study the transmission of the morphological traits in various plant species (e.g. *Lychnis*, *Papaver*, and *Solanum*), while the scepticism arose from the unfavourable results of the application (e.g. Whitman 1904; McCracken 1905, 1906, 1907; Reid 1905; Prout 1907; Saunders 1907; Hart 1909; Holmes and Loomis 1909). Problem-solving promise has been widely acknowledged as a virtue of scientific practice (e.g. Kuhn 1970b; Laudan 1977; Nersessian 2008), while problem-defining novelty is also recently highlighted (Shan 2020b, 2020a). In addition, interdisciplinarity is another important virtue in scientific practice. Science advances with so many interactions of different disciplines, for example, astrophysics, biochemistry, and bioinformatics.[8] The interdisciplinarity of an exemplary practice helps to widen the scope and explore the novel lines of scientific inquiry. Mendel's work on pea hybrid development (1866) is a good example of an useful exemplary practice. As I have shown in greater detail (Shan 2020a), Mendel's exemplary practice introduced in his study of pea hybrid development was useful in the sense that it was repeatable in practice and provided the foundation for the twentieth century study of heredity to solve the problems of transmission of the morphological traits of other species and to generate more potential testable research problems across the areas like cytology, evolution, and heredity.

In addition, it should be noted that usefulness is community-dependent. A particular exemplary practice might be taken as useful by some scientific communities but not others. The Mendelian-Biometrician controversy in the first decade of the twentieth century well illustrates this point. The Mendelians, led by William Bateson, were optimistic on the future of the Mendelian approach to the study of heredity, while sceptics, including Weldon and Karl Pearson, doubted the usefulness of Mendel's approach (especially its conceptualisation, hypothesisation, and experimentation) to study the phenomena of heredity. In other words, Weldon and Pearson overlooked the 'progressive' element of the Mendelian approach due to their failure of the recognition of its usefulness. Thus, the usefulness of a given exemplary practice might not be obvious to a scientific community. The progress thus achieved is not judged or known by the community. It is in this sense that my approach is not internalist.

As I have argued in Shan (2019), this new functional approach well resolves the four main problems of the Kuhn-Laudan functional approach. Firstly, I have argued that whether science progresses depends on whether more useful exemplary practices are proposed. In order to determine whether there is progress in science, one has to examine whether there is a new exemplary practice which provides a more reliable framework to solve unsolved problems, and whether it proposes more novel and testable problems across more areas. Given such a qualitative notion of usefulness, there is no need to look for a quantitative framework to calculate and weigh the significance and amount of the problems. Thus, the problem of quantitative weighing is inapplicable to my functional approach. Secondly, my functional approach is not internalist. Bird construes the Kuhn-Laudan approach as internalist in the sense that scientific progress is only judged and known by a community, independent of any features unknown to them. However, this does not apply to my approach. The usefulness of an exemplary practice is not straightforwardly recognisable by the scientific community, as I just illustrated in the Mendelian-Biometrician controversy. Thirdly, I contend that by highlighting the significance of problem-defining, the problem of sufficiency is resolved. Fourthly, the problem of intuition can also be solved. As I mentioned, the functional approach is somehow neglected in the recent debate for its conflict with the intuition that scientific progress is about knowledge and/or truth. However, I argue that my functional approach can be compatible with this intuition. Usefulness of an exemplary practice could be *somehow* interpreted in terms of knowledge if knowledge is not merely construed as something propositional or theoretical. Knowledge is traditionally classified into know-that and know-how.[9] A particular way of problem-defining and problem-solving can be argued as a case of know-how. Thus, that more useful exemplary practices are proposed could be understood in the sense that more useful know-how is obtained. Moreover, usefulness of exemplary practices can be understood in terms of truth *to some extent*. In particular, it is well explained by the 'contextualist' theory of truth (Chang 2012; Massimi 2018). Michela Massimi (2018), for example, proposes that truth in the context of scientific practice should be defined in a perspectival way.

> Knowledge claims in science are [perspective-dependent] when their *truth-conditions* (understood as rules for determining truth-values based on features of the context of use) depend on the scientific perspective in which such claims are made. Yet such knowledge claims must also be assessable from the point of view of other (subsequent or rival) scientific perspectives.
>
> (Massimi 2018, 354)

If truth is defined in this perspectival way, then the increase of the usefulness of an exemplar practice implies a reliable framework with more novel problems and more confirmable hypotheses. The Mendelian approach to the study of heredity, for example, generated more confirmable hypotheses (e.g. the law of segregation) and factual knowledge (e.g. the summary of the transmission of morphological traits of various plants). All these hypotheses and factual knowledge are 'true' according to its perspective (i.e. the Mendelian approach) by means of experiments, while they are also assessable from the point of view of the subsequent scientific perspective (e.g. the Morgan approach) by new ways of experimentation. Therefore, that more useful exemplary practices are proposed could be interpreted as that more perspective-dependent true knowledge claims are attained.[10]

To sum up, I argue that new functional approach is better than both the Kuhn-Laudan and Popper-Lakatos functional approaches to respond the four main objections.

4.   Transcending Knowledge, Truth, and Understanding

In this section, I argue that my new functional approach is better than its rival approaches. There are three other main approaches to scientific progress: the epistemic approach (e.g. Bird 2007), the semantic approach (e.g. Niiniluoto 1980, 2014), and the noetic approach (e.g. Dellsén 2016, 2021). The epistemic approach defines scientific progress in terms of knowledge. The semantic approach construes scientific progress in terms of truthlikeness. The noetic approach characterises scientific progress in terms of understanding. According to the epistemic approach, science progresses if scientific knowledge accumulates. According to the semantic approach, science progresses if more scientific truths are obtained or scientific theories are approximating truths. According to the noetic approach, science progresses if there is an increased understanding of some phenomena, where understanding is defined in terms of accuracy and comprehensiveness of dependency models.[11] I will argue that the new functional approach has an important advantage over these approaches: it provides a fuller picture of progress in the history of science.

It has been shown that the new functional approach better accounts for progress in genetics (Shan 2019, 2020a) and conservation biology (Justus and Wakil 2021) than the epistemic and semantic approaches.[12] In particular, I argue that the new functional approach is better than the epistemic, semantic, and noetic approaches in accounting for the non-theoretical aspect of scientific progress. All of the epistemic, semantic, and noetic approaches to scientific progress pay too much attention to theoretical achievements: is there more propositional knowledge than before? Are our current best scientific theories more truthlike? Do we have more accurate and comprehensive dependency models? However, such approaches downplay the significance of the non-theoretical aspect of science. Scientific practice is much more than theorising or modelling.

It would be surprising if the introduction of new research problems and the improvement of the experimental methods and devices are excluded from the constituents of scientific progress. Reconsider the case of the origins of genetics. It seems plausible to argue that the progress made by Mendel was to propose the law of composition of hybrid fertilising cells to advance our knowledge of the mechanism of heredity. Similarly, de Vries' law of segregation, Correns' Mendelian rule, and Bateson's Mendelian principles provide a better knowledge of heredity than Mendel's law. Accordingly, it can be argued that we knew more and more about the mechanism of heredity with the theoretical development from Darwin to Bateson. However, it is definitely not the only aspect of the progress achieved in the study of heredity in that period. We learnt more and more about how to define and refine good research problems, how to design and undertake good experiments, and how to use these problems and experiments to study the mechanism of heredity in a better way. In a word, the non-theoretical aspect of scientific progress should be taken into account as well as the theoretical aspect. As Heather Douglas (2014, 56) points out, science is not just about theory. Hence, scientific progress should be examined in both theoretical and non-theoretical aspects. In particular, not all of these non-theoretical activities can easily be accounted for in terms of propositional knowledge, theories, or dependency models. Thus, I contend that my functional approach provides a better account of the non-theoretical aspect of scientific progress.

In addition, the epistemic, semantic, and noetic approaches are even more problematic when they are applied to some sciences in which theorising is not a key task. As James Jutus and Samatha Wakil argue,

> [The epistemic and semantic approaches] seem utterly ill-equipped to account for progress in applied, ethically-driven sciences. These sciences don't deliver anything resembling justified true beliefs about a mind-independent cosmos, at least as that idea is usually philosophically expressed about, say, particle physics. Instead, they supply data-driven, evidence-based, and in the present instance algorithmically-rigorous means for achieving ethical goals.
>
> (Justus and Wakil 2021, 189)

Such an objection is also applied to the noetic approach, since understanding is determined by accuracy and comprehensiveness of dependency models. In contrast, the new functional approach better accounts for progress in these sciences. It can well characterise the main activities in applied, ethically driven sciences (i.e. 'delivering scientific insights and tools that promote achieving ethical goals') in terms of problem-defining and problem-solving.

Moreover, all the epistemic, semantic, and noetic approaches focus too much on the explanatory activities in scientific practice. This is rooted in a theory-centric view widely received in philosophy of science. As C. Kenneth Waters summarises:

> Philosophers (perhaps I should say we) typically analyze [science] by identifying central explanatory theories. Then for each theory, we analyze its central concepts and principles (or laws), detail how it can be applied to explain the phenomena, reconstruct how it is justified, explore how it might be further developed or how its explanatory range might be extended (the so-called 'research program'), and consider how it should be interpreted (for example, instrumentally or realistically).
>
> (Waters 2004, 784)

However, this kind of philosophical analysis overlooks the significance of investigative or exploratory activities in scientific practice. Science not only aims to explain puzzling phenomena, but also aims to investigate them 'towards open-ended research' (Waters 2004, 786). As Waters (2004, 786) insightfully points out, scientific practice often 'aimed towards developing knowledge about phenomena which fall outside the domain, even the potential explanatory domain, of any existing theory'. It is evident that none of the epistemic, semantic, and noetic approaches captures this investigative aspect of scientific progress. In contrast, by highlighting the significance of problem-defining, the new functional approach sheds light on the investigative or exploratory feature of scientific practice.

5. Objections and Responses

In this section, I address two objections to my new functional approach.

5.1. Problem of Compromise

One objection arises from a concern that the new functional approach seems to be too friendly to the epistemic and semantic approaches. As my reply to the problem of counter-intuition involves the notions of truth and knowledge, one may wonder whether my approach can be still classified as 'functional' rather than epistemic or semantic. It seems that my new functional approach can be reinterpreted in the way that science progresses if more and more know-how is attained, or that science progresses if more perspective-dependent truths are obtained. As Ilkka Niiniluoto complains:

> [Shan] is open to the introduction of the notions of know-how and perspectival truth, so that his "new functional approach" is a compromise with what Bird (2007) calls the "epistemic view" of progress. (Niiniluoto 2019)

In response, I argue that my functional approach is not reducible to the epistemic or the semantic approach. As I have pointed out, the solution to a research problem is not something purely theoretical or propositional. There are some indispensable non-theoretical aspects. For example, problem specification and experimentation cannot be reduced to knowledge, truth, or other theoretical elements. Accordingly, I do not see that there is any true or correct solution to a research problem, given its practical nature. For example, it is implausible to claim that there is a true or correct way of experimentation or problem-refining. Moreover, I would like to highlight that my approach aims to capture the multiple facets of scientific progress, but it does not imply that the functional, epistemic, and semantic aspects constitute the nature of scientific progress in equal shares. Rather more know-how (the epistemic aspect) and more well-corroborated hypotheses (the semantic aspect) may only partially constitute usefulness of problem-defining and problem-solving (the functional aspect). In short, both the epistemic and semantic accounts of scientific progress can be explained by the new functional approach, but the new functional account cannot be fully accounted for in terms of knowledge or truthlikeness. Therefore, there is no compromise between the epistemic and the new functional approaches. Nor is the distinction between the epistemic and new functional approaches blurred. As I have argued in Section 4, the new functional approach is better than both the epistemic and semantic approaches.

5.2. *Problem of Problem-Solving Centrism*

Another objection is that the new functional approach still assumes the central role of problem-solving in scientific practice, so it will be challenged by the similar problems that the Kuhn-Laudan approach encounters.

> Yafeng Shan (2019) has recently offered a version of the functional-internalist approach that seems to me to be an improvement on the Kuhn – Laudan version. Nonetheless, I think it still fails on the point raised in this paragraph, that not all progress involves solving a problem.
>
> (Bird 2022, 42 f2)

> In Shan's new functional approach to scientific progress, too, the notion of the *solved problem*, which is the basic unit of scientific progress according to Laudan's problem-solving model of progress, is central.
>
> (Mizrahi 2022, 3)

However, as I have highlighted in Section 3, problem-defining and problem-solving are two mutually intertwined activities. They cannot be analysed in isolation in the examination of scientific progress. Unlike the Kuhn-Laudan approach, my new functional approach does not view science as an essentially problem-solving enterprise. Accordingly, problem-solving success is not central to the new functional approach, while usefulness of exemplary practices is.

6. Conclusion

In this chapter, I have argued that the functional approach should not be conflated with the Kuhn-Laudan functional approach. There are other versions of the functional approaches, such as the Popper-Lakatos approach and my new functional approach. I have also argued that my new functional approach is the most promising version of the functional approach. I have shown that this new functional approach is immune to the main objections to old functional approaches. Moreover, I have argued that the new functional approach is better than the epistemic, semantic, and noetic approaches by providing a fuller picture of scientific progress. In a word, scientific progress is best characterised in terms of usefulness of exemplary practices.

Notes

1. The differences between Kuhn's and Laudan's approaches are marginal, whereas their similarities are quite fundamental. Thus, it is plausible to regard Kuhn's and Laudan's approaches as the same version of the functional approach.
2. A typical example is the neglect of Mendel's work. For an in-depth analysis, see Shan (2020a Chapter 7).
3. It should be noted that Popper (1963) also developed the concept of verisimilitude which later played a central role in the semantic approach to scientific progress (see Chapter 3). Thus, it can be argued that Popper pioneered both the semantic and functional approaches.
4. The Popper-Lakatos approach is largely neglected in the recent discussion on scientific progress. Only Bird (2007) and Shan (2019, 2020a) briefly mention it as a version of the functional approach. What is worse, there is little detailed examination of it. Bird (2007, 67) simply dismisses it without argument: 'Much of what I have to say will apply to Lakatos's methodology of scientific research programmes also'. However, as I shall argue, the Popper-Lakatos approach is immune to some objections to the Kuhn-Laudan approach.
5. It seems appropriate for Bird (2007, 67) to call these approaches 'functional-internalist'.
6. For more discussion on the distinction between macro- and micro-scientific consensus, see Shan (forthcoming).
7. For an overview of the Mendelian-Biometrical controversy, see Mackenzie and Barnes (1975) and Shan (2021).
8. See more discussion on interdisciplinary progress in Chapter 19.
9. Jason Stanley and Timothy Williamson (2001) famously reject this distinction by arguing that know-how is reducible to know-that. Whether there is a genuine distinction between how-that and know-how, my point still holds.

Science does not only tell us something theoretical which can be formulated in the propositions, but also tell us something practical, whether which can be reformulated in the propositions or not.

10. It should be highlighted that the notion of usefulness can be explicated by the contextualist theory of truth does not imply that my functional approach assumes a contextualist theory of truth. It does not eliminate the possibility that it can also be explicated by other theories of truth.

11. This is Finnur Dellsén's most recent formulation of his noetic approach (Dellsén 2021), which is different from his early formulation (Dellsén 2016). According to the early formulation, 'Science makes (cognitive) progress precisely when scientists grasp how to correctly explain or predict more aspects of the natural world than they did before' (Dellsén 2016, 75). Dellsén's early version is more similar to the functional approach, while his recent one is closer to the semantic approach.

12. It is also argued that my functional approach well characterises progress in other scientific disciplines, such as economics (see Chapter 12), seismology (see Chapter 9), and interdisciplinary sciences (see Chapter 19).

References

Bird, Alexander. 2007. "What Is Scientific Progress?" *Noûs* 41 (1): 64–89.

———. 2016. "Scientific Progress." In *The Oxford Handbook of Philosophy of Science*, edited by Paul Humphreys, 544–563. New York: Oxford University Press.

———. 2022. *Knowing Science*. Oxford: Oxford University Press.

Bury, J. B. 1920. *The Idea of Progress*. New York: Macmillan.

Chang, Hasok. 2012. *Is Water H2O? Evidence, Realism and Pluralism*. Dordrecht: Springer.

Correns, Carl. 1900. "G. Mendels Regel Über Das Verhalten Der Nachkommenschaft Der Rassenbastarde." *Berichte Der Deutschen Botanischen Gesellschaft* 18 (4): 158–168.

Dellsén, Finnur. 2016. "Scientific Progress: Knowledge versus Understanding." *Studies in History and Philosophy of Science* 56: 72–83.

———. 2018. "Scientific Progress: Four Accounts." *Philosophy Compass* 13 (11): e12525.

———. 2021. "Understanding Scientific Progress: The Noetic Account." *Synthese* 199: 11249–11278.

Douglas, Heather. 2014. "Pure Science and the Problem of Progress." *Studies in History and Philosophy of Science* 46: 55–63.

Hart, D. Berry. 1909. "Mendelian Action on Differentiated Sex." *Transactions of the Edinburgh Obstetrical Society* 34: 303–357.

Holmes, S. J., and H. M. Loomis. 1909. "The Heredity of Eye Color and Hair Color in Man." *Biological Bulletin* 18: 50–56.

Justus, James, and Samantha Wakil. 2021. "The Algorithmic Turn in Conservation Biology: Characterizing Progress Inethically-Driven Sciences." *Studies in History and Philosophy of Science* 88: 181–192.

Kuhn, Thomas Samuel. 1962. *The Structure of Scientific Revolutions*. 1st ed. Chicago, IL: The University of Chicago Press.

———. 1970a. "Logic of Discovery or Psychology of Research?" In *Criticism and the Growth of Knowledge*, edited by Imre Lakatos and Alan Musgrave, 1–23. Cambridge: Cambridge University Press.

———. 1970b. *The Structure of Scientific Revolutions*. 2nd ed. Chicago, IL: University of Chicago Press.

Lakatos, Imre. 1978. "Falsification and the Methodology of Scientific Research Programmes." In *The Methodology of Scientific Research Programme*, edited by John Worrall and Gregory Currie, 8–101. Cambridge: Cambridge University Press.

Laudan, Larry. 1977. *Progress and Its Problems: Toward a Theory of Scientific Growth*. Berkeley and Los Angeles, CA: University of California Press.

———. 1981. "A Problem-Solving Approach to Scientific Progress." In *Scientific Revolutions*, edited by Ian Hacking, 144–155. Oxford: Oxford University Press.

MacKenzie, Donald A., and S. Barry Barnes. 1975. "Biometriker versus Mendelianer. Eine Kontroverse Und Ihre Erklärung." In *Wissenschaftssoziologie: Studien Und Materialien*, edited by Nico Stehr and René König, 165–196. Wiesbaden: VS Verlag für Sozialwissenschaften.

Massimi, Michela. 2018. "Four Kinds of Perspectival Truth." *Philosophy and Phenomenological Research* 96 (2): 342–359.

McCracken, Isabel. 1905. "A Study of the Inheritance of Dichromatism in Lina Lapponica." *Journal of Experimental Zoology* 2: 117–136.

———. 1906. "Inheritance of Dichromatism in Lina and Gastroidea." *Journal of Experimental Zoology* 3: 321–336.

———. 1907. "Occurrence of a Sport in Melasoma (Lina) Scripta and Its Behavior in Heredity." *Journal of Experimental Zoology* 4: 221–238.

Mendel, Gregor. 1866. "Versuche Über Pflanzenhybriden." *Verhandlungen Des Naturforschenden Vereins Brünn* IV (1865) (Abhandlungen): 3–47.

Mizrahi, Moti. 2022. "What Is the Basic Unit of Scientific Progress? A Quantitative, Corpus-Based Study." *Journal for General Philosophy of Science*. https://doi.org/10.1007/s10838-021-09576-0.

Nersessian, Nancy J. 2008. *Creating Scientific Concepts*. Cambridge, MA: MIT Press.

Niiniluoto, Ilkka. 1980. "Scientific Progress." *Synthese* 45 (3): 427–462.

———. 2014. "Scientific Progress as Increasing Verisimilitude." *Studies in History and Philosophy of Science Part A* 46: 73–77.

———. 2019. "Scientific Progress." In *Stanford Encyclopedia of Philosophy*, edited by Edward N. Zalta. Winter 2019 ed. https://plato.stanford.edu/archives/win2019/entries/scientific-progress/.

Park, Seungbae. 2017. "Does Scientific Progress Consist in Increasing Knowledge or Understanding?" *Journal for General Philosophy of Science* 48: 569–579.

Popper, Karl. 1959. *The Logic of Scientific Discovery*. 1st ed. London: Hutchinson & Co.

———. 1963. "Truth, Rationality, and the Growth of Scientific Knowledge." In *Conjectures and Refutations: The Growth of Scientific Knowledge*, 215–250. London: Routledge and Kegan Paul.

Prout, Louis B. 1907. "Xanthorhoe Ferrugata (Clark) and the Mendelian Hypothesis." *Transactions of the Entomological Society of London for the Year 1906*: 525–531.

Reid, George Archdall. 1905. *The Principles of Heredity with Some Applications*. New York: E. P. Dutton & Co.

Saunders, Charles E. 1907. "The Inheritance of Awns in Wheat." In *Report of the Third International Congress on Genetics 1906*, 370–372. London: The Royal Horticultural Society.

Shan, Yafeng. 2019. "A New Functional Approach to Scientific Progress." *Philosophy of Science* 86 (4): 739–758.

———. 2020a. *Doing Integrated History and Philosophy of Science: A Case Study of the Origin of Genetics.* 1st ed. Boston Studies in the Philosophy and History of Science. Cham: Springer.

———. 2020b. "Kuhn's 'Wrong Turning' and Legacy Today." *Synthese* 197 (1): 381–406.

———. 2021. "Beyond Mendelism and Biometry." *Studies in History and Philosophy of Science* 89: 155–163.

———. forthcoming. "The Historiography of Scientific Revolutions." In *Handbook of the Historiography of Science*, edited by M. L. Condé & M. Salomon. Cham: Springer.

Stanley, Jason, and Timothy Williamson. 2001. "Knowing How." *The Journal of Philosophy* 98 (8): 411–444.

Vries, Hugo de. 1900a. "Das Spaltungsgesetz Der Bastarde (Vorlaufige Mittheilung)." *Berichte Der Deutschen Botanischen Gesellschaft* 18 (3): 83–90.

———. 1900b. "Sur La Loi de Disjonction Des Hybrides." *Comptes Rendus de l'Academie Des Sciences (Paris)* 130: 845–847.

Waters, C. Kenneth. 2004. "What Was Classical Genetics?" *Studies in History and Philosophy of Science Part A* 35 (4): 783–809.

Weldon, Walter Frank Rapheal. 1902a. "Mendel's Laws of Alternative Inheritance in Peas." *Biometrika* 1 (2): 228–254.

———. 1902b. "On the Ambiguity of Mendel's Categories." *Biometrika* 2 (1): 44–55.

Whitman, C. O. 1904. "Hybrids from Wild Species of Pigeons, Crossed Inter Se and with Domestic Races." *Biological Bulletin* 6: 315–316.

4 The Noetic Approach

Scientific Progress as Enabling Understanding

Finnur Dellsén

1. Introduction

What is it for science to make progress? When does a cognitive change in science, such as the replacement of one theory with another, constitute an improvement on what came before?

Consider, for example, the five successive models of the atom proposed by John Dalton, J.J. Thomson, Ernest Rutherford, Niels Bohr, and Erwin Schrödinger, respectively. According to Dalton's original atomism, atoms were considered the smallest possible units of matter, and thus indivisible. This was revised by Thomson, whose work on electricity led him to adopt a version of Lord Kelvin's hypothesis that atoms contain negatively charged electrons as well as a positively charged substrate that pervaded the atom. Rutherford then suggested that the positively charged part of the atom was highly concentrated in the atom's centre, that is in its nucleus. Bohr, with his quasi-quantum model of the atom, subsequently conjectured that the electrons were located on various orbits at fixed distances from the nucleus. Schrödinger revised Bohr's model by suggesting that although electrons are most likely to be found around Bohr's orbits, their locations are undefined prior to being observed.

Even if only Schrödinger's model is considered close to fully accurate at this point, each one of these models was an improvement on the previous one. But why? What exactly makes each development an instance of scientific progress? This question should be of interest to philosophers of science in part because it is almost universally agreed that science frequently makes progress, and that significant regress (the opposite of progress) rarely occurs in science. In many other human endeavours, for example arts and politics, many of us are less optimistic that humankind is generally making progress and very rarely regressing.[1] So the question 'What is scientific progress?' is in part an attempt to understand what makes science so successful as compared to other human endeavours.

However, while it is widely agreed that scientific progress is prevalent, there is much less agreement on what scientific progress consists in. Until relatively recently, philosophers had mostly given three main types of

DOI: 10.4324/9781003165859-6

answers to this question.[2] One of these answers is that progress amounts to getting closer to the truth, that is in increasing the truthlikeness (verisimilitude) of accepted theories. This is *the truthlikeness account*, a.k.a. the verisimilitudinarian account (Popper 1963; Niiniluoto 1984, 2014). Another answer is that progress amounts to having fewer or less important unsolved scientific problems, either by solving problems that arise within science or by eliminating or downgrading the importance of problems that perhaps cannot be solved. This is *the problem-solving account* (Kuhn 1970; Laudan 1977; see also Shan 2019).[3] The third answer is that progress consists in accumulating knowledge, where 'knowledge' – as per epistemological orthodoxy – minimally requires truth, belief, and some sort of epistemic justification. This is *the epistemic account* (Bird 2007b, 2016).

Although all of these accounts contain valuable insights into the nature of scientific progress, I maintain that they are only plausible in so far as they approximate a fourth account of scientific progress. On this account, scientific progress is defined in terms of increasing *understanding*: roughly, scientific progress regarding some phenomenon occurs when, and to the extent that, scientific research enables relevant members of society to better understand that phenomenon. Since the Greek word 'nous' is sometimes translated as 'understanding' (just as 'episteme' is often translated as 'knowledge'), I have called this view *the noetic account* of scientific progress.[4]

The main aim of this chapter is to articulate what I consider to be the most plausible version of the noetic account, and an associated view of scientific understanding. I will focus here on the positive project of building the account, as opposed to arguing against rival accounts – in part because I have done enough of the latter elsewhere (Dellsén 2016, 2018c, 2021b, 2022), in part because this will make it easier to rigorously evaluate the account (giving my opponents a better chance of refuting it!), and in part because I hope that discussions of various more concrete aspects of scientific progress (such as those considered in much of this volume) would benefit from being able to draw upon the details of a relatively fleshed-out account of scientific progress.

2. What We (Ought to) Talk About When We Talk About Scientific Progress

The truthlikeness, functional, epistemic, and noetic accounts are competing accounts. But what is that these accounts are meant to be accounts of? What is the question to which an account of scientific progress is meant to be the answer? There is some unfortunate confusion on this point in the literature, which in turn fuels an even more unfortunate scepticism about the topic itself. We are therefore forced to precisify what exactly accounts of scientific progress are meant to address before proceeding to build such an account. Accordingly, I will start my discussion with several preliminary points about the nature of our topic (see Dellsén 2021b for a fuller discussion of these).

First, accounts of scientific progress are meant to capture the ways in which various *mental states* (e.g. beliefs) and/or *representational devices* (e.g. theories) improve over time. It is difficult to be more precise than this, since different accounts of scientific progress will to some extent take quite different types of mental states/representational devices to be driving scientific progress. For example, while the epistemic account focuses on accumulation of a specific kind of mental state, namely knowledge, the truthlikeness account is normally formulated in terms of improvement in certain representational devices, namely theories. Even so, it should be clear that there are many ways in which science might improve that are not meant to be captured by extant accounts of scientific progress. Note, for example, that a scientific discipline might improve – and thus make progress, in a perfectly legitimate sense of the term – by becoming better funded than it was before, or by increasing opportunities for minorities and underprivileged groups. Although these types of improvements in science are certainly forms of progress in science, they are not our topic here since they don't concern changes in mental states and/or representational devices. For short, we may say that we are concerned only with *cognitive* scientific progress.

Second, it's important to keep in mind that progress is a partially evaluative concept – it refers to *improvement* over time, as opposed to mere change. It follows that accounts of scientific progress are at bottom normative claims, and that accepting a given account has normative implications for how science should proceed. To illustrate with a simple example, if an account of scientific progress implies that a given research project would not add at all to scientific progress even if it were successful, then, other things being equal, progress-seeking scientists should pursue other projects according to that account. More generally, since scientific progress is clearly something that comes in degrees (i.e. there can be *more* and *less* progress over a given episode), an account of scientific progress will deliver normative verdicts about how much progress would be made by a given research project if successful. These normative verdicts can then be used, in conjunction with the scientists' (degrees of) beliefs about whether a given project will be successful, as grounds for choosing which research projects to pursue.

Third, a related point is that accounts of scientific progress are not accounts of the meaning or use of the *words* 'scientific progress' or any of their cognates, in English or any other language. Due presumably to the influence of ordinary language philosophy, philosophers have an unfortunate tendency to test philosophical views against their linguistic intuitions, for example (in the current case) by asking whether we would be inclined or disinclined to describe a given episode with the words 'scientific progress'. Although there may be a role for pre-theoretic judgements (or 'intuitions') in debates about scientific progress, the relevant subset of these will not be concerned with meaning or usage of specific words. To see this most clearly, note that even if there were no such words as 'scientific progress' (in English or some other language), we could still debate (in that language) which kinds of cognitive changes in science are genuinely improvements on what came before.

Fourth, any account of scientific progress should be consistent with the truism that scientific progress can be made in many ways. This might seem to automatically invalidate all four accounts mentioned earlier, since each such account identifies a single type of cognitive change as scientifically progressive. Not so. To see why, let us introduce a distinction between promoting and constituting progress. A cognitive change *promotes* progress when the change is an improvement only in so far it brings about, or raises the probability of, progress at some later time.[5] By contrast, a cognitive change *constitutes* progress when the change is an improvement regardless of what other changes are thereby brought about, or are made more likely to be brought about, at some later time. This distinction mirrors that between *instrumental* and *intrinsic* value, where instrumentally valuable things are valuable in so far as they (eventually) lead to something of intrinsic value.[6]

Finally, it is often helpful to distinguish a general notion of *overall progress*, that is progress with respect to all topics or phenomena studied in science, from a more specific notion of *progress-on-X*, that is progress with respect to some particular topic or phenomenon X.[7] Although many authors sometimes write as if they are concerned with the former,[8] the examples and arguments they give often betray that it is the latter that is their immediate concern. In particular, when they put forward a particular episode of cognitive change as a putative instance of scientific progress, these authors do not usually (if indeed ever) consider the possibility that science has regressed on other topics or phenomena, leading to an overall regress or flatlining of science during the period in question. In my opinion, this is just as well, since we should indeed be most directly concerned with accounting for progress on a particular topic or phenomenon (progress-on-X) rather than progress on all topics or phenomena (overall progress). This is partly because discussing scientific progress would otherwise be rather unmanageable, since we would have to consider all scientific changes that occur at over a given time period in order to say whether the episode is progressive.

Another reason to focus on progress-on-X rather than overall progress is that it allows us to set aside, for the time being at least, difficult questions about how to balance progress on one topic or phenomenon, with progress on another, in an account of the overall progress during some period. For example, making scientific progress on COVID-19 is (at the time of writing) presumably much more significant than making progress on *misophonia*, the obsessive preoccupation with noises made by other humans (e.g. chewing sounds). Thus, it seems plausible that a given amount of progress on COVID-19 should be taken to contribute more to overall progress than would the same amount of progress made on misophonia, simply because COVID-19 is (currently) the much more significant topic.[9] But why exactly is the former more significant than the latter, and how much more does the former therefore contribute to overall progress? There may be no universally correct answer to such questions, and I suggest we set them aside by focusing – at least for now – on the narrower issue of what constitutes progress on a particular topic or phenomenon.

In brief, then, I suggest that the question 'What is scientific progress?' can be reformulated in a clearer and more precise manner as follows:

(Q) What type of cognitive change with respect to a given topic or phenomenon X constitutes a (greater or lesser degree of) scientific improvement with respect to X?

If I am right, this captures in a nutshell what debates about the nature of scientific progress are about. To be sure, there are various related questions that could be asked as well, for example regarding what promotes progress on each account, and what constitutes overall progress (rather than progress-on-X). But (Q), I suggest, is the primary question that any adequate account of scientific progress must address.

3.   The Noetic Account of Scientific Progress

The noetic account of scientific progress can be divided into two parts. The first part specifies, roughly, *what type of* cognitive achievement is relevant for progress (see Section 3.1). The second part specifies, roughly, *whose* cognitive achievements are relevant for progress (see Section 3.2). Jointly, these two claims constitute a revamped noetic account of scientific progress (see Section 3.3).

3.1.   *Understanding as Dependency Modelling*

As I have noted, the noetic account takes scientific progress to be explicable in terms of *understanding*. But what, exactly, is understanding?[10]

Understanding is closely related to *explanation*: when we understand something to a reasonably high degree we are normally in a position to explain it or several aspects of it. But, as I argue in detail elsewhere (Dellsén 2020), understanding and explanation can also come apart in important ways. For example, we can increase our understanding of phenomena by learning things that *rule out* explanations of those phenomena. For a case in point, consider the fact that in classical physics the gravitational acceleration of a given object is, somewhat surprisingly, independent of the object's mass. At the surface of the Earth, for example, the gravitational acceleration of any object is $g \approx 9.8$ m/s^2 regardless of the object's mass. When we learn this fact, our understanding of gravitational acceleration increases, and yet it does not supply us with any explanation of the object's gravitational acceleration. On the contrary, we now know that no explanation of an object's gravitational acceleration could possibly appeal to the object's mass.

Another way in which understanding can come apart from explanation concerns the *direction* of explanation. A typical causal explanation of some fact or event X cites the (salient) *causes*, rather than the *effects*, of X. For example, the explanation of the car's crashing into the tree cites the driver's mistake, the bald tires, the icy road, and so forth, as opposed to citing the injuries sustained by the driver or the fact that she was unable to attend a meeting scheduled later that day. By contrast, a relatively complete understanding of a fact or event might partly consist in a representation of its (salient) effects.[11] For example, grasping how the car crash caused the driver's injuries, or how it made her unable to attend the meeting, might well form part of one's understanding of the crash.[12] In sum, while (causal) explanation is arguably backward-looking, understanding looks both ways.[13]

In order to capture these ways in which understanding differs from explanation, and yet respect the idea that the two notions are tightly connected, I have developed a view of understanding that directly refers to the *dependence relations*, such as causation, constitution, or grounding, that are often associated with explanation. More specifically, this view defines understanding in terms of *modelling* such dependence relations, where the relevant kind of a 'model' is just an information structure that represents the dependence relations, or the lack thereof, in which the phenomenon and its aspects stand to other aspects or phenomena. Someone's degree of understanding at a given time is then determined by the accuracy (the extent to which its claims are correct) and comprehensiveness (the extent to which it is informative) of their dependency model (Dellsén 2020: 1268):

> An agent *understands* X if and only if they grasp a sufficiently accurate and comprehensive dependency model of X; and the agent's *degree of understanding* of X is proportional to the accuracy and comprehensiveness of their dependency model of X.

To avoid circularity, the notion of 'grasping' used in this definition cannot itself be defined in terms of the same notion of understanding as the definiendum. Fortunately, if 'grasping' is a type of understanding, it is so in a quite different sense of 'understanding' (see Strevens 2013: 511–512).[14]

In many sciences, the most efficient way to convey understanding is through equations or sets thereof. To illustrate with a familiar example, consider a simple gravity pendulum swinging from side to side, where the pendulum's length is much greater than its swing and its mass is concentrated at the bottom of the pendulum. Christiaan Huygens discovered that the period of such a pendulum is approximately given by $T \approx 2\pi\sqrt{L/g}$, where L is the length of the pendulum and g is the gravitational acceleration at its location. This tells us not just *what* the pendulum's period depends on, namely its length and the gravitational acceleration at its location; it also tells us *how* the pendulum's period depends on these variables, that is what T would be for other possible values of L and g. Furthermore, Huygens' law also tells us a great deal about what T does *not* depend on – for example, that the period does not (significantly) depend on the pendulum's mass or the amplitude of its swings. So an equation like Huygens' law conveys a great deal of understanding in a condensed form.[15]

As noted, the concept of understanding thus defined is closely connected to explanation. If, as many believe, explanation always works by somehow citing or exploiting dependence relations, such as causal relations, then discovering or proposing (approximately) correct explanations always increases our understanding of the explained phenomena. But there are also ways of having (some degree of) understanding that don't directly involve explanation – including, importantly, reliable prediction. This important feature of scientific practice also relies on us latching onto dependence relations in one way or another. For example, when I use the barometer to successfully predict upcoming storms, I rely on a representation of the disjunctive fact that there is *either* a chain of dependence relations between barometer readings and the occurrences of storms *or* between each of these and some third variable that explains them both (i.e. a common cause). By contrast, if I thought there was no dependence relations whatsoever between barometer readings and occurrences of storms, or between each of these and some common cause, then I would have no reason whatsoever to think that barometer readings would help to predict storms.

In sum, then, explanation and prediction are both activities that require us to have at least somewhat accurate and comprehensive dependency models. Thus, on the account of understanding sketched here, gaining an ability to correctly explain or reliably predict some phenomenon inevitably involves increasing one's understanding. This in turn implies that on accounts of scientific progress that appeal to understanding (in this sense of the term), correctly explaining and reliably predicting a phenomenon X are both ways of contributing to scientific progress with respect to X.[16]

3.2. *Enabling Understanding With Public Information*

So far I have said that scientific progress can be 'defined in terms of' increased understanding. But *whose* understanding must increase when scientific progress is made? This question is rather more difficult to answer than one might think. In this section, I will address this question as it arises for understanding-based accounts, but it's worth noting that one could ask analogous questions regarding rival accounts. For example, *whose* knowledge must accumulate on the epistemic account in order for scientific progress to occur?

Since scientific progress is made *by scientists,* it might seem obvious that an understanding-based account of scientific progress should define progress in terms of increasing the understanding *of scientists.* However, there are two problems with this answer. The first is that it is not clear what exactly 'the understanding of scientists' would be. To spell that out would require us to delineate the extension of 'scientists' in some way, which seemingly requires both a solution to the notoriously impenetrable demarcation problem (which disciplines count as 'science'?), and some principled way of drawing a line between genuine members of a given science and those who play various auxiliary roles (e.g. lab assistants, graduate students, janitors). It would also require an account of what it would be for *a collection of scientists,* rather than individual scientists, to increase their understanding. After all, it is surely insufficient for scientific progress that a single scientist increases her understanding, even significantly, since other scientists might simultaneously decrease their understanding.

The other problem with defining progress as increasing the understanding of scientists is deeper – and darker. Note that a group of scientists who gain understanding of some phenomenon, for example by obtaining some experimental results, may for whatever reason decide to keep the relevant result secret, temporarily or even permanently. This happens frequently in 'research and development' at private companies, where making information publicly available would blunt one's own company's competitive edge. It happens occasionally in science as well, although it is usually met with considerable resistance.[17] Now, when scientific results are kept secret in this way, does obtaining them nevertheless constitute scientific progress? According to the conception of scientific progress on which it is determined by the increasing understanding of scientists themselves, the answer appears to be 'yes' – at least if sufficiently many scientists are in on the secret. This seems to me to be the wrong thing to say. Not because it would be counterintuitive or even unusual to describe such episodes with the words 'scientific progress',[18] but because our account of scientific progress should not encourage secrecy of this kind by counting such episodes as progressive. Results that are kept secret in this way are not (better: should not be classified as) genuine improvements in or of science.

In my view, the fundamental issue underlying both problems is that we are in the grips of a framework that focuses on the *producers* rather than the *users* of scientific results. Scientific progress is not (better: ought not be)

determined exclusively by the cognitive benefits it brings to scientists themselves, for example in terms of their increased understanding; rather, it is
(ought to be) determined also by the cognitive benefits it brings to various
groups of non-scientists who consume scientific information. Accordingly,
I suggest that progress should be defined in terms of the information that
is made publicly available in scientific research, for example in journal
articles and research repositories, on the basis of which various relevant
members of society – including scientists themselves but not excluding
various non-scientists – may increase their understanding (see Dellsén ms).

This is emphatically not to deny that in many cases (especially in 'basic'
or 'pure' research), it is appropriate for scientific results to be primarily communicated to other scientists. After all, those other scientists are normally the
people who have the most use for the relevant information. The lay public,
by contrast, may not care much at all, or have much direct use for, such
information. In cases of this sort, scientists are both the users and the producers of scientific information. In other cases, however, scientific results are
highly relevant to various groups of non-scientists, for example engineers,
medical professionals, and policy-makers, whose role in society could not
be successfully carried out without such information. In those cases, the
conception of scientific progress as determined by making information publicly available for the benefit of the potential consumers of such information
requires that those non-scientists have access to the relevant results as well.

To summarise, I am suggesting that scientific progress should be defined
not in terms of the increased understanding of those *by whom* scientific
progress is made, but in terms of enabling increased understanding among
those *for whom* scientific progress is made. In concrete terms, this means
that progress is determined not by the actual cognitive states of some particular group of people, namely 'scientists', but by changes or additions
to public information that has been made available by scientific research.

3.3.   A (Re-)Statement of the Noetic Account

We are now finally in a position to formulate a more precise version of
the idea that progress can be defined in terms of increasing understanding:

> *The noetic account (restated):* Scientific progress is made with respect
> to a given phenomenon X just in case, and to the extent that, changes
> in publicly available scientific information enable relevant members of
> society to increase their understanding of X, i.e. to increase the accu-
> racy and/or comprehensiveness of their dependency models of X.[19]

More colloquially, the noetic account thus reformulated holds that
scientific progress on some phenomenon consists in making scientific
information publicly available so as to enable relevant members of society
to better understand that phenomenon.

As emphasised here (see Section 2), any account of scientific progress, including the noetic account, is in the first instance an attempt to say what *constitutes* scientific progress. But any such account also has direct implications for what *promotes* scientific progress. In particular, the noetic account implies that to promote progress on some phenomenon X is to cause or probabilify scientific activities that ultimately help us understand X. Consider, for example, the quintessentially scientific activity of collecting empirical data to serve as evidence for or against scientific theories. Since an uninterpreted dataset arguably does not by itself convey any understanding, the mere collection of data evidently does not *constitute* progress on the noetic account. However, on the noetic account, collecting data strongly *promotes* progress if and in so far as it leads us to accept true hypotheses and reject false ones, since that in turn enables us to increase our understanding via developing more accurate and/or more comprehensive dependency models than we would otherwise have had.

This might seem harsh on those scientists whose main contribution to scientific progress is empirical as opposed to theoretical. I have two things to say in response. First, this objection conflates the distinction between constituting and promoting progress with different extents to which something contributes to progress. A purely progress-promoting episode may promote much more progress than a purely progress-constituting episode constitutes, in which case the latter contributes much more to progress than the former. Second, the alternative position – according to which collecting data constitutes rather than promotes progress – is, on reflection, quite implausible. To see why, just consider cases in which collecting data would not promote progress, for example because the relevant data, although accurate, was collected in such an unsystematic manner that no inferences can be drawn from it.[20] If collecting data constitutes progress, then such cases would have to be classified as progressive.

4. Connections to Rival Accounts

As noted in the introduction, I will not be defending the noetic account by arguing against rival accounts (for that, see Dellsén 2016, 2018c, 2021b, 2022). In this final section, I will instead explore certain points of similarity and dissimilarity between the noetic account and its most prominent rivals, namely the truthlikeness account, the problem-solving account (including its recent reincarnation in Shan's new functional account), and finally the epistemic account.

4.1. *The Truthlikeness Account*

Of the three rival accounts described earlier, the noetic account is arguably most similar to the truthlikeness account. The truthlikeness account identifies scientific progress with increases in the truthlikeness of accepted theories, where truthlikeness measures the extent to which a given theory captures the whole truth about some topic or phenomenon (or indeed the entire world). The noetic account identifies scientific progress with enabling increased understanding, where (degrees of) understanding is determined by a combination of *accuracy* and *comprehensiveness* in one's dependency models. These two criteria for understanding are closely related to truthlikeness, for a theory is more truthlike to the extent that the claims that it does make approximate the truth (≈accuracy) and to the extent that those claims cover more of the whole truth (≈comprehensiveness).

There are, however, two important ways in which the noetic account comes apart from the truthlikeness account (see Dellsén 2021b for a more detailed discussion). Firstly, the truthlikeness account counts all increases in truthlikeness as progressive, whereas the noetic account counts some of them as non-progressive. In particular, note that accepting a sufficiently truthlike correlation, even if entirely spurious, would increase the truthlikeness of accepted theories and thus constitute progress on the truthlikeness account. For example, the discovery of a close correlation between margarine consumption and divorce rates in Maine in 2000–2009 (Vigen 2015: 18–20) amounts to an increase in the truthlikeness of our theories. But this discovery conveys no understanding whatsoever, since it fails to tell us anything about what either of the correlated phenomena do or don't depend on. Thus it would not count as progressive on the noetic account.[21]

On the other hand, the noetic account counts as progressive some episodes in which there is no change at all in the truthlikeness of accepted theories. For example, consider how a set of theories (including theories about the state of various initial conditions) is often used to explain a familiar phenomenon by a simple derivation from theories to phenomena, as in Newton's explanation of Kepler's laws of planetary motion. In cases where the theories were already accepted prior to the explanation, such developments cannot constitute progress on the truthlikeness account. This is so for the simple logical reason that if $T_1, \ldots, T_n$ logically entail C, then the conjunction $T_1 \& \ldots \& T_n \& C$ is logically equivalent to $T_1 \& \ldots \& T_n$, so the addition of C cannot increase the truthlikeness of accepted theories. By contrast, an explanation of this type would increase understanding in so far as it reveals, or helps to reveal, what the explained phenomenon depends on – which is what ordinary, garden-variety explanations do. Thus the noetic account straightforwardly counts as progressive developments that don't increase the truthlikeness of accepted theories, but rather involve the application of (previously accepted) theories to explain (previously known) phenomena.

4.2.   *The Problem-Solving and New Functional Accounts*

Problem-solving accounts of scientific progress define progress in one way or another in terms of scientific problems or puzzles and their solutions.

According to the most influential version of this account, namely, Laudan's problem-solving account (1977, 1981; see also Kuhn 1970), scientific progress amounts to decreasing the number or importance of unsolved problems, either by solving such problems, or by eliminating or downgrading the importance of the problems that perhaps cannot be solved. But what exactly is a *problem*, and what is involved in *solving* it? On Laudan's view, these terms can only be understood with reference to a particular *research tradition* (corresponding roughly to Kuhn's notion of a 'paradigm'/'disciplinary matrix'), which Laudan defines as a set of assumptions about the entities and processes in some domain and the appropriate methods for studying them (Laudan 1977: 81–95). These assumptions entirely determine not only what counts as a 'problem' at a given time (and which problems are more and less 'important'), but also what counts as a 'solution' to such a problem.

There is a certain relativism inherent in Laudan's problem-solving account. Given two distinct research traditions, for example successive traditions within some discipline, the questions and answers that count as problems and solutions relative to the first tradition may not count as problems or solutions relative to the second. Moreover, there are no factive standards whatsoever for what counts as a problem, or a solution, on the problem-solving account. If the assumptions of a given research tradition are completely misguided, for example in implying that a radically false theory would provide a solution to an entirely spurious problem, then this 'solution' will still count as progressive according to the problem-solving account. For example, medieval theories of the various medical benefits of bloodletting, although now known to be false, were progressive at the time according to the problem-solving account. After all, the research traditions of the day considered them to be 'solutions' to genuine 'problems', such as why bloodletting (allegedly!) cures most diseases (Laudan 1977: 16).

The noetic account, by contrast, does not relativise scientific progress to 'research traditions' – or, indeed, to the assumptions or beliefs of scientists at a given time. Whether some historical development enables increased understanding is an objective matter, in that understanding requires one to grasp dependency models that are in fact *accurate* (rather than, say, models that are merely *assumed* or *believed* to be accurate).[22] For example, since most diseases are not in fact cured or even alleviated by bloodletting, that is since there is no causal relation (or other dependence relation) between bloodletting and the termination or alleviation of most diseases, bloodletting theories do not convey any understanding of those diseases. Accordingly, medieval bloodletting theories do not contribute to scientific progress regarding these diseases.[23]

Let me end my discussion of Laudan's account on a conciliatory note. In my view, there is an important kernel of truth in Laudan's emphasis on problem-solving in science. To wit, much of scientific progress consists not in discovering, formulating, or accepting *new theories*; rather, it consists in *applying theories* to specific phenomena in various concrete situations. Now, the problem-solving account refers to this process of applying theories to phenomena as 'problem-solving', and analyses problem-solving in terms of what counts as a problem/solution relative to a research tradition. The noetic account, by contrast, focuses on how theories are used to *understand* phenomena, which is an achievement for which there are objective success conditions. With that said, the two accounts share the basic conviction that theories, by themselves, do not constitute scientific progress; rather, progress occurs when theories are applied to various specific targets of scientific interest.

With Laudan's account out of the way, let me also briefly address a more recent development of the idea that scientific progress has something to do with problems, namely Shan's *new functional account* (2019, 2022). Shan's account identifies progress with proposing more useful exemplary practices, where an exemplary practice is 'useful' just in case it (i) provides a repeatable way of solving and defining research problems, (ii) provides a reliable framework for solving problems, and (iii) generates more testable research problems across different areas or disciplines. If I understand Shan's account correctly,[24] I suspect it is both too demanding and not demanding enough.

I suspect it is too demanding because there are surely ways of making progress that don't involve any exemplary practices being proposed at all. In particular, it seems to me that progress can be made by simply utilising previously proposed exemplary practices in various ways. For example, it seems to me that progress was made each time Newtonian mechanics and its associated methodology was applied to increase our understanding of yet another physical phenomenon during the eighteenth and nineteenth centuries. Such applications of Newtonian mechanics seemingly did not involve proposing any new exemplary practices since the relevant exemplary practice had already been proposed years earlier. Hence they would not count as progressive on Shan's account.

To see why I suspect Shan's account is also not demanding enough, note that Shan's account resembles Laudan's problem-solving account in that the there is no requirement that the problem definitions and problem solutions that a useful exemplary practice is meant to provide must be factive, for example in that the 'problems' must be based on correct assumptions about which things actually exist. Accordingly, it seems to me that Shan's account opens the door for all sorts of deeply mistaken practices to constitute contributions to scientific progress. Indeed, returning to our earlier example of bloodletting, it seems to me that the medieval practices of bloodletting would count as contributions to scientific progress on Shan's account, since these practices provided an ambitious framework for 'defining' and 'solving' various 'problems'. At the very least, a proponent of Shan's account cannot say that the problem with these practices is that bloodletting theories were radically false/inaccurate, since truth(likeness)/accuracy is not a requirement for progress on Shan's account.[25]

4.3. The Epistemic Account

The epistemic account identifies scientific progress with accumulation of knowledge, where 'knowledge' is understood to require truth, belief, and some form of justification (e.g. reliability, safety, or evidential support). This is not to say that knowledge can be *identified* with justified true belief, since the infamous Gettier cases suggest that truth, belief, and justification are not jointly sufficient for knowledge.[26] Nevertheless, knowledge uncontroversially *requires* truth, belief, and some form of justification, in the sense that S cannot know that P if P is false, not believed by S, or epistemically unjustified for S. It follows that, on the epistemic account, science makes progress only when there is accumulation of justified true beliefs. Put differently, whenever some scientific claim is put forward, it cannot constitute progress on the epistemic account unless it is (i) true, (ii) believed, and (iii) epistemically justified. Whenever one of the requirements (i)–(iii) fails, there is no progress on the epistemic account.

There are some similarities between the epistemic and noetic accounts. In particular, the truth requirement of the epistemic account resembles the accuracy criterion of the noetic account. We might say that the epistemic and noetic accounts are both broadly *veritistic* – like the truthlikeness account, but unlike the problem-solving account. In addition, the belief requirement of the epistemic account is in some ways mirrored in the grasping requirement on understanding, in that both define progress with respect to a specific mental state. With that said, Bird's epistemic account seems to require that scientists themselves, or perhaps communities thereof (see Bird 2019), *be in* the relevant mental state (i.e. knowledge/belief). The noetic account, by contrast, instead requires that scientists *enable* various agents – some but not all of which are themselves scientists – to come to be in the relevant mental state (i.e. understanding/grasping). So the noetic account, unlike the epistemic account, does not really concern itself with what goes on in the minds of scientists, or communities thereof; rather, what matters are the *products* of scientific research, such as published journal articles.

The main difference between the epistemic and noetic accounts concerns the epistemic account's requirement that scientific claims be epistemically justified in order for their accumulation to constitute progress. The noetic account, by contrast, does not require epistemic justification for progress, since justification is not required for understanding according to the definition with which it operates (see Section 3.1).[27] What difference does this make?

One might think that imposing a justification requirement on progress would help to explain the value of various practices in which scientific claims are justified, for example through experimentation, observation, and theoretical argument. Indeed, an account of scientific progress that does not impose a justification requirement might seem to imply that scientists would have no reason to engage in such justificatory activities, since making scientific progress would be perfectly possible without it. But this line of thought is too quick. Any veritistic account of scientific progress implies that justificatory activities *promote* scientific progress in so far as they cause, or raise the probability of, scientific progress. After all, the whole point of justificatory activities is to separate fact from fiction, accuracy from error. In so far as these justificatory activities are successful, they ensure that the cognitive changes that are made in science (e.g. replacing one theory with another) constitute progress rather than regress or flatlining. Given this straightforward instrumental value to justificatory activities on any veritistic account, there is no need to also impose a justification requirement on scientific progress in order to explain the scientific value of justificatory activities.[28]

Moreover, as I argue in much more detail elsewhere (Dellsén 2022), there are several independent reasons not to impose a justification condition on scientific progress. Consider, for example, the fact that in a gradual transition from a scientific consensus on a theory T_1 to a consensus on a rival theory T_2, there will inevitably be a period of time during which those scientists who are most knowledgeable with respect to T_1 and T_2 disagree about which theory is more accurate (or more likely to be accurate). Given widely accepted views in the epistemology of disagreement, at least part of this period will be such that very few, if any, of the relevant scientists are epistemically justified in believing either T_1 or T_2, roughly because the justification they would normally have for T_1/T_2 is undercut by their awareness of the widespread disagreement on T_1/T_2 among the relevant experts. Thus the epistemic account, and indeed any account that imposes a similar justification-requirement, implies that the progress which had previously been made with the introduction of T_1 disappears, or is at least sharply reduced, during this period of widespread disagreement, only to re-emerge once the disagreement abates in favour of T_2.

Awkward consequences of this kind suggest that we shouldn't *require* justification for progress. This does not imply that scientific claims are typically unjustified, or that justification is unimportant for the progress of science. It implies only that it is *possible* to make scientific progress in the absence of the type of justification required for knowledge. This is no doubt a rare occurrence, because the ethos of science dictates that one shouldn't present something as true unless one has evidence to back it up. But rare things do happen occasionally (see e.g. Dellsén 2016: 76–77).

5. Conclusion

I have articulated a version of the idea that scientific progress can be defined in terms of increasing understanding. The account I favour, the noetic account, holds that scientific progress on some phenomenon consists in making scientific information publicly available so as to enable relevant members of society to increase their understanding of that phenomenon. Within this account, 'understanding' is defined in terms of the accuracy and the comprehensiveness of one's representation of the dependence relations, for example the causal relations, in which the target of understanding stands to other things. Since dependence relations undergird correct explanation and reliable prediction (among other things), explanation and prediction are some of the central ways in which scientific progress manifests itself on the noetic account.[29]

Notes

1. This sentiment is expressed by authors such as Sarton (1936: 5) and Kuhn (1970: 160).
2. For an opinionated overview of the following accounts, see Dellsén (2018b).
3. A note on terminology: Bird labels the problem-solving account 'the functional-internalist account', while Shan (2019, 2022) uses 'functional account' to refer to a more general class of accounts which include Kuhn's and Laudan's problem-solving account as well as falsificationist ideas about scientific progress put forward by Popper (1963) and Lakatos (1978). I am happy to reserve the term 'problem-solving account' for Kuhn and Laudan's accounts specifically, as per Shan's suggestion. I am less happy to use the term 'functional account' in the way suggested by Shan, since it seems to me that just about *any* account of scientific progress identifies progress with 'the fulfilment of a certain function' (Shan 2019: 739). (These accounts just differ in what they take the relevant function of science to be; for example, the epistemic account can be said to identify scientific progress with the fulfilment of the function of accumulating knowledge.)
4. The noetic account is not the only account of scientific progress that appeals to 'understanding' in some sense of the term. For example, Rowbottom (2019: 19–21, 110–127) argues that science progresses, *inter alia*, through increases in 'empirical understanding', and McCoy (2022) develops a version of the problem-solving account in which 'understanding' features centrally. So, in a sense, the noetic account developed here is just one among many understanding-based accounts of scientific progress. Note also that the noetic account *of scientific progress* should not be confused with Bengson's (2015) noetic account *of understanding*, which is an alternative to the dependency modeling account of understanding discussed here (see Section 3.1).
5. See Sharadin and Dellsén (2019) for a detailed formal account of the promotion-relation that can be applied, *mutatis mutandis*, to flesh out the notion of promoting progress.
6. Indeed, were it not for the fact that the most prominent accounts in the literature are clearly only meant to account for what constitutes progress, I would have been inclined to mark the distinction between promoting and constituting progress with the terms 'instrumental progress' and 'intrinsic progress'.

7. In addition, one might distinguish these two notions of progress from *progress-in-a-discipline*, that is the progress of a particular scientific field. However, the reasons I give for not focusing on overall progress apply to progress-in-a-discipline as well, so I do not consider this notion specifically.

8. For example, it is common for authors in the debate to state their accounts using phrases which refer to science as a whole, for example 'an episode in science is progressive when . . . ' (Bird 2007b: 64) and 'science makes progress by . . . ' (Niiniluoto 2014: 75).

9. See Kitcher (2001, 2011) for an influential discussion of this issue and an attempt at defining the notion of 'significant truths'.

10. The notion of understanding that I discuss here is canonically ascribed to a subject S with a sentence of the form 'S understands X', where X is an object or phenomenon. In the philosophical literature, this is often referred to as 'objectual understanding', to distinguish it from 'understanding-why' (a.k.a. 'explanatory understanding'), which is canonically ascribed with 'S understands why P', where P is a proposition.

11. Note that this is a claim about (objectual) understanding, not a claim about understanding-why (explanatory understanding).

12. Of course, it will be a highly contextual matter *which* effects are relevant to understanding a given fact or event. But the same is true of its causes.

13. Similarly, a constitutive explanation is 'downward-looking': such an explanation of X cites the grounds of X as opposed to what X grounds. By contrast, understanding looks both up and down: we gain understand of H_2O by considering its molecular structure and subatomic composition (i.e. H_2O's grounds) as well as by considering its macro-level properties such as its transparency and freezing point (i.e. what H_2O grounds).

14. See also Bourget (2017) for an extended discussion of how to define 'grasping' in a way that avoids this type of circularity.

15. See Rowbottom (2019: 8–16) for an account of how Huygens' law provides us with a different type of understanding, namely what he calls 'empirical understanding'.

16. In previous work on scientific progress, I defined understanding operationally roughly as being able to correctly explain and reliably predict the understood phenomenon (Dellsén 2016: 74–75). I now prefer to define understanding as the underlying state that normally undergirds such abilities. Although this shift makes little extensional difference, it highlights the underlying unity of understanding, and thus the unity of scientific progress, according to the noetic account.

17. For example, Merton (1973: 273) notes that secrecy violates 'communism', that is the common ownership of scientific results, which is one of his four norms constituting the ethos of modern science.

18. Recall (from Section 2) that the central question of scientific progress is not concerned with the meaning or usage of particular words.

19. As noted earlier (footnote 16), this is a restatement – or, development, if you will – of the original noetic account (Dellsén 2016).

20. See Dellsén (2016: 78) for some concrete examples of this kind.

21. At most, the discovery of a spurious correlation of this type might *promote* progress, if and in so far as it leads to progress at a later time. For example, if awareness of this correlation were to lead social scientists to discover that consuming margarine and filing for divorce has a common (probabilistic) cause, for example suffering from depression, then discovering the correlation would have promoted progress on the noetic account.

22. Indeed, whether or not some model or theory is *believed* to be accurate or true is quite irrelevant to whether it enables increased understanding in my view,

since understanding is compatible with lack of belief in the propositions on the basis of which one understands (see Dellsén 2017: 247–251).

23. There are some rare cases of diseases which bloodletting does in fact help to alleviate, for example *hemochromatosis* (a condition that causes excess iron to be stored in one's organs). A theory that accurately explains this would of course count as progressive on the noetic account, since it would reveal the causal relationship between bloodletting and alleviation of hemochromatosis.

24. Shan's account is quite complex, appealing to a number of elusive terms that are left undefined (e.g. 'framework', 'exemplary', 'practices', and 'problem'). So my objections may be based on a misunderstanding of Shan's intentions; if so, I apologise and hope that the current discussion may prompt clarifications of the account going forward.

25. In response to an objection that resembles the one pushed here, Shan says that 'usefulness of exemplary practices is also explicable in terms of truth *to some extent*. In particular, it is well explained by the 'contextualist' theory of truth . . .' (Shan 2022: 7). Shan's point seems to be that one could adopt a notion of truth that counts certain theories as false *in our context* while counting the same theories as true *in another context*, presumably in such a way that now-discarded theories that are false in our current context were true in the context in which they were accepted. I find this 'contextualist' notion of truth problematic for the usual reasons, but I won't argue the point here. Instead, I'll just note that adopting this theory of truth fails to address the current objection. After all, *in our context*, bloodletting theories are simply false. Thus, bloodletting theories arguably do not contribute to progress *in our context*. Shan's account, however, is forced to say that they do contribute to progress *in our context*. (Whether or not bloodletting theories might count as true in another context is neither here nor there as far as this argument is concerned, since the argument does not hinge on whether something counts as progress in some context other than the current context.)

26. Indeed, Williamson (2000) argues that knowledge is not analysable at all in terms of necessary and sufficient conditions. Bird (2007a) endorses this view.

27. Some epistemologists suggest that justification is required for understanding as well as for knowledge, for example on the grounds that understanding is a species of knowledge (e.g. Sliwa 2015; Khalifa 2017). I argue otherwise (Dellsén 2017, 2018a, 2021a; see also Hills 2016). But more importantly, we can set these views aside here since our current concern is not with explicating the concept of *understanding*, but that of *scientific progress*. For these purposes, we can simply stipulate that the term 'understanding', as it is used here, does not require justification. (If this sounds incoherent to some readers, they may mentally replace 'understanding' with 'understanding-*sans*-justification'.)

28. Here I am in agreement with Rowbottom (2008), who was the first to argue, against Bird's epistemic account, that justification is instrumental for, rather than constitutive of, scientific progress.

29. Many thanks to an anonymous reviewer for helpful comments on a previous draft of this chapter.

References

Bengson, J. (2015). A Noetic Theory of Understanding and Intuition as Sense-Maker. *Inquiry* 58: 633–668.

Bird, A. (2007a). Justified Judging. *Philosophy and Phenomenological Research* 74: 81–110.

Bird, A. (2007b). What Is Scientific Progress? *Nous* 41: 64–89.

Bird, A. (2016). Scientific Progress. In P. Humphreys (ed.), *The Oxford Handbook in Philosophy of Science*. Oxford: Oxford University Press, pp. 544–565.

Bird, A. (2019). The Aim of Belief and the Aim of Science. *Theoria: An International Journal for Theory, History and Foundations of Science* 34: 171–193.

Bourget, D. (2017). The Role of Consciousness in Grasping and Understanding. *Philosophy and Phenomenological Research* 95: 285–318.

Dellsén, F. (2016). Scientific Progress: Knowledge versus Understanding. *Studies in History and Philosophy of Science* 56: 72–83.

Dellsén, F. (2017). Understanding without Justification or Belief. *Ratio* 30: 239–254.

Dellsén, F. (2018a). Deductive Cogency, Understanding, and Acceptance. *Synthese* 195: 3121–3141.

Dellsén, F. (2018b). Scientific Progress: Four Accounts. *Philosophy Compass* 13: e12525.

Dellsén, F. (2018c). Scientific Progress, Understanding, and Knowledge: Reply to Park. *Journal for General Philosophy of Science* 49: 451–459.

Dellsén, F. (2020). Beyond Explanation: Understanding as Dependency Modeling. *The British Journal for the Philosophy of Science* 71: 1261–1286.

Dellsén, F. (2021a). Rational Understanding: Toward. Probabilistic Epistemology of Acceptability. *Synthese* 198: 2475–2494.

Dellsén, F. (2021b). Understanding Scientific Progress: The Noetic Account. *Synthese* 199: 11249–11278.

Dellsén, F. (2022). Scientific Progress Without Justification. To appear in K. Khalifa, I. Lawler, and E. Shech (eds.), *Scientific Understanding and Representation: Modeling in the Physical Sciences*. London: Routledge.

Dellsén, F. (ms). *Scientific Progress: By-Whom or For-Whom?*

Hills, A. (2016). Understanding Why. *Nous* 50: 661–688.

Khalifa, K. (2017). *Understanding, Explanation, and Scientific Knowledge*. Cambridge: Cambridge University Press.

Kitcher, P. (2001). *Science, Truth, and Democracy*. Oxford: Oxford University Press.

Kitcher, P. (2011). *Science in a Democratic Society*. Oxford: Oxford University Press.

Kuhn, T. S. (1970). *The Structure of Scientific Revolutions*, 2nd edition. Chicago: University of Chicago Press.

Lakatos, I. (1978). "Falsification and the Methodology of Scientific Research Programmes." In J. Worrall and G. Currie (eds.), *The Methodology of Scientific Research Programme*, edited by. Cambridge: Cambridge University Press, pp. 8–101.

Laudan, L. (1977). *Progress and Its Problems: Towards a Theory of Scientific Growth*. London: Routledge and Kegan Paul.

Laudan, L. (1981). A Problem-solving Approach to Scientific Progress. In I. Hacking (ed.), *Scientific Revolutions*. Oxford: Oxford University Press, pp. 144–155.

McCoy, C. D. (2022). Understanding the Progress of Science. To appear in K. Khalifa, I. Lawler, and E. Shech (eds.), *Scientific Understanding and Representation: Modeling in the Physical Sciences*. London: Routledge.

Merton, R. K. (1973). The Normative Structure of Science. In R. K. Merton and N. W. Storer (eds.), *The Sociology of Science: Theoretical and Empirical Investigations*. Chicago and London: University of Chicago Press, pp. 267–278.

Niiniluoto, I. (1984). *Is Science Progressive?* Dordrecht: D. Reidel.

Niiniluoto, I. (2014). Scientific Progress as Increasing Verisimilitude. *Studies in History and Philosophy of Science* 46: 72–77.

Popper, K. R. (1963). *Conjectures and Refutations: The Growth of Scientific Knowledge*. London: Hutchinson.

Rowbottom, D. P. (2008). N-rays and the Semantic View of Progress. *Studies in History and Philosophy of Science* 39: 277–278.

Rowbottom, D. P. (2019). *The Instrument of Science: Scientific Anti-Realism Revitalised*. London: Routledge.

Sarton, G. 1936. *The Study of the History of Science*. Cambridge, MA: Harvard University Press.

Shan, Y. (2019). A New Functional Approach to Scientific Progress. *Philosophy of Science* 86: 739–758.

Shan, Y. (2022). The Functional Approach: Scientific Progress as Increased Usefulness. To appear in Y. Shan (ed.), *Scientific Progress*. London: Routledge.

Sharadin, N., and F. Dellsén (2019). Promotion as Contrastive Increase in Expected Fit. *Philosophical Studies* 176: 1263–1290.

Sliwa, P. (2015). Understanding and Knowing. *Proceedings of the Aristotelian Society* 115: 57–74.

Strevens, M. (2013). No Understanding without Explanation. *Studies in History and Philosophy of Science* 44: 510–515.

Vigen, T. (2015). *Spurious Correlations: Correlation Does Not Entail Causation*. New York: Hachette Books.

Williamson, T. (2000). *Knowledge and Its Limits*. Oxford: Oxford University Press.

Part II

Historical Case Studies

5   Progress in Physics
A Modular Approach

Olivier Darrigol

The progress of science seems to have endangered the best established principles, the ones that were indeed regarded as fundamental. Yet there is no reason to think that we will not be able to save them. Even if we do not quite succeed in so doing, they will still subsist in a transformed guise. We should not compare the advance of Science to the changes of a city in which old edifices are pitilessly torn down to make room for new ones, but to the continuous evolution of zoological types which constantly develop and end up being unrecognizable in a layman's eye, although a trained eye can always detect traces of the work of centuries past. We should not believe that outmoded theories were vain and sterile.[1]

(Henri Poincaré, 1905)

Through a continuous tradition, every physical theory passes on to the next theory the share of natural classification it was able to construct, as in some antic games each runner handed on the burning torch to the next runner. This continuous tradition ensures for science a perpetuity of life and progress.[2]

(Pierre Duhem, 1906)

There are several kinds of progress physicists may perceive in their field. The socio-institutional kind concerns institutions, management, funding, teaching, and applications. The experimental kind concerns apparatus, measurement techniques, and metrological background. The theoretical kind involves new theories, improved theories, better theoretical understanding of a given physical object, cross-theoretical principles, and physico-mathematical techniques. The participative kind consists in an increased role of physical theories in the general progress of science and technology, for instance when physical apparatus and concepts are used in biology. Historians should not separate these three components, because they are intimately and symbiotically related. For instance, the assessment of theoretical progress depends on experimental techniques, which rely on metrological institutions and decisions, which in turn involve the societal

DOI: 10.4324/9781003165859-8

implications of applied physics. It would be especially harmful to ignore instrumental progress, for it conditions the very object of the physicists' inquiries, most evidently so in domains which, like modern cosmology and particle physics, strongly depend on the limits of observation. Philosophers nevertheless tend to focus on theoretical progress, and this is the reason why I will somewhat artificially restrict the following discussion to this kind of progress.

Theoretical progress may be assessed from different perspectives with different results. Firstly, it may be judged internally by the physicists themselves, or externally by philosophers developing their own criteria. I will primarily discuss the internal perception of progress, and secondarily compare with the philosophers' criteria. The physicists' assessments of progress strongly depend on the time scale. Typically, few are those who judge a new theory progressive in its infancy, and their criteria differ from those which may later justify unanimity in favour of this theory.[3] One should expect more rationality and more objectivity to be found at the larger time scale, although time may also serve to preserve notions doomed to be someday abandoned, as for instance happened with the ether in the nineteenth century. Individual, short-term judgements tend to be more subjective. This is why in the following I will be mostly concerned with long-term, collective assessments of progress.

Section 1 of this chapter briefly describes a few cases of alleged theoretical progress in the history of post-Aristotelian physics, from Descartes to quantum physics. Section 2 summarises the various criteria that were used in these cases. Section 3 defends the rationality of some of these criteria by means of a modular conception of physical theory, enabling the comparison of successive theories and suggesting a partially cumulative view of progress. Section 4 briefly recalls the competing ideas of progress that are currently discussed in philosophy of science, considers their implicit reliance on forms of holism, and indicates how the modular conception may help to improve on them.

1.   A Few Cases of Progress

Newton's Principia

The theory expounded in Newton's *Principia mathematica* covers the entire universe, from terrestrial physics to astronomical motions. Although there is no single obvious choice of an earlier theory to be compared to Newton's in assessing progress, Descartes's system is a most relevant candidate, because it shared the scope and the mechanical underpinning of Newton's theory, and because for Newton himself and for most natural philosophers in the years to come, Descartes's natural philosophy was most frequently opposed to Newton's.[4]

In the long run, there was a consensus, even among the remaining Cartesians, that Newton's theory best accounted for the observed motion of celestial bodies and for the mechanics of terrestrial bodies. From an empirical point of view, there was a clear method of comparison based on shared geometric, optical, and instrumental knowledge. From a formal point of view, Descartes's and Newton's systems enjoyed similar unity and simplicity: they both proceeded deductively from a few premises and purported to cover all physical experience.[5] Newton's had the advantage of mathematical rigor, while Descartes's offered a fuller mechanical explanation: Descartes explained gravitation through a complete vortex-based mechanism, whereas Newton assumed an unexplained action at a distance. This is why Christiaan Huygens and many French scholars questioned the superiority of Newton's system over Descartes's and tried to improve on the latter's mechanism, although they rarely followed Descartes's theological and metaphysical arguments in favour of a purely geometric concept of matter. Yet, all along the eighteenth century, the empirical advantage of Newton's theory kept growing through new calculations and matching observations. Late in that century direct action at a distance acquired philosophical legitimacy in the systems of Ruđer Bošković, Immanuel Kant, and Pierre Simon de Laplace. Altogether, empirical adequacy decided Newton's victory, and metaphysical arguments against his theory dwindled.[6]

Wave Optics

In the first third of the nineteenth century, the wave optics of Thomas Young and Augustin Fresnel gradually displaced Newtonian corpuscular optics, which had dominated the previous century. Both theories were mechanistic in some sense, and there was therefore no sharp metaphysical criterion to decide between them. The comparison mostly rested on empirical adequacy, although it also depended on the power of an established tradition of Newtonian molecular physics, which culminated in the early nineteenth century under Laplace's lead. On the empirical side, the comparison long remained undecided or controversial. Reflection, simple refraction, diffraction, and dispersion were equally well explained in both systems. Rectilinear propagation and polarisation challenged the wave system; double refraction and interference challenged the corpuscular system. Fresnel's exquisite work on diffraction and his explanation of rectilinear propagation through interference did not suffice to shake the Laplacian confidence in the corpuscular theory. But very few resisted the appeal of his impressively accurate theory for the propagation of light in anisotropic media.[7]

All along this history, experimental comparison between the competing theories rested on simple observations of patterns of luminous intensity, and also on these theories' compatibility with the ontologically neutral theory of geometrical optics. As was mentioned, Fresnel offered the first convincing proof that wave optics contained geometrical optics as a limit. This result, later confirmed through more rigorous mathematics, implies that wave optics contains a substructure of the older corpuscular theory, and may beat it in the explanation of observed departures from rays' optics (interference, diffraction, double refraction).

Maxwell's Theory

When Maxwell's electromagnetic field theory appeared in the second half of the nineteenth century, its main competitors were the German theories of direct action at a distance by Franz Neumann and Wilhelm Weber. Initially, there was no empirical way to decide between the competing theories because they gave identical predictions for the closed or quasi-closed currents that were accessible in contemporary experiments. British belief in the superiority of Maxwell's theory rested on a metaphysical preference for field-mediated actions and also on the electromagnetic theory of light. Maxwell had indeed reduced light to electromagnetic waves and thereby offered a credible alternative to the elastic-solid theories then in force for the luminiferous medium.[8]

We here encounter a new aspect of comparison not present in the two previous cases: the domain of the new theory exceeds those of earlier theories, and its ability to unify the domains of the earlier theories plays an important role in assessing progress. In particular, the electromagnetic theory of light enabled Maxwell to derive a satisfactory value of the velocity of light from electrically known quantities. Hermann Helmholtz, George FitzGerald, and Hendrik Lorentz later proved that this theory contained the correct laws for the reflection and refraction of light, with none of the artificial assumptions needed in the elastic-ether theories.

In Germany in the 1870s, Helmholtz proved the superiority of Maxwell's theory over the German theories for open currents. He did this firstly by showing that Weber's theory implied the instability of electric equilibrium in conductors, and secondly by designing new crucial experiments, which ended up favouring Maxwell's theory. The experiments were genuinely crucial, for their interpretation depended on concepts and methods that were shared by the competing theories. Interestingly, Helmholtz's Kantian espousal of action at a distance did not interfere with his appraisal of Maxwell's theory, for he was able to reformulate it as a theory of action at a distance in an indefinitely polarisable medium. As is well known, for most continental physicists, the turning point in favour of Maxwell's theory was Heinrich Hertz's demonstration, in 1887–1888, that electromagnetic waves propagating at the velocity of light could be produced by an electric oscillator. Preconceptions in favour of action at a distance vanished, and field-based physics became the new norm. Yet the superiority of Maxwell's theory over Weber's theory of electricity and optical ether theories was not entirely evident. These older theories better accounted for magnetism, electrolysis, magneto-optics, optical dispersion, and the optics of moving bodies. In later years, Lorentz deeply altered Maxwell's theory to integrate aspects of these older theories. So to speak, this was progress enhanced by regress.[9]

Boltzmann's Theory

Ludwig Boltzmann's original purpose, in the last third of the nineteenth century, was to retrieve the laws of gases and the laws of thermodynamics in the molecular-mechanical picture of matter. He therefore emphasised his ability to retrieve thermodynamic behaviour in a statistico-mechanical manner while he belittled any predicted departure from this behaviour. At the same time, along with Maxwell and other supporters of the kinetic theory of gases, he insisted that the new theory predicted new relations between transport coefficients (viscosity, diffusion, heat conduction) and determined their temperature and density dependence in conformity with experimental measurements. He also believed that the molecular picture offered a better intelligibility of thermal physics, based on clearly constructed mechanical pictures.[10]

The detractors of Boltzmann's theory rejected his statistical interpretation of thermodynamic irreversibility, denounced the theory's incompatibility with the measured values of the specific heats of gases, and contrasted Boltzmann's highly mathematical deductions from unwarranted molecular assumptions with the direct derivations offered by thermodynamics from evident macroscopic principles. Boltzmann did not regard any of these objections as fatal, and he reproached the 'phenomenologists' with tacitly assuming a kind of atomism through the discrete, arithmetic foundation of their differential equations. Yet he refrained from declaring victory, and instead pleaded for a peaceful cohabitation of molecular and phenomenological theories until further experiments would come to verify specific predictions of his theory regarding the low-density behaviour of gases and fluctuations around equilibrium (Brownian motion).[11]

Even before such experiments came to the fore in the early twentieth century, physicists obtained experimental proofs of a first atomistic entity, the electron. Lorentz obtained the best verified electromagnetic theory of the time by integrating electrons, ions, atoms, and molecules in the basic picture. Atomic and molecular theories gained traction, both in chemistry and in physics. There were successful hybrids of Lorentz's and Boltzmann's theories, and many former adversaries of the latter theory changed their mind, a few years before Jean Perrin's famous experiments on Brownian motion. To summarise, the progressive character of Boltzmann's theory long remained undecided, until the basic entities of this theory, the molecules, became less hypothetical and until experiments were imagined and realised to verify specific departures from thermodynamic behaviour. What used to be a matter of taste became a matter of fact.[12]

With Boltzmann's theory, we have a case of a theory employing invisible, hypothetical entities to be compared with a theory much more directly related to observed phenomena. The comparison is based on molecular models of macroscopic objects, devices, and quantities. Thanks to these models we can predict the observable evolution of measurable quantities in humanly controllable processes. Thermodynamics thus becomes a kind of approximation of statistical mechanics, and we can imagine the conditions of observation of the effects neglected in this approximation.

Special Relativity

Special relativity has no clear, single antecedent to be compared with, because it is a frame theory, that is, a set of principles meant to constrain other theories, rather than a full theory by itself. The same could be said for thermodynamics compared to earlier theories. In Einstein's parlance, we have a principles theory instead of a constructive theory. From a historical point of view, the most relevant antecedent of special relativity is Lorentz's electrodynamics, since relativity theory resulted from a criticism of Lorentz's theory. Another important antecedent or competitor was Max Abraham's electromagnetic theory based on a strictly rigid electron model in an electromagnetic world picture. Compared with Abraham's theory, Einstein's theory had the merit of strict compatibility with the relativity principle, and it was the only one correctly predicting the deflection of fast electrons in electric and magnetic fields. But this is retrospective judgement. In the early twentieth century, one could still reasonably doubt the strict validity of the relativity principle, the electron-deviation experiments were subject to criticism, while Abraham's electromagnetic worldview offered the hope of a new grand unification of physics.[13]

The comparison with Lorentz's theory was not at all to be decided by experiments. Indeed, the predictions of Einstein's and Lorentz's theories were strictly identical once a few errors in Lorentz's original formulation were corrected in the manner indicated by Henri Poincaré. The difference between the two theories was entirely formal: whereas the Lorentz-Poincaré theory was a constructive theory based on a dynamical ether and distinguishing between the *true* space, time, and fields in the ether frame from the *apparent* ones measured in other inertial frames, Einstein's theory assumed the perfect equivalence of all inertial frames in direct conformity with the relativity principle. In Einstein's hands and eyes, special relativity had the advantage of eliminating theoretical distinctions without any empirical counterpart. In Lorentz's and Poincaré's eyes, the ether theory was preferable because it offered a better intuition of physical phenomena and preserved the ancestral notions of space and time. Yet Einstein's theory more naturally implied typical relativistic effects such as the dilation of time, the contraction of lengths, and the inertia of energy. In Minkowski's hands, this theory acquired an appealing mathematical simplicity and harboured a tensor calculus that eased the construction of future relativistic theories (including gravitation theory). Minkowski's work was instrumental in convincing the concerned physicists that special relativity marked a decisive progress over earlier electron theories.[14]

Since, historically, the superiority of Einstein's theory over other theories was not decided by crucial experiments, it seems preposterous to discuss the possibility of such experiments. It may still be good to know that many years after their deduction by Lorentz and Einstein, the departure of fast-particles dynamics from Newtonian dynamics, the relativistic dilation of time, and the inertia of energy were directly confirmed. All these effects can be seen as departures from the predictions of Newtonian mechanics, since electromagnetism here plays at most an accessory role. The comparison is possible despite the incompatibility of the basic concepts of space and time used in the two compared theories. Indeed, it is enabled by observational devices whose working does not involve the earlier theory more than is allowed by its domain of validity.[15]

Quantum Mechanics

The name quantum mechanics may refer to two different theories. It originally meant a new mechanics of atomic motion, supplemented with rules giving the relation between this motion and radiation emitted or absorbed by the atoms. After the invention of quantum electrodynamics, it also meant a general way of associating a quantum theory to any classical theory (in Hamiltonian form). I will here consider only the first meaning, which was historically sufficient to establish the superiority of quantum mechanics over classical theory regarding the constitution of atoms and their interaction with radiation. The basis of comparison is classical electrodynamics, or the Bohr-Sommerfeld theory that covered the same domain in intermediate years.

In the 1910s already, the elite of theoretical physics came to agree that classical electrodynamics broke down at the atomic scale, and the Bohr-Sommerfeld theory succeeded in explaining a few regularities of atomic spectra and constitution. However, this early quantum theory was usually regarded as an open, incomplete theory, to be someday replaced by a fully consistent generalisation of classical electrodynamics. So to speak, the move was empirically progressive but formally regressive. In 1923–1925, this theory failed in dealing with the helium atom, the anomalous Zeeman effect, and radiation scattering. Quantum mechanics arose in 1925–1926 from lessons of this multiple failure and from the deep formal analogy Bohr anticipated between the new theory and the received electrodynamics. Originally, this new theory only duplicated the confirmed results of the older quantum theory (the hydrogen spectrum, for instance). Yet it was immediately judged superior to its predecessor for offering an internally consistent 'generalisation' of classical mechanics and electrodynamics (in Bohr's sense of generalisation, which accommodates a radical difference in the basic concepts). When, a few months later, its promoters successfully solved the very problems that had precipitated the fall of the previous theory, many came to believe that the true quantum mechanics had been found. This hope has since then been abundantly confirmed.[16]

The comparison between the empirical predictions of classical electrodynamics, the old quantum theory, and quantum mechanics requires a common language to express the design and the result of experiments, and the compatibility of this language with each of the compared theories. According to Bohr, this language is the language of classical physics. The rationality of the comparison then rests on the compatibility of quantum mechanics, when applied to the measuring apparatus, with the classical behaviour of some aspects of this apparatus.[17]

Hydrodynamics

In the previously discussed cases, the experimental comparison of two theories varied in time not only because of the evolution of experimental techniques but also because the compared theories, at least one of them, evolved in time. The kind of evolution I have in mind is not any adjustment in the foundations of the theory, but an advancement of the means through which these foundations are brought to bear on conceivable experiments. Typically, the answer of a given theory to a given experimental question is not an analytically solvable problem, and some strategy is needed to fill the gap between foundations and applications. For instance, Newton's theory of gravitation would not be of much help without a perturbation theory that allows us to go beyond the two-body problem. Also, the spectrum of conceivable applications may widen in time, as happened with Boltzmann's theory. In general, theoretical progress is not limited to the transition from the current to a better theory, it also occurs within the received theories. Thomas Kuhn acknowledges this kind of progress, since for him progress is an increase in the number of puzzles solved, and since this increase may occur within a given paradigm as 'normal science'. Even though he admits that puzzle-solving can be 'fascinating work' for the practitioners, he also writes that 'the aim of normal science is not major substantive novelties'. It is mere 'mopping up', whereas the truly deep innovations, those affecting the conceptual apparatus of theories, occur only in scientific revolutions. *Pace* Kuhn, the history of hydrodynamics gives a splendid case in which the internal evolution of a theory on fixed foundations involves highly creative conceptual work that completely alters the significance of the theory.[18]

The fundamental equations and principles of hydrodynamics were known in the 1750s for inviscid fluids (Euler's equations), and in the

1820s for viscous fluids (Navier-Stokes equations). In the eighteenth and nineteenth centuries, a few problems of practical interest could be solved, including the variations of hydraulic head for irrotational, inviscid flow; the propagation of waves of small amplitude on a water surface; creeping (low Reynolds-number) flow; and vortex motion in perfect liquids, with striking physical consequences. These developments should not be underestimated, for they brought new concepts and methods that redefined the key physical quantities in specific kinds of fluid motion.

However, two practically most important problems, fluid retardation (in pipes) and fluid resistance (to the motion of immersed bodies), remained essentially unsolved in the most common case of slightly viscous fluids (high Reynolds number). Progress in this domain awaited the boundary-layer theory through which Ludwig Prandtl, in the early twentieth century, distinguished two regions of the flow: one, next to the solid walls, in which viscous shear is important, another, at sufficient distance from the walls, in which the flow may be regarded as irrotational. Solutions in these two regions could be handled by specific approximations and methods, and then matched in the overlapping zone. This refined approximation method combined asymptotics (asymptotic behaviour in the limit of vanishing viscosity) and regionalisation of the flow patterns. After generalisation to turbulent boundary layers, it produced universal solutions of the retardation and resistance problems most frequently encountered by hydraulic and aeronautic engineers.

The moral is that the Navier-Stokes equations, although they implicitly cover most cases of flow encountered in nature, remain useless for an enormous variety of concrete problems until proper conceptual furniture is brought in through specialisation, approximation, regionalisation, and asymptotics. There are evident reasons for this complexity and richness: the infinite number of degrees of freedom, the drastic non-linearity of the fundamental equations, and the ensuing instability of simple solutions. In the course of time physicists gradually learned how to harness this complexity, although there are still many open questions especially regarding turbulence. It was much easier to discover the fundamental equations – no offense to d'Alembert, Euler, and Navier – than to discover their physical content in the application game.

2. The Historical Criteria of Progress

Empirical Adequacy

In the cases discussed earlier, the assessment of progress generally depended on two kinds of criteria, empirical and formal. The empirical criteria refer to empirical adequacy à la van Fraassen: a better theory ought to account for the results of a larger number of observations and experiments.[19] Empirical adequacy thus defined does not presume any concept of independent reality. This is a serious advantage when investigating historical assessments of progress, because judgements on the reality of the basic entities of a theory and judgements on its empirical adequacy do not necessarily overlap. The increase of empirical adequacy may occur within the same theory, as was explained in the case of hydrodynamics; or it may involve the comparison of two or more theories, under the condition that the domain of the new theory embraces the conjunction of the domains of the older theories.

Empirical adequacy is not just about how much empirical ground a given theory may cover in principle. It is also about how many experimental predictions we can effectively make, considering the finiteness of our cognitive means. In this pragmatic view, computation techniques, approximation methods, and interpretive substructures play an essential role, as was shown in the case of hydrodynamics. As Marion Vorms has argued, the (re)formulations of a theory, for instance the Newtonian, Lagrangian, and Hamiltonian formulations in the case of mechanics, also play an important role because the new formulations are better adapted to specific kinds of application.[20] Even when the foundations of a theory are fully settled, its empirical adequacy varies considerably in time, not only for the trivial reason that more experimental results become available but also because the empirical exploitability of the theory grows in time. Therefore, there can be important progress within a given theory, and the empirical comparison of two theories varies in time.

Formal Criteria

In the aforementioned examples, empirical adequacy played the strongest role in the long-term, retrospective, and collective assessment of the superiority of a given theory. But a few other criteria also played a role in short-term or medium-term assessments of progress. I call these criteria formal, because they concern the way in which a theory is built, not its empirical content. A first criterion, for modern physical theories, is mathematical, *deductive rigor*. Descartes's theory did not pass this test for it relied on too many loose analogies and suffered from internal contradictions. Nor did the Bohr-Sommerfeld theory because of its open, heuristic nature. Another criterion is the perceived *security* or plausibility of the foundations. This security may be of various kinds: empirical for the principles of thermodynamics, theological and metaphysical in Descartes's system; metaphysical for action at a distance versus medium-based interaction; traditional and methodological for nineteenth-century mechanical reduction. A third important criterion is *unity*/simplicity: the theory should start with a minimum of premises, with a pyramidal deductive structure.[21] This gives it the advantage when compared to a theory that is the mere conjunction of two other theories (for instance, Maxwell's theory is superior to the conjunction of earlier optical and electrodynamic theories). A fourth criterion is *directness*: the distance between the foundational premises and experimental predictions should be as small as possible. As we saw, this was an argument in favour of phenomenological theories and against molecular theories. A typical counterargument was based on another criterion: *intuitiveness (Anschaulichkeit)* in the foundations and in the deductions. Still another important criterion is *workability*: how much we can derive from the premises.[22] This criterion was used to favour semi-empirical hydraulics over hydrodynamics, or to reject otherwise attractive theories such as vortex-sponge theories of a universal ether in the late nineteenth century or string theories of everything in recent times. It is not purely formal, since it conditions empirical adequacy as earlier defined. A last criterion is *heuristic power*, meaning the ability of a theory to serve as a springboard to future generalisation. For instance, Einstein's special relativity was superior to Lorentz's theory in its ability to suggest higher generalisations.

3. A Modular Basis for Comparing Theories

The various formal criteria can be mutually conflicting (for instance intuitiveness versus directness); most of them, except perhaps for deductive rigor, have a subjective component; and none of them could weigh much against a persisting difference in empirical adequacy.[23] It is only when such difference is negligible or contestable that formal criteria may have the last word. Even in such cases, physicists welcome belated confirmations of original predictions of the preferred theory. Granted that empirical adequacy remains the most decisive criterion for the long-term comparison of theories, the question is: can there truly be, at a given time, a clear-cut difference of empirical adequacy between two competing theories? This is a broader version of the question: can there be crucial experiments? Famously, the heirs of Duhemian holism, including Karl Popper, Imre Lakatos, Thomas Kuhn, and many contemporary philosophers, have answered these questions negatively, against the intuition of many physicists. It is of course true that a single experiment, no matter how well conceived and executed, cannot by itself decide the fate of a theory. But there may be extremely high confidence that an experiment contradicts a theory. There are even cases in which the refuting experiment was designed by the promoter of the theory.[24] The protective strategies imagined by Poincaré, Lakatos, Kuhn, and many others often exaggerate the pliability in the theoretical interpretation of experimental results. They are legitimate when they involve (yet undetected) entities that still belong to the original theory (for instance Neptune in celestial mechanics or the neutrino in quantum field theory). They are not when they involve entities or processes of a kind foreign to the original theory,[25] or when they question aspects of the experimental setups that physicists have good reasons to trust.

The common source of these misconceptions is the holistic view of physical theory that has been dominating the philosophy of physics from Duhem to van Fraassen. A few years ago, I proposed a definition of physical theory in which interpretive substructures and theoretical 'modules' play an essential role, in conformity with the observed practice in theoretical physics from Newton to the present. In this definition, a theory comprehends a *symbolic universe*, a growing set of *interpretive schemes*, and various techniques for determining the behaviour of the schemes under the laws of the symbolic universe. The symbolic universe roughly corresponds to what physicists call the formalism of a theory: it gives a purely mathematical description of systems, states, and transformations in abstract spaces, together with laws restricting the states and their transformations. An interpretive scheme is a mathematical blueprint for a conceivable experiment. It involves a subclass of systems of the symbolic universe, as well as formal prescriptions for measuring some of the attached quantities. These prescriptions may be describable in the symbolic universe of the theory under consideration, or they may also rely on external theories (modules). The laws of the symbolic universe in principle imply relations between the schematic quantities. Proper machinery, including techniques of approximation, is needed to effectively determine these relations and test them experimentally. I include this evolving machinery in the definition of a theory, because the theory otherwise remains a mostly ineffective construction.[26]

In almost every physical theory, the construction of the symbolic universe appeals to earlier theories. The interpretive schemes also involve external theories. Furthermore, physicists often focus on subclasses of interpretive schemes for which the induced schematic laws take a simpler form. This means that the theory under consideration is variously and intimately connected to *modules* defined as theories that connect in various ways to the given theory although they have a different domain of application. There are *reducing modules* that serve the construction of the symbolic universe: for instance, a mechanical module is used to construct the universe of Boltzmann's theory. There are *defining modules* that serve to define basic quantities in the symbolic universe: for instance, a geometric module is used in almost every theory of physics, and a mechanical module is used in electrodynamics to define the implied forces. There are *schematic modules* that serve to conceive some of the measurements implied in interpretive schemes: for instance, an optical module may be used to perform measurements in a physical geometry. There are *specialising modules* obtained by setting a parameter of the interpretive schemes to zero, for instance electrostatics as a module of electrodynamics, and there are *approximating modules* obtained by considering the limit of schematic laws when some schematic quantities become very small or very large: for instance, Newtonian mechanics is an approximating module of relativistic mechanics when all velocities in the schemes become small compared to the velocity of light.

The modular structure of a theory evolves together with the theory, and it plays an essential role in the construction, verification, comparison, and communication of theories. Let us focus on verification and comparison. Defining modules and schematic modules are essential in verification because they dictate the construction of some of the measuring devices used in the concrete realisation of the schemes. Typically, these modules are already well-known theories used in their trusted domain of validity. We can thereby profit from earlier knowledge acquired in applying these theories. The comparison between two theories is possible when they can be made to share some of their interpretive schemes. This is possible when all schematic quantities are defined by shared modules that are used in their specific domains of validity. In this case, all knowledge implied in the process of comparison is solid, and it is not compromised by the more hypothetical aspects of the compared theories. For instance, Hertz's famous experiments could decide between the predictions of continental electrodynamics and those of Maxwell's electrodynamics because the quantities measured in these experiments were completely defined by a few shared modules or geometry, mechanics, and quasi-stationary electrodynamics.[27] Contrary to the holistic view, the modular view severely limits the aspects of an experimental setup about which we may reasonably have doubts.

Given the modular approach to verification and comparison, there is a clear sense in which a theory can be verified: its predictions for a given class of interpretive schemes are confirmed at the precision of the experiments. Of course, as is already the case for simple empirical laws, the confirmation has an inductive nature: from a finite but large and diverse number of conclusive tests we come to trust the predictions of the theory for interpretive schemes of the same kind. This inductive success does not mean that the symbolic world acquires ontological weight. It only means that the implied schematic laws are verified. The extent of this verification is not precisely known until the theory becomes an approximating module of a superseding theory. Knowledge captured by the former theory is not suddenly lost in this process. On the contrary, it is often advantageous to keep using the older theories in restricted domains: aeronautical engineers do not appeal to relativistic or quantum mechanics to study the mechanical components of planes, they use good old Newtonian mechanics. There is some ambiguity in what we here mean by Newtonian mechanics: it is not Newton's original theory with the attached ontology of absolute space and time in geometric guise, it is a modernised mechanics that carries the same schematic predictions for a well-identified class of interpretive schemes. Progress is cumulative in this sense, despite the obsolescence of the ontologies carried by older theories. In Poincaré's parlance, 'les rapports vrais' of older theories remain. In Duhem's parlance, 'la part de classification naturelle' is passed between generations.[28] The modular conception of theories thus justifies the old cumulative view of scientific progress.

As we saw in our historical cases, empirical adequacy was rarely the sole criterion used to decide between competing theories. Formal criteria such as unity, simplicity, workability, intuitiveness, and heuristic power also played a major role. In the modular conception of theories, it becomes clear that these criteria are not independent of the criterion of empirical adequacy, because modular structure is involved in the application of both kinds of criteria. This is obvious for workability, since this quality strongly depends on the development of a modular substructure that drives the applications of the theory. The unity, simplicity, and intuitiveness of a theory often depend on recourse to a reducing module in the construction of its symbolic universe. Conversely, heuristic power has to do with a theory's ability to serve as a reducing module for a new theory. Increased unity often means that the new theory contains earlier distinct theories as specialising modules. Globally, physics becomes more unified when the strength, diversity, and multiplicity of modular relations between the available theories increase, even if we never reach the dream of a central module of which all theories would be approximating modules. At the same time, modular structure conditions our ability to apply any given theory.

As the modular possibility of verifying, comparing, and preserving theories plays an important role in the forthcoming discussion of various definitions of progress, I will now illustrate this possibility through the case of rays and wave optics. In the conception of physical theories sketched earlier, the symbolic universe of rays optics is defined by a continuous distribution of rays (continuous lines) in Euclidean space, by a first scalar field representing the optical index, a second scalar field representing the absorptivity of the medium, and an angle-dependent distribution of emitted rays at the surface of sources. For the sake of simplicity, I ignore polarisation and dispersion, although it would not be difficult to take them into account. The principal laws of the symbolic universe are Fermat's principle for the rays, and an absorption law for the density of the rays.

An interpretive scheme is a system of the symbolic universe in which the index, absorptivity, and emissivity are meant to be realised by concrete sources, lenses, mirrors, screens, absorbers, etc. The relevant quantities in these schemes are the density of rays; the shape and position of the components just mentioned; and the relevant index, absorptivity, and emissivity values. A typical optical experiment verifies relations between these various quantities. The measurement of these quantities is based on a geometrical module that rests on a well-established physical geometry; it also depends on a photometric module for comparing and measuring light intensities (densities of rays). The photometric module can be a specialising module of the optical theory, for instance based on the additivity of the intensities from two sources (or the inverse squared law for the light from a point source) and relying on the eye for the comparison of the luminosities of two contiguous surfaces. Or the module could imply external theories, for instance heat theory in a bolometric measurement.

Now consider the wave theory of light. The symbolic universe is given by a wave function in Euclidean space, a scalar field for the propagation velocity, another field for absorptivity, and proper distributions for the sources. The principal law of the symbolic universe is the field equation that connects all these functions and distributions. An interpretive scheme is again a system of the symbolic universe that can be realised through luminous sources, lenses, mirrors, etc. The only significant differences are that the quantity representing the luminous intensity is now given by the time average of the squared modulus of the wave function, and that the optical index is now given by the inverse of the wave velocity. Again, a geometric module is needed to measure the position and shape of the optical apparatus, and a photometric module is needed to measure the light intensity.

In order to compare the predictions of the two theories, we need to make sure that the employed interpretive schemes imply the same modules in both cases. This is obvious for the geometric module. For the photometric module, this is also obvious in the bolometric approach for which the measurement does not rely on optical theory, but no so obvious in the case of measurement based on the additivity of intensities. In general, there is no such additivity in the wave theory: the linearity of the wave equation implies the additivity of amplitudes, not of intensities. But the additivity of intensities still holds for separate, mutually incoherent sources. That is to say, the ray theory and the wave theory can be made to share the same specialising module of intensity measurement.

We are now equipped to compare the predictions of the two theories through fully matching interpretive schemes. As we know, the wave theory wins since diffraction and interference phenomena elude the ray theory. Yet, the predictions of the ray theory remain valid in many cases, as all optical instrument makers know. The conditions for this approximate validity can be precisely derived in the wave theory. In particular, there should be no superposition of mutually coherent beams, and the wavelength of the light should be much smaller than the characteristic dimensions of the optical apparatus (size of diaphragms, diameter of lenses, etc.). The latter condition was vaguely anticipated since Francesco Maria Grimaldi's discovery of diffraction. But the former condition did not occur to anyone before Thomas Young's discovery of optical interference.

To summarise, matching interpretive schemes based on shared modules principally permit (and actually permitted) the comparison of rays optics and wave optics. The former turns out to be an approximating module of the latter, which is exceedingly well verified as along as subtle quantum effects are ignored. This is why the knowledge acquired in rays optics never vanished, why this theory is still in use, and why the invention of wave optics marked a significant progress in the history of physics.

4.   The Philosophers' Criteria of Progress

As was mentioned, Poincaré and Duhem both defended a cumulative view of progress in physics, with nuances. For Poincaré, it is only the *rapports vrais* of a theory, not the specific way in which the theory is expressed and understood at a given time, that remain and keep accumulating. For Duhem, one can cleanly separate the *classification naturelle* brought by the symbolic system of a theory from the metaphysical assumptions and the fanciful illustrations that may encumber it at a given time of history. The progress of science is like a rising tide, with successive waves of metaphysical belief that come and crush, and a global continuous advance of the true phenomenal content of the theories.[29] In the mid-twentieth century, the cumulative view was multiply attacked, most notoriously by Popper and Kuhn, who both denied that a theory could ever be said to be true, even after the filtering imagined by Duhem and Poincaré.[30] This attack induced a series of proposals and counter-proposals in the definition of scientific progress.[31]

Increased Problem-Solving Power

Based on the observed collapse of all past theories and on the alleged incommensurability between successive theories, Kuhn denied any truth value to scientific theories and proposed to gauge progress through the number and quality of puzzles solved. So did too Larry Laudan, based on a slightly different notion of problem-solving. These authors seem to have regarded an increased ability at problem-solving as a sufficient condition of progress, even in the absence of empirical verification. There is little doubt, however, that Kuhn acknowledged increases of empirical adequacy both within a paradigm and during scientific revolution. The notion of problem is vague, and it may include experimental questions and anomalies. Unfortunately, it is not clear why for Kuhn puzzles can easily traverse the barrier of incommensurability whereas no theoretical statement can do so, and why for Laudan some problems can be shared by different research traditions even though problems are generally context-dependent.[32]

The modular answer to this difficulty is to restrict incommensurability to the higher concepts of the theory (involved in the linguistic description of the symbolic universe) and acknowledge the module-based comparability of theories. Then the increase in problem-solving power (for a proper class of problems) may translate into an increase of empirical adequacy, and the empirical truths of the newer theory simply add to those of the older theory, no matter how disparate the basic entities of the two theories may be, and no matter how much ontological weight these entities may carry in their inventors' heads.

Another weakness of the Kuhn-Laudan definition of progress is the presumed stability of the stock of problems to be solved in a progressive episode; what counts is an evolution in the proportion of solved and unsolved problems, not so much the number and significance of problems. As Yafeng Shan recently pointed out, in some cases (especially in biology), progress has to do with ability to generate qualitatively new problems. This is also true in physics: Boltzmann's theory generated fluctuation problems that were not conceivable in earlier thermodynamics; relativistic dynamics brought an unexpected relation between mass and energy content. In my terminology, this means that the novel theory may suggest interpretive schemes that were not imaginable in the older theory and for which the two theories yield different predictions. That said, Shan's purpose is broader than the comparison of theories. He means to cover progress in the non-theoretical aspects of science, which remain understudied in the literature on scientific progress.[33]

Increased Truthlikeness

Another definition of progress, first proposed by Popper and later elaborated by Ilkka Niiniluoto, is based on the idea that we can judge not the truth of theories, but their proximity to truth. Niiniluoto's trick is to assume a universal language L in which a complete description C can be given for any possible world. The language is chosen to support a definition of the distance $d(C_i, C_j)$ between two possibilities C_i and C_j. A theory T is defined as the disjunction of a subset $C_1, C_2, C_i, \ldots$ of possibilities. Its truthlikeness is the minimum value $\text{Tr}(T, C^*)$ of the distance $d(C_i, C^*)$ between the theoretical possibilities C_i compatible with T and the description C^* of the real world. Niiniluoto further introduces the *expected verisimilitude* under a given empirical evidence E as

$$\text{ver}\,(T) = \sum_i P(C_i \,/\, E)\text{Tr}(T, C_i),$$

in which $P(C_i \,/\, E)$ represents the probability of the possibility C_i under the evidence E.[34]

The true description C^* being forever unknown, the truthlikeness of a theory can never be determined. In contrast, its expected verisimilitude can be determined, provided the language L is given and the probabilities $P(C_i \,/\, E)$ can be estimated in this language. This proposal remains highly abstract. In particular, it is not clear how we should pick the language L to allow for the needed determinations, and the function ver evidently depends on the chosen language.[35] One appealing feature of this proposal is the idea that a common basis – here a language – is employed for the comparison of two theories. I believe the modular structure of theories offers a much more effective basis for this comparison, namely, the notion of shared interpretive schemes in which all quantities are defined through shared modules. Once this substructure is admitted, the need to distinguish between truth and truthlikeness disappears, for there is a clear sense in which a theory can be true with respect to a given class of interpretive schemes.

Accumulated Knowledge

In 2007, Alexander Bird criticised the verisimilitude approach for allowing truth or truthlikeness without justification, in which case one would not intuitively see progress.[36] Whether or not this criticism is acceptable, it prompted Bird to replace truthlikeness with *knowledge* in the assessment of progress. Knowledge entails justified true belief, and more (to exclude the Gettier cases). While Bird does not precisely define knowledge, he assumes that verified theories deliver genuine knowledge, not in the sense that a verified theory T should be regarded as true (no theory is ever true in a strict sense) but in the sense that the theory holds approximately, to a degree depending on the extent of the domain of application. In Bird's symbols, $Ap(T)$ is true. In a progressive sequence of theories $T_1, T_2 ...$, knowledge accumulates in the sense that $Ap(T_1) < Ap(T_2) < ...$. Bird's idea of verification and comparison thereby resembles the modular conception, except that it does not spell out the actual means of verification and comparison.[37]

Increased Understanding

A last notion of progress, most recently propounded by Finnur Dellsén, is based on understanding defined as our capacity to explain and predict phenomena. Unlike Bird's knowledge, Dellsén's understanding can do without justification, and sometimes even without belief. For instance, no credible justification is given when the theory used in a true prediction is highly speculative at the time of a verified prediction.[38] One merit of this view is that it allows progress within a theory, and even progress without a theory, as was potentially the case in Kuhn's puzzle-solving approach. Once restricted to the comparison of theories, this view does not truly contradict Bird's epistemic view, because a good theory may be regarded, by definition, as entailing some sort of justification at least in the long run.

*

There are difficulties in comparing the competing views of scientific progress because, even though they share a focus on theoretical progress, they deal with different varieties of progress that may not be captured by the same criteria: progress in the comparison between two theories, progress within a given theory, progress in applying theoretical knowledge to a given object or class of objects (for instance the sun or the atom), progress without a fully articulated theory. In addition, the perspective varies: it can be short-term or long-term; and it can be internal or external. In order to have a common denominator best fitting the case of physics, let us focus on fully articulated theories, compared in a long term for which internal and external judgments of progress hopefully converge.

The modular definition of progress is based on empirical adequacy, controlled by modular verification and comparison. The four definitions recalled in this section also have to do with empirical adequacy, but in looser ways. The problem-solving approach does so by including empirical problems, for instance the length dependence of pendulum oscillations or the existence of characteristic spectra, in the problems to be solved; but it does not stipulate why and how different theories may address such problems. The truthlikeness-based approach defines degrees of empirical adequacy through a formal notion of verisimilitude; but it does not give the tools for effectively computing this quantity in a given theory. The knowledge-based approach defines the empirical content of a theory as the totality of its verified predictions in a given approximation and defines progress through the increase of this content; but it does not tell how a common measure of empirical content may apply to different theories. The understanding-based approach intuitively appeals to empirical adequacy through a broad idea of the explanation and prediction of phenomena; but it does not give a precise basis for comparing the explanatory and predictive merits of two different theories.

Over these four approaches, the modular approach has the advantage of giving effective means for evaluating and comparing the empirical adequacy of theories, in conformity with the physicists' practice since the beginnings of modern physics. But it is narrower than the other approaches, which go beyond judging the empirical adequacy of theories in various ways: the problem-solving approach by including non-empirical problems and untested progress in theoretical development, the truthlikeness-based approach by assuming a realist ontology as a gauge of truth, the knowledge-based approach by requiring a good dose of justified belief, the understanding-based approach by allowing non-theoretical explanations. These deviations have merits, and there is ample room for a pluralist vision of progress in agreement with the variety of science. The modular approach best captures the incremental progress Poincaré and Duhem saw in the history of physics, and it most faithfully represents the means by which physicists evaluate and compare the empirical adequacies of their theories.

Acknowledgements

I am thankful to Yafeng Shan and to an anonymous reviewer for useful comments and suggestions.

Notes

1. Poincaré (1905, p. 8).
2. Duhem (1906, p. 48).
3. Larry Laudan (1977, pp. 108–111) similarly distinguishes between the *context of pursuit* and *the context of acceptance* of theories (or research traditions).
4. Newton (1687); Descartes (1644). For lucid introductions to Newton's and Descartes' systems, see Smeenk and Schliesser (2013); Schuster (2013).
5. However, Newton believed his premises to be of inductive origin, whereas Descartes believed his to be purely rational.
6. For a competent survey of the competition between Cartesian and Newtonian views, see Heilbron (1982, ch. 1). The implied notion of 'empirical adequacy' is similar to Bas van Fraassen's: see Sec. 2.
7. On early-nineteenth century optics, see Buchwald (1989).
8. An old but still informative source on this history is Whittaker (1910).
9. On this integration of microphysics into Maxwell's theory, see Buchwald (1985).
10. On the nature and contents of Boltzmann's project, see Darrigol (2018) and further reference there.
11. On Boltzmann's philosophical strategy of defense, see de Courtenay (1999).
12. On the contemporary rise of atomism, see Nye (1972).
13. On the genesis and early life of special relativity, see Miller (1981).
14. On the circumstances that ultimately favoured Einstein's theory, see Darrigol (2022, Chap. 7), and further reference there.
15. On the tests of special relativity, see Will (2005). Friedrich Paschen's confirmation of Sommerfeld's fine-structure formula for hydrogen was used to argue the superiority of Einstein's theory over Abraham's. This happened in 1917, well after most electron theorists had decided in favor of Einstein's theory; and Paschen's results were repeatedly contested in later years: see Kragh (1985, pp. 82–102).
16. These broad historical features can already be found described in Jammer (1966).
17. The earliest expression of this view is in Bohr's Como lecture of 1927 (Bohr 1928).
18. On these developments, see Darrigol (2013) and further reference there. 'Mopping up' and 'fascinating' are in Kuhn (1962, p. 24); 'not major substantive novelties' ibid. p. 36.
19. See van Fraassen (1980).
20. See Vorms (2011).
21. The unity criterion of progress has been emphasised by many authors, including Maxwell (1876, p. 101), Poincaré (1905, p. 134), Ernst Cassirer (1910), Michael Friedman (1974), and Margaret Morrison (2000). I write 'unity/simplicity' instead of just 'unity' because unity may also be achieved in a non-pyramidal manner through the interconnectivity of theoretical modules: see the discussion of unity versus disunity in Darrigol (2008, pp. 195–196, 221–222).
22. Workability is not just fruitfulness, defined as the ability to predict new phenomena. It also covers the ability to cover already known phenomena. It implies the evolving substructures and approximation methods that relate the premises of the theory to the empirical world.
23. The first two points are made in Kuhn (1977), though with different criteria and different intentions.
24. For example, Helmholtz (1875) refuted his own electrodynamic-potential theory by detecting the electrification of a metallic blade rotating in a

magnetic field. An adept of the Newtonian theory of light, Siméon Denis Poisson, suggested the diffraction experiment through which Arago (1819, p. 16) vindicated Fresnel's theory.

25. For instance, in Helmholtz's aforementioned refutation of his own potential theory, it could be tempting to say that his assumption of an indefinitely polarizable ether saved the theory from refutation. In reality, this assumption altered the basic ontology of the theory, and it led to a theory that was empirically equivalent to Maxwell's.
26. See Darrigol (2008; 2014, Chap. 9). For connections with model-based conceptions of physical theory, especially with Sneedian structuralism, see Darrigol (2008, p. 200).
27. There surely were problems with Hertz's experiments, but these problems were not caused by an improper definition of interpretive schemes. They were caused by a defective realisation of these schemes, and they were properly corrected in later experiments: see Buchwald et al. (2021).
28. Poincaré (1902, pp. 192–193, 205); Duhem (1906, p. 48). John Worral and Steven French have recently defined a similar persistence of structures in their defense of a structural realism. In their defense of a selective realism, Philip Kitcher and Stathis Psillos have advocated a somewhat different sort of continuity in which the 'idle components' of theories are filtered out (Psillos) or the 'working posits' are separated from unnecessary 'presuppositional posits' (Kitcher). On these various arguments for cumulative progress, see Psillos (2018).
29. Duhem (1906, p. 58).
30. Popper (1935); Kuhn (1962).
31. For a clear overview of these proposals, cf. Dellsén (2016).
32. Kuhn (1962); Laudan (1977). On Laudan's position with regards to commensurability, cf. Pearce (1984).
33. Shan (2019).
34. Niiniluoto (1980, 2014).
35. The latter criticism is developed in Miller (2006, Chap. 11).
36. Bird (2007, pp. 65–67). This criticism is deflected in Niiniluoto (2014, p. 76).
37. Bird (2007, pp. 76–78).
38. Dellsén (2016). In the main example given by Dellsén, which is Einstein's theory of Brownian motion, the employed theory, statistical mechanics, was much more justified in 1905 than Dellsén thinks: it was known to contain thermodynamics in a certain approximation, it already had an impressive number of verified specific predictions, and there already was empirical evidence for atomistic entities. At any rate, the assessment of progress and the justification of a theory are time dependent. The best we can hope is their convergence in the long run, as happened in the case of statistical mechanics.

References

Arago, François. 1819. "Rapport fait par M. Arago à l'Académie des Sciences au nom de la commission qui avait été chargée d'examiner les mémoires envoyés au concours pour le prix de la diffraction." *Annales de chimie et de physique*, 11: 5–30.

Bird, Alexander. 2007. "What is scientific progress?" *Noûs*, 41: 64–89.

Bohr, Niels. 1928. "The quantum postulate and the recent development of atomic theory." *Nature*, 121: 580–590.

Buchwald, Jed. 1985. *From Maxwell to microphysics: Aspects of electromagnetic theory in the last quarter of the nineteenth century*. Chicago: The University of Chicago Press.

————. 1989. *The rise of the wave theory of light: Optical theory and experiment in the early nineteenth century*. Chicago: The University of Chicago Press.

Buchwald, Jed, Chen-Pang Yeang, Noah Stemeroff, Jenifer Barton, and Quinn Harrington. 2021. "What Heinrich Hertz discovered about electric waves in 1887–1888." *Archive for History of Exact Sciences*, 75: 125–171.

Cassirer, Ernst. 1910. *Substanzbegriff und Funktionsbegriff: Untersuchungen über die Grundfragen der Erkenntniskritik*. Berlin: Bruno Cassirer.

Darrigol, Olivier. 2008. "The modular structure of physical theories." *Synthese*, 162: 195–223.

————. 2013. "For a philosophy of hydrodynamics." In Robert Batterman (ed.), *The Oxford handbook of philosophy of physics*. Oxford: Oxford University Press, pp. 12–42.

————. 2014. *Physics and necessity: Rational pursuits from the Cartesian past to the quantum present*. Oxford: Oxford University Press.

————. 2018. *Atoms, mechanics, and probability: Ludwig Boltzmann's statistico-mechanical writings – an exegesis*. Oxford: Oxford University Press.

————. 2022. *Relativity principles and theories from Galileo to Einstein*. Oxford: Oxford University Press.

de Courtenay, Nadine. 1999. *Science et philosophie chez Ludwig Boltzmann: la liberté des images par les signes*. Thèse de doctorat, Université de Paris 4.

Dellsén, Finnur. 2016. "Scientific progress: Knowledge versus understanding." *Studies in History and Philosophy of Science*, 56: 72–83.

Descartes, René. 1644. *Principia philosophiae*. Amsterdam: Ludovicus Elzevirius.

Duhem, Pierre. 1906. *La théorie physique, son objet et sa structure*. Paris: Chevalier & Rivière.

Friedman, Michael. 1974. "Explanation and scientific understanding." *Journal of philosophy*, 71: 5–19.

Heilbron, John. 1982. *Elements of early modern physics*. Berkeley: University of California Press.

Helmholtz, Hermann. 1875. "Versuche über die im ungeschlossenen Kreise durch Bewegung inducirten elektromotorischen Kräfte." *Annalen der Physik*, 158: 87–105.

Jammer, Max. 1966. *The conceptual development of quantum mechanics*. New York: McGraw-Hill.

Kragh, Helge. 1985. "The fine structure of hydrogen and the gross structure of the physics community, 1916–26." *Historical Studies in the Physical Sciences*, 15: 67–125.

Kuhn, Thomas. 1962. *The structure of scientific revolutions*. Chicago: The University of Chicago Press.

————. 1977. "Objectivity, value judgment, and theory choice." In *The essential tension*. Chicago: The University of Chicago Press, Chap. 13, pp. 320–339.

Laudan, Larry. 1977. *Progress and its problems*. London: Routledge.

Maxwell, James Clerk. 1876. *Matter and motion*. London: Society for Promoting Christian Knowledge.

Miller, Arthur. 1981. *Albert Einstein's special theory of relativity: Emergence (1905) and early interpretation, 1905–1911*. Reading, MA: Addison-Wesley.

Miller, David. 2006. *Out of error: Further essays on critical rationalism*. Burlington: Ashgate.

Morrison, Margaret. 2000. *Unifying scientific theories: Physical concepts and mathematical structures*. Cambridge: Cambridge University Press.

Newton, Isaac. 1687. *Philosophiae naturalis principia mathematica*. London: Streater & Smith.

Niiniluoto, Ilkka. 1980. "Scientific progress." *Synthese*, 45: 427–462.

———. 2014. "Scientific progress as increasing verisimilitude." *Studies in History and Philosophy of Science*, 46: 73–77.

Nye, Mary Jo. 1972. *Molecular reality: A perspective on the scientific work of Jean Perrin*. London: MacDonald.

Pearce, David. 1984. "Research traditions, incommensurability and scientific progress." *Journal for General Philosophy of Science*, 15: 261–271.

Poincaré, Henri. 1902. *La science et l'hypothèse*. Paris: Flammarion.

Poincaré, Henri. 1905. *La valeur de la science*. Paris: Flammarion.

Popper, Karl. 1935. *Logik der Forschung; zur Erkenntnistheorie der modernen Naturwissenschaft*. Wien: Springer.

Psillos, Stathis. 2018. "Realism and theory change in science." In Edward Zalta (ed.), *The Stanford encyclopedia of philosophy* (Summer 2018 edition). https://plato.stanford.edu/archives/sum2018/entries/realism-theory-change/

Schuster, John. 2013. *Descartes-Agonistes: Physico-mathematics, method and corpuscular-mechanism 1618–33*. Dordrecht: Springer.

Shan, Yafeng. 2019. "A new functional approach to scientific progress." *Philosophy of Science*, 86: 739–758.

Smeenk, Christopher, and Eric Schliesser. 2013. "Newton's principia." In Jed Z. Buchwald and Robert Fox (eds.), *The Oxford Handbook of the History of Physics*. Oxford: Oxford University Press, pp. 109–165.

van Fraassen, Bas. 1980. *The scientific image*. Oxford: Clarendon Press.

Vorms, Marion. 2011. *Qu'est-ce qu'une théorie scientifique?* Paris: Vuibert.

Whittaker, Edmund Taylor. 1910. *A history of the theories of aether and electricity from the age of Descartes to the close of the nineteenth century*. London: Longmans, Green and Co.

Will, Clifford. 2005. "Special relativity: A centenary perspective." In Thibault Damour, Olivier Darrigol, Bertrand Duplantier, and Vincent Rivasseau (eds.), *Einstein 1905–2005. Poincaré seminar 2005*. Basel: Birkhäuser, pp. 33–58.

6 Progress in Chemistry

A Cumulative and Pluralist View

Robin Findlay Hendry

1. Introduction

Among historians, philosophers, and sociologists of science, there is a widespread view that the inevitable fate of scientific theories is to be swept away in favour of radical new reinterpretations of nature. I think this is a deeply mistaken view of the development of chemistry which, I will argue, has built up a system of theoretical and practical knowledge since the late eighteenth century, with new developments typically building on and enriching what went before, rather than erasing it. I would endorse the view of Nevil Sidgwick, President of the Chemical Society (later to become the Royal Society of Chemistry):

> There are two ideas about the progress of science which are widely prevalent, and which appear to me to be both false and pernicious. The first is the notion that when a new discovery is made, it shows the previous conceptions on the subject to be untrue; from which unscientific people draw the very natural conclusion that if to-day's scientific ideas show those of yesterday to be wrong, we need not trouble about them, because they may themselves be shown to be wrong to-morrow. It is difficult to imagine any statement more opposed to the facts. The progress of knowledge does indeed correct certain details in our ideas, but the main result of every great discovery is to establish the existing doctrines on a firmer foundation, and give them a deeper meaning.
>
> (Sidgwick 1936, 533)

Chemistry might be regarded as different from other sciences because many of its achievements are practical and material: known substances are analysed into their constituent elements, some previously unknown, and new compound substances are synthesised. One way to demonstrate chemistry's cumulative development would therefore be to list elements by their date of discovery (see for instance Scerri 2007, 7) or to track the seemingly inexorable rise in the number of known chemical substances

DOI: 10.4324/9781003165859-9

(see Schummer 1997). Maybe so, but that will not be my focus. I will argue for cumulativity in chemistry's theoretical understanding of its subject matter. One aspect of this is that knowledge of the composition and structure of individual substances can be secure enough to resist revision. Water was discovered to be a compound of hydrogen and oxygen in the eighteenth century, its composition understood quantitatively as H_2O in the nineteenth century, and its structure at the molecular scale investigated in the twentieth century. Benzene was a conundrum for the earliest theories of molecular structure in the 1860s, the accepted solution being that benzene is a regular hexagon with alternating double and single bonds (Brock 1992, Chapter 7; Rocke 1985). The alternating bonds were later dropped, but the symbol suggesting them lives on, as does benzene's hexagonal structure, demonstrated by x-ray crystallographer Kathleen Lonsdale in the 1920s (Lonsdale 1928, 1929).

There is cumulativity not just in the longevity of theoretical claims about particular substances, but also in the broad frameworks for understanding composition and structure. These frameworks themselves are long-lived, and newer frameworks emerge from the problems faced by older ones. Not only that: *pace* Kuhn, the emerging awareness of new kinds of structure does not typically force chemists, physicists, and crystallographers into a radical reinterpretation of previously well-established structural claims. Good examples of this kind of conservative development include the extension of structure theory to explain optical isomerism in the 1860s and 1870s, and the discovery of quasi-crystals in the 1980s.

In the next section I will set out the historical case, as I see it, for Sidgwick's view of the development of chemistry. Now if chemistry has developed cumulatively, just what has it accumulated? I will conclude the discussion by briefly assessing two conceptions of progress in science, involving problem-solving and the accumulation of knowledge respectively.

2.   The Development of Chemistry

In what follows I will provide brief discussions of seven key episodes in the history of chemistry, from the 1790s to the 1980s, each of which transformed chemists' understanding of the composition and structure of chemical substances. The central theme running through my discussion is that even though these developments were transformational, they did not undermine previous scientific achievements. This cumulativity is expressed through the longevity of compositional and structural claims about particular substances, that these compositional and structural claims have been deepened and enriched over time rather than overturned or revised radically, and by the fact that each major development has been possible only given the prior developments.

2.1. Lavoisier and the 'Chemical Revolution'

Antoine Lavoisier played the central role in what has come to be known as the chemical revolution. He is credited with establishing that oxygen is an element and water its compound with hydrogen, refining experimental methods in chemistry, reforming chemical nomenclature along systematic lines, providing an operational definition of what it is to be a chemical element, and permanently denying phlogiston a place in accepted chemical explanation (for an overview, see Brock 1992, Chapter 3). Before Lavoisier there was no universally accepted list of elements tied closely to laboratory practice, but after it there was a stable list that underwrote subsequent compositional research, and a clearly identifiable research programme involving the analysis of compound substances and the identification of new elements (see Hendry 2021). The results of that programme, and the empirical and conceptual problems it generated, are key to understanding the later development of chemistry.

Lavoisier identified oxygen as the key agent of combustion and calcination and a small number of substances – oxygen, hydrogen, and nitrogen – as components of a very wide range of other substances. These claims are still accepted by modern chemistry. Lavoisier also argued for compositional claims that modern chemists would reject: gases are no longer thought to contain a 'principle of expansibility' which explains their shared gaseous nature; heat is no longer thought of as anything like a chemical substance that is able to combine with other substances to form compounds; and it is now acknowledged that acids need not contain oxygen, which cannot therefore be the principle of acidity (for discussion, see Hendry 2005, 2010, 2021b).

Did these achievements constitute a revolution in chemistry? Lavoisier's name appears throughout *The Structure of Scientific Revolutions* and his work is a key example for Kuhn's model of scientific change (Kuhn 1970). In this technical sense, the defeat of the phlogiston theory has been called a scientific revolution because: (i) it involved wholesale reinterpretation of experimental evidence concerning chemical change; (ii) it involved wholesale revision of accepted views about the composition of important classes of substances, including metals and their calxes, acids, and gases; (iii) it was accompanied by an influential reform of chemical nomenclature that embedded oxygen into the basic compositional language of chemistry and (perhaps more importantly) denied phlogiston such a place; (iv) it involved the introduction of new concepts and methods into chemistry (for philosophers' arguments along these lines, see for instance Thagard 1990; Kitcher 1993).

Ursula Klein (2015) has challenged the term 'revolution', emphasising the conceptual continuity of Lavoisier's system with what went before and downplaying the empirical innovation. Lavoisier's *Traité Elementaire de Chimie* (Lavoisier 1790) drew on experimental knowledge that had been developed by phlogistonists; furthermore (*pace* Chang 2012), the conceptual underpinnings of his system are not importantly different.[1] However, taking on board Klein's caveats, one is left with the following facts: working with largely the same set of conceptual tools and empirical methods as were available to various versions of the phlogiston theory, Lavoisier and his collaborators reinterpreted, and in some cases refined, the experimental evidence. Crucially, their system used a different compositional basis of 'simple substances' to the phlogistonists, and their compositional claims were highly revisionary, reversing compositional relationships between whole classes of substances, including metals and their calxes. They undertook an accompanying project of nomenclature reform. Although controversial at first, both their compositional claims and their nomenclature came quite quickly to be widely accepted. Not every item on Lavoisier's table of simple substances has lasted, but most of them did, and from the 1790s onwards there was a broad research program involving the identification of new chemical elements – most of them metals – by separating them out from minerals and other naturally occurring substances, and then distinguishing them from other elements by virtue of their chemical and physical properties (see Hendry 2021). The need to systematise the growing list of elements uncovered by this project, and also the similarity relationships between them, led eventually to the periodic table. The important point is that one can defend the anti-phlogistic chemistry as a turning point in the development of chemistry without needing to think that it involved significant conceptual innovation.

2.2. *Chemical Atomism*

It is well known to philosophers that, in 1808, John Dalton published *A New System of Chemical Philosophy*, in which he presented an initial version of a theory that Alan Rocke has called 'chemical atomism'.[2] Dalton's theory offered the sketch of an explanation of the laws of constant and multiple proportions: the fact that typically, when elements combine, they do so with fixed proportions between their weights, and the fact that the same two elements may combine in different proportions to form distinct compounds. Chemical atomism assumed that the elements are composed of distinct kinds of atom that combine with atoms of other elements in fixed ratios to form 'compound atoms' of which the substance is composed.

When I say that Dalton offered a *sketch* of an explanation of the laws of constant and multiple proportions I mean just that: the explanation was sketchy, offering no clue about how atoms can bind together. His theory simply assumed that they do, and was effectively a projection down to the atomic scale of the elemental composition of a substance. For this reason, Paul Needham (2004a, 2004b, 2020) has argued that Daltonian atomism provides no explanation at all. I have argued elsewhere that this is too strong (Hendry 2020; see also Zwier 2011). Dalton does offer us an explanation, though a weak one. If one agrees with Alexander Bird (1998, 153) that atomism was vindicated via inference to the best explanation, one must conclude that the original Daltonian explanations provided very little in the way of epistemic justification. The detailed chronology for its justification is controversial: Bird attributes the required evidence to the success of structural theory in the 1860s and 1870s, while Needham (2004a, 2004b) delays the required justification to the 1930s (see Hendry 2020 for a reply to Needham). One can see why so many chemists were thoroughly sceptical of Dalton's theory, and why many chemists interpreted structural theory instrumentally.

Whether one agrees with Bird or with Needham, one cannot regard the chemical atomic theory as having anything like a solid epistemic justification until after it had been worked on for some decades. Yet Dmitri Mendeleev regarded the theory as being counterfactually necessary to many key empirical discoveries of the nineteenth century:

> The law of combining weights is formulated with great ease, and is an immediate consequence of the atomic theory. [W]ithout it, it is difficult even to understand. Data for its evolution existed previously, but it was not seen until those data were interpreted by the atomic theory. Such is the property of hypotheses. They are indispensable to science; they bestow an order and simplicity which are difficultly attainable without their aid. The whole history of science is a proof of this. And therefore one may boldly say that it is better to hold to an hypothesis which may afterwards prove untrue than to have none at all.
>
> (Mendeleev 1891, vol. 1, 220)

It is also true that without chemical atomism, structure theory would probably never have emerged. The theory therefore seems to have an anomalous place in the development of nineteenth-century chemistry: with some justification it was regarded as highly speculative by many scientists. Yet without it, many of chemistry's central scientific achievements would not have happened.

2.3.   *The Periodic System*

It is sometimes said that the growth in the number of the elements during the first half of the nineteenth century made a system of the elements necessary. The Karlsruhe Congress of 1860, which forged an international consensus to adopt Stanislao Cannizzaro's method for determining the atomic weights of elements, made such a system possible. It is only with respect to this agreement that Mendeleev and others could embark on the project of investigating what connection there is, if any, between the atomic weights of the elements and their chemical properties. Mendeleev used relationships of chemical similarity as a guide to the structure of the table, crucially leaving gaps for as-yet undiscovered elements.[3]

The periodic table is another scientific achievement whose longevity within chemistry would be surprising if the widespread scepticism about cumulativity in science among historians, philosophers, and sociologists of science was justified. Changes to the periodic table have occurred, but as we have seen Lavoisier provided the first list of elements containing a significant number of recognisable modern elements. With a few exceptions (light and heat have dropped off the list), the modern table is an extension of Lavoisier's list, with some added structure in the form of spatial relationships indicating similarities between elements. There are still debates about how, in detail, the elements should be arranged – which arrangement best reflects various important similarities between elements – but one standard table is close enough to being universal to have become a scientific icon. In any case, if one distinguishes between the table and the periodic system that underlies it (see for instance Scerri 2007, xiii), then debates about different tabulations come to seem much less significant. If they involve substantive disagreements at all, then they are proxies for other issues concerning, for instance, which candidate groups of elements carve chemical reality at its joints.

There are, however, two major changes to the periodic *system* that might challenge the view that its development has been continuous. One is the discovery of the noble gases, beginning with argon in the 1890s (see Brock 1992, Chapter 9). Here was a whole *group* of elements for which Mendeleev had failed to leave gaps. Yet the noble gases did not really disrupt the rest of the periodic system, or existing knowledge of other elements, so adding them was a conservative extension. A second important change is the discovery of isotopy and the replacement of atomic weight by atomic number (or charge) as the property that defined places in the periodic table. K. Brad Wray has argued that the replacement of atomic weight by atomic number should therefore count as a Kuhnian revolution (Wray 2018). A profound shift in chemists' understanding of the structure and composition of chemical substances clearly did occur during the late nineteenth century and the first half of the twentieth century, drawing on theory and experiment from both chemistry and physics.[4]

According to Wray, the discovery of isotopy and the consequent replacement of atomic weight by atomic number involved the adoption of a new concept of element because 'the concept of an isotope is a conceptual impossibility as long as chemists assumed that atomic weight defined a chemical element. After all, if two samples had different atomic weights they were, *by definition*, different elements' (2018, 215). Now I would agree that some nineteenth-century chemists, including Mendeleev, *thought* that an element's atomic weight is what makes it the element that it is, but that is not the same thing. An alternative interpretation of the historical facts is as follows: Mendeleev and others formulated (false) hypotheses about the nature of the elements. The discovery of isotopy led to the rejection of those hypotheses and the formulation of a new one identifying atomic number as what individuates the elements. The very possibility that defenders of these two hypotheses could *disagree* requires a single subject of for the two hypotheses, namely, the elements themselves. How is that possible? Conceptual and semantic continuity of zthe kind required is secured by (i) the fact that chemists' reasoning about the elements has applied a single compositional concept of element since the eighteenth century and (ii) semantic continuity in the names of individual elements, which is secured in turn by chemists' laboratory access to samples of the elements in the context of the conceptual continuity (for arguments see Hendry 2005, 2010, 2021b). Atomic number was always what individuated the elements, because it is the physical property that gathered groups of atoms together.[5] The twentieth-century discoveries merely made this explicit.

In the 1960s, Kuhn rejected the idea of a theory-independent reality that is robust enough to commensurate the terms of competing theories. At that time, he was able to draw tacitly on widely shared assumptions about the meaning of theoretical terms, but those assumptions have long been challenged by realist philosophers of science. Even if one accepts Wray's argument that the concept of element changed in the early twentieth century, it does not follow that the elements themselves changed. It remains open to us to regard them as constant objects of study throughout.

2.4.   *Structure Theory*[6]

During the 1860s there emerged within organic chemistry a theory of the structure of substances at the molecular scale whose main tenets live on in modern chemistry. Agreement on how to determine atomic or equivalent weights had stabilised the elemental compositions that were assigned to particular substances on the basis of chemical analysis. On this firm empirical basis there emerged a puzzle: isomerism, whereby two chemical substances that are known to be distinct, on account of their separability and their different chemical and physical behaviour, are found to be composed of the same elements in the same proportions. Isomerism is widespread in organic chemistry because in that domain many thousands of substances are composed from just a few elements (mostly, but not exclusively, carbon, hydrogen, oxygen, and nitrogen). Since isomers share their elemental composition, they must differ in *how* the elements are combined in the different substances. This is the information that a successful theory of structure will provide. But how would such a theory be tested, given that, in the 1860s, chemists had no direct experimental access to reality at the molecular scale? The answer comes in what Alan Rocke has called 'isomer counting' (2010, Chapter 7): a good theory of structure will be able to predict how many distinct chemical substances there are with a given elemental composition, and this is just what structural theory began doing successfully. Each new case of isomerism presents a challenge to structure theory, or perhaps a Kuhnian 'puzzle': to provide the structures of the isomers, showing how they are different. Sometimes these puzzles resisted solution, in that a particular puzzle could not be solved in the same way as previous cases. This would force chemists to imagine new ways for substances to differ from each other at the molecular scale, leading to a growing menu of kinds of isomerism. In the decades after 1860, chemists extended structure theory to explain structural isomerism, geometrical isomerism, and optical isomerism.

In organic chemistry, structures were represented using a diverse range of representational media, including August Kekulé's sausage formulae, A.W. Hofmann's glyptic formulae and the chemical graphs of Alexander Crum Brown and Edward Frankland (see Meinel 2004; Rocke 2010, Chapter 5). Underlying this diversity of presentation was a single body of theory, based on the core idea of fixed atomic valences generating a fixed number of ways a given collection of atoms of different kinds can be connected together via bonds (see Rocke 1984, 2010). The theory was developed amid considerable controversy concerning whether hypotheses about structure at the atomic scale should have any legitimate role in chemistry (see Brock and Knight 1967). The resulting interpretative caution among the chemists of the time has been understood by historians and philosophers as a form of instrumentalism. Perhaps it was that, but that label obscures finer interpretative distinctions in the way the theories were being used. Like many chemists at the time, Frankland was careful to point out that structural formulae were intended 'to represent neither the shape of the molecules, nor the relative position of the constituent atoms' (see Biggs et al. 1976, 59). William Brock and David Knight have described Frankland as one of the 'moderate conventionalists' about structural formulae, people who were 'prepared to employ them as useful fictions' (1967, 21). Frankland's interpretative stance requires careful handling, however. Rather than a blanket conventionalism or instrumentalism about structure, I would argue that it reflects a more selective combination of commitment and epistemic caution: although the bonds must be shown pointing in some direction or other, only the bond topology matters:

> The lines connecting the different atoms of a compound, and which might with equal propriety be drawn in any other direction, provided they connected together the same elements, serve only to show the definite disposal of the bonds: thus the formula for nitric acid indicates that two of the three constituent atoms of oxygen are combined

with nitrogen alone, whilst the third oxygen atom is combined both
with nitrogen and hydrogen.

(Frankland, quoted in Biggs et al. 1976, 59)

Structural formulae are understood to represent only the order of the connections between the atoms. Outside of this, the shapes of the molecules and
the relative positions of the atoms played no more part in the representation
than the particular colours chosen to represent the different kinds of atoms
in Hofmann's glyptic formulae.[7] All of this reflects the conventionalist's
awareness that a given 'fiction' will be useful for some things and not for
others. The explanatory scope of the structural formulae was soon extended
so that inferences could be made about the relative positions of the atoms
in a molecule. The original theory made no commitment on the geometry
of tetravalent carbon, between square planar and tetrahedral arrangements.
Chemists had not worked out how to make any testable inferences on
that point, and so refrained from doing so. Then, in 1874, Jacobus van
't Hoff explained why there are two isomers of compounds in which four
different groups are attached to a single carbon atom by supposing that the
valences are arranged tetrahedrally (with the two isomers conceived of as
mirror images of each other). Adolf von Baeyer explained the instability and
reactivity of some organic compounds by reference to strain in their molecules, which meant their distortion away from some preferred geometry
(see Ramberg 2003, Chapters 3 and 4). These stereochemical theories were
intrinsically spatial, because their explanatory power depended precisely on
their describing the arrangement of atoms in space.

When a theory is extended conservatively, its inferential content
before the extension (that which determines its explanatory and predictive power) is a proper subset of its inferential content afterwards. This
will happen, for instance, when a theory is extended or reinterpreted
so as to allow inferences to be made about a new range of phenomena.
Mauricio Suárez (2004) presents a deflationary account of representation:
inferential use exhausts what can usefully be said about representation.
Whether or not one agrees with that, inference is certainly a very useful
way to track the interpretation of a representation by its users. We can
track the extension of structure theory through the interpretation of its
visual representations. This extension did not require a revision of the
structures that had previously been assigned to organic substances: they
were simply embedded in space, allowing new classes of inference to be
made. Hence stereochemistry constituted a conservative extension of the
earlier structural theory of the 1860s.

2.5.   *The Instrumental Revolution*

During the twentieth century, chemistry developed an ever-closer relationship with physics. Chemists and physicists identified the physical

structures of chemical substances at the molecular scale, a development reflected in the International Union of Pure and Applied Chemistry's systematic nomenclature for chemical substances. This allowed them to apply physical theories to the motions and interactions of these structures. These applications underpinned the emergence of physical methods for investigating and individuating chemical substances and their reactions, including X-ray crystallography, and infrared, ultraviolet, mass, and NMR spectrometry. By the end of the twentieth century, these methods had been fully integrated into chemical practice, alongside traditional 'chemical' methods. One might say that, from a scientific point of view, the distinction between chemical and physical theories and methods gradually becomes obsolete.

The theoretical integration involved a further extension of structural theory to accommodate motions at the molecular level. The structural formulae of the nineteenth century are sometimes described as static, though this is misleading. It would have been obvious to nineteenth-century chemists that matter might be dynamic at the atomic scale, but they had no way of using structure theory to make reliable inferences about its motions; therefore they did not make any. Although the structural formulae were static, they were not interpreted as representing *molecules* to be static. In 1911, Niels Bjerrum applied the 'old' quantum theory to the motions of molecules and their interaction with radiation, potentially explaining their spectroscopic behaviour (see Assmus 1992). From the 1920s onwards, Christopher Ingold and others applied G.N. Lewis's understanding of the covalent bond as deriving from the sharing of electrons to give detailed insight into reaction mechanisms. These explained how, when chemical reactions occur, the structure of the reagents transforms into the structure of the products (see Brock 1992, Chapter 14; Goodwin 2007). These developments involved further conservative extensions of the structural theories that had emerged in the nineteenth century, allowing inferences to be made about their motions, and providing explanations of various spectroscopic and kinetic phenomena that had not previously been within their scope.

The development of the new physical methods for investigating structure has been called the 'instrumental revolution', which encompasses both the development of the new methods in the early twentieth century, and their widespread integration into chemical practice from the 1950s onwards (see Morris and Travis 2002). Instruments that had initially been expensive and exotic became much cheaper and more common, available in every chemistry laboratory rather than only in specialised facilities. Because the new methods offered new modes of epistemic access to reality at the molecular scale, and because they did not disrupt structural theory, they had the overall effect of increasing the epistemic justification of the structures attributed to particular substances. A flavour of this process can be captured through the example of X-ray crystallography. In her Nobel lecture, Dorothy Crowfoot Hodgkin relates how she 'first met the subject of X-ray diffraction of crystals in the pages of the book W.H. Bragg wrote for school children' (1964, 71). Bragg had observed that 'the discovery of X-rays has increased the keenness of our vision over ten thousand times' (1964, 71), exploiting the physical continuity between X-rays and visible light to extend the scope of 'observation', just as philosopher of science Grover Maxwell did later (1962). Hodgkin saw the promise that this technique would allow direct access to previously hypothetical structures at the molecular scale as part of what drew her into a career in X-ray crystallography:

> At Oxford, seriously studying chemistry, with Robinson and Hin-shelwood among my professors, I became captivated by the edifices chemists had raised through experiment and imagination – but still I had a lurking question. Would it not be better if one could really 'see' whether molecules as complicated as the sterols, or strychnine were just as experiment suggested?
>
> (Hodgkin 1964, 71)

Although Hodgkin acknowledged that 'the process of "seeing" was clearly more difficult to apply to such systems than my early reading of Bragg had suggested' (1964, 71), it is quite clear that in some important sense molecular structure became more 'visible' to scientists during the first half of the twentieth century. Philosophers might see the route between evidence and structure in X-ray crystallography as less like perception and more like eliminative induction: X-ray crystallographers use diffraction patterns to choose between candidate structures. Nevertheless, it is clear that in the twentieth century the development of these new techniques caused a signifi-cant shift in the degree of trust that physical scientists were willing to place in structural theory and in the structures attributed to particular substances. In the 1860s, molecular structures had been widely regarded as specula-tive explanations of observed chemical behaviour. In this context chemists rightly regarded the hexagonality of benzene as a hypothesis, something that explained benzene's chemical behaviour but could not be confirmed directly. Crystallographers regard Kathleen Lonsdale (1928, 1929) as pro-viding that confirmation through her study of hexamethylbenzene.

2.6. *Quantum Mechanics*

What prospects are there for extending my tale of cumulative develop-ment into the era of quantum mechanics? This might seem like a hopeless task: surely the arrival of quantum mechanics swept away the Victorian assumptions that atoms in molecules would be classical. The quantum-mechanical molecule must surely be radically different from the classical one! But this is too quick. Firstly, 'classical' chemical structures are not classical in any way that has a direct connection with physics. They involve no application of (for instance) classical mechanics or Boltzmann statistics. Secondly, when quantum mechanics did come to be applied to molecules in the 1920s and 1930s, the resultant Schrödinger equations were insoluble. To make progress, chemists and physicists developed semi-empirical methods that effectively inserted the classical structures into quantum calculations by hand. This is why Linus Pauling, for instance, regarded quantum chemistry as a synthesis of 'classical' structure and quantum mechanics (for details see Hendry 2008). The development of hybrid semi-empirical methods characterised the development of quantum chemistry from the very beginning. This approach could be surprisingly successful: Kenichi Fukui and Roald Hoffmann were jointly awarded the 1981 Nobel Prize in Chemistry for their development[8] of the frontier molecular orbital methods for explaining and, more importantly, predicting how some very complex chemical reactions proceed (see Brush 2015, Chapter 10).

By the 1980s, greatly increased computing power and the development of density-functional theory allowed theoretical chemists to compute accurate electron-density distributions for molecules, which were used by Richard Bader (1990) to define structures which could be interpreted as corresponding to the traditional atoms and bond topologies of molecules. The details are quite interesting: the electron-density distribution of a given molecule can be used to define 'bond paths' between atoms. These bond paths generate a 'molecular graph' for the molecule. In many cases the resulting molecular graph is strikingly close to the classical structure for the molecule. As Bader puts it, 'The recovery of a chemical structure in terms of a property of the system's charge density is a most remarkable and important result' (1990, 33). But the correspondence between bond path and chemical bond is not perfect. The main problems concern repulsive (rather than attractive) interactions between neighbouring atoms in a molecule. Bader's algorithm finds bond paths corresponding to these repulsive interactions, even though chemists would not normally regard the mutually repelling pairs of atoms as bonded to each other.

Despite these issues about understanding molecular reality in the light of quantum mechanics (on which see Hendry 2022), key claims about the structure of substances at the molecular level remain unaltered and indispensable in understanding the chemical and physical behaviour of matter. Methane is tetrahedral. Carbon dioxide is cylindrical. Benzene is hexagonal. Even if quantum mechanics has brought about a radical change in scientists' conceptions of the nature of reality at the molecular scale, these facts are still part of it.

2.7. *Quasi-crystals*

I will close this historical section with an example from crystallography. From the origins of modern crystallography in the nineteenth century until the 1980s, models of crystal structure were built around the idea that a crystal can be understood as a three-dimensional array of regularly repeated unit cells of atoms (see Glazer 2017, Chapter 1). One way to think about this is that the structure of a crystal is generated as a three-dimensional tessellation, which builds in a requirement that the structure of any crystal be translationally symmetrical. This requirement of translational symmetry is equivalent to a requirement that unit cells display only twofold, threefold, fourfold, or sixfold symmetry. Fivefold symmetry and higher than sixfold symmetries are barred.

In 1984, Dan Shechtman detected an electron diffraction pattern that suggested tenfold symmetry. Shechtman understood this to signal the discovery of a new type of structure: a quasiperiodic crystal or quasicrystal (see Glazer 2017; Shechtman 2019). These new structures bear a relation to standard crystals that is akin to that of Penrose tilings to standard tilings. When we think of a two-dimensional tiling, we may think of something that has translational symmetry. Penrose tilings, in contrast, never repeat, though they do exhibit short-range (for instance fivefold) symmetry. Shechtman's work was originally rejected for publication, and its validity challenged by prominent scientists including double-Nobel Laureate Linus Pauling (see Shechtman 2019). It came to be accepted however, and the discovery led to the exploration of a whole new class of materials, a significant enough achievement for Shechtman to be awarded the 2011 Nobel Prize in Chemistry.

Shechtman (2019) has characterised his discovery as bringing about a 'paradigm shift in crystallography', a description that seems justified on the basis of the following considerations: Firstly, the assumption of translational symmetry is a substantive restriction on what can count as a crystal that seems to have arisen *a priori* in early thinking about crystal structure during the nineteenth century, before X-ray diffraction methods provided direct empirical access to the structure of crystals at the atomic scale at the beginning of the twentieth century. Secondly, the structures of a great many common crystalline materials can be understood within the space of possibility afforded by this restrictive assumption. Thirdly, there are alternative explanations of the kind of diffraction patterns that quasi-crystals give rise to, which could be appealed to by critics of Shechtman's work: Pauling argued that Shechtman's results could be explained as a form of multiple twinning, arising from crystals in different orientations (see Glazer 2017, 3–56). Fourthly, the discovery of quasi-crystals enforced a revision to the previously accepted definition of what counts as a crystal, which assumed a periodic structure. However, it had no tendency to enforce a revision to previously accepted crystal structures. The paradigm might have changed, in the sense that the space of structural possibility had been expanded, but this expansion did not change the nature of the structures assigned to traditional crystals. Once again, previous discoveries were left untouched.

3. Progress in Science

I have argued that chemistry has developed cumulatively, but just what is it that has accumulated? Philosophers have articulated different conceptions of progress that see science as an activity that is, or should be, directed towards the following:[9]

1. Solving empirical problems, in the sense of applying a fixed theoretical framework to describing the world (for the classic version see Kuhn 1970); or
2. Increasing the amount of truth in theoretical representations of the world (see for instance Niiniluoto 2014); or
3. Gathering knowledge about the world (see Bird 2007, 2022); or
4. Increasing understanding of the world (see Dellsén 2016).

This is not the place to embark on a survey of these four accounts, and how they relate to the historical development of chemistry. Instead I will concentrate on progress as problem-solving, and as the accumulation of knowledge.

Kuhn's account of science and progress centres on the idea that scientists are engaged with solving empirical problems (Kuhn 1970). They work within frameworks of metaphysical assumptions, concepts, principles, and (theoretical and experimental) methods, using them to develop growing libraries of successful treatments of empirical phenomena. The resulting account of scientific change is historiographically rich, but its motivations are as much philosophical as they are historiographical. Kuhn resisted general attempts to commensurate the achievements of different frameworks and his arguments on this are philosophical, involving substantive claims about perception and its role in scientific inference, conceptual change, the meaning of theoretical terms, and the values that motivate choice among theories.[10] At the time, Kuhn's assumptions about the meaning of theoretical terms seem to have been shared widely enough to be invisible, but realist philosophers of science such as Richard Boyd and Hilary Putnam later challenged them (for discussion in the context of chemistry, see Hendry 2005, 2010).

One might say that Kuhn's broadly anti-rationalist and anti-realist philosophical stance motivates a mutually reinforcing historiographical approach. If that's right, then scepticism about framework-transcendent progress in science is not something that is forced on us by the historical phenomena. Although this is far from being a get-out-of-jail-free card for scientific realists, it does mean that arguments for semantic incommensurability need to be made out in careful historical detail. This is particularly salient in the case of the chemical elements: since the 1790s, the list of known elements has grown cumulatively and there has been a great deal of continuity in how chemists reason about the elements. Whole new groups of elements were added to the periodic table without disrupting what was previously known, and in the twentieth century, the elements were reinterpreted from within a new theoretical setting. It is sometimes argued that the twentieth century brought a new concept of chemical element (see for instance, Wray 2018), but then it needs to be shown how in detail chemists' reasoning about elements changed. My own view is that it did not, except that new information about them was added (see Hendry 2005, 2010, 2021b).

The problem-solving conception of progress in science is strictly compatible with theoretical frameworks being extended conservatively. Such extensions seem pervasive in chemistry, however: in addition to the chemical elements, we have conceptions of structure from the 1860s onwards, and the discovery of quasi-crystals in the 1980s. This surely goes against the internalist spirit of the problem-solving conception.[11] If so, one might seek alternative ways of understanding progress in science, appealing (for instance) to the world's natural-kind structure. These points notwithstanding, a focus on problem-solving is historiographically valuable. Each of the extensions to structure theory arose as a new kind of solution to a recalcitrant empirical problem, and then served as the basis for something like Kuhnian normal science, in which the new theoretical template is applied to a range of similar problems.

Neither the problem-solving nor the semantic conceptions of progress are set up to accommodate the accumulation of specific items of knowledge as a central kind of scientific achievement. In my account of the historical development of chemistry, there was cumulativity at the level of both the particular and the general. The epistemic view of progress in science, as developed by Alexander Bird (2007, 2022), addresses this shortcoming by conceiving of scientific progress simply as the accumulation of knowledge. This makes the view an attractive one: chemistry has progressed primarily in the sense that twenty-first century chemists know much more than eighteenth-century chemists about the composition and structure of chemical substances. Water and benzene demonstrate how successive generations of chemists have developed deeper and more detailed knowledge of the composition and structure of particular substances.

One particular advantage of the epistemic account is that it shows how scientific activity directed towards increasing the justification of scientific beliefs can promote progress. The 'instrumental revolution', in which physical methods were added to so-called 'wet' chemical methods for determining structure, provides an excellent example of this. Hodgkin's comments about X-ray crystallography were clear: here was a new and more direct way of accessing structure at the molecular scale that enabled chemists, crystallographers, and physicists to cut through the complex inferences that chemists had previously made about structure on the basis of chemical reactivity. Similarly, although scientific debate about the structure of benzene had petered out before the end of the nineteenth century, Lonsdale's crystallographic work on hexamethyl benzene established in a more direct way that it is a regular hexagon. The perceived change in the epistemic status of benzene's structure was significant enough for Lonsdale to be able to pre-publish her conclusions in *Nature* (Lonsdale 1928).

However, one of Bird's criticisms of the semantic view can be turned round, forming an objection to the epistemic view itself (see Bird 2007, Section 2.1). Bird imagines a series of beliefs that display increasing truth content, but lack sufficient evidential support to count as knowledge. The semantic view would count this as progress, he argues, but according to both intuition and the epistemic view, it is not progress. Now consider the development of Dalton's atomic theory. As we have seen, when Dalton first proposed his version of atomic theory, it was widely viewed as lacking sufficient justification to count as knowledge. Chemists acknowledged its explanatory power, but most did not see this as sufficient justification for accepting the theory. There is an argument to be had about just when enough evidence was gathered for the truths embodied in Dalton's theory to become known, but whatever date one picks (e.g. the 1860s or the 1930s), chemists would have been working on the theory for decades in advance of that. As we have seen it was extremely fruitful scientific activity, from a heuristic point of view. How was it contributing to the growth of knowledge? Some important scientific activities, it seems, contribute to the growth of knowledge only indirectly, being counterfactually indispensable to growth in knowledge without themselves constituting it. To understand such activities, we should probably turn to one of the other conceptions of progress.

4. Conclusion

I have argued that, from the 1790s to the present day, chemistry has developed cumulatively, establishing broad and stable frameworks for understanding the composition and structure of chemical substances. Chemistry provides little evidence for the pessimistic thoughts about scientific knowledge that motivate anti-realist interpretations of science. I will close with some pluralistic comments about progress in science, which I do not claim to have argued for rigorously, but which are in keeping with everything I have established.[12] Discussion of the different conceptions of progress in science is sometimes framed as a debate between competing accounts, but it is not clear why. Perhaps they are better understood as mutually complementary perspectives on scientific activity, for three broad kinds of reason. First, different conceptions of progress might have something distinctive to offer in helping to understand the historical development of science, something that is not offered by other conceptions. Second, I suspect that none of these conceptions of progress offers a *complete* understanding of the historical development of science in the sense that, for each conception of progress there are important developments for which it does not adequately account. Both these points are illustrated by the brief comments I have made on the problem-solving and epistemic conceptions of progress in science in relation to the history of chemistry. Seeing science as a problem-solving activity nicely captures some aspects of the development of structure theory, but not the cumulative growth of knowledge about the composition and structure of particular substances. The epistemic conception deals much better with that, but not so well with the important work that scientists do in developing speculative theories that do (as things turn out) embody significant truths, before the supporting evidence is strong enough for them to count as knowledge. Lastly, attempts to reduce the aim motivating one conception of progress to that motivating another are unedifying. In so far as knowledge, truth and understanding are distinct, attempts to reduce one to the other simply fudge the differences between them. Knowledge, truth and understanding are all honourable goals for science, and each might well motivate different kinds of scientific activity.

Acknowledgements

I would like to thank Yafeng Shan for his comments on an earlier version of this chapter, and the John Templeton Foundation for funding the project 'The Successes and Failures of Science through the Lens of Intellectual Humility: Perspectives from the History and Philosophy of Science' (Grant ID 62247).

Notes

1. For the background to this point, see Klein 1994; see also Hendry 2021 for further discussion.
2. See Brock 1992, Chapter 4. Rocke 1984 provides the authoritative analysis of the structure of the theory and its reception; Rocke 2005 provides a detailed account of its origins. It is important to distinguish Dalton's chemical atomism from early-modern corpuscularianism (see Klein 1994; Chalmers 2009; Hendry 2020).
3. Karoliina Pulkkinen analyses the role of scientific values in this process (Pulkkinen 2019).
4. These developments came to fruition in the twentieth century with the 'instrumental revolution' and the application of quantum mechanics to the explanation of chemical bonding (see later).
5. Whether or not anyone knew it, samples of silver has always been roughly equal mixtures of two different isotopes.
6. This section draws on Hendry (2018 and 2020).
7. Some of the colour choices made by chemists at that time live on: in the molecular models of today, oxygen atoms are still red, carbon black, and hydrogen white. See Meinel (2004) for details.

8. R.B. Woodward, with whom Hoffmann developed the Woodward-Hoffmann rules encapsulating this approach, had died in 1979. Woodward had been awarded the 1965 Nobel Prize in Chemistry for work on organic synthesis.
9. For surveys, see Dellsén 2018 and Bird (2022, Chapter 3).
10. For analyses of his changing views on incommensurability, see Sankey 1993 and Bird 2000.
11. This internalist spirit is explicit in Yafeng Shan's presentation of his version of the view (Shan 2019).
12. Darrell Rowbottom (forthcoming) is also developing a pluralist approach to progress in science. Kuhn's influential analysis of epistemic values in science (Kuhn 1977) is another source of support.

References

Assmus, Alexi 1992 'The molecular tradition in early quantum theory' *Historical Studies in the Physical and Biological Sciences* 22, 209–231

Bader, Richard F.W. 1990 *Atoms in Molecules: A Quantum Theory* (Oxford: Oxford University Press)

Biggs, Norman L., E. Keith Lloyd and Robin J. Wilson 1976 *Graph Theory, 1736–1936* (Oxford: Clarendon Press)

Bird, Alexander 1998 *Philosophy of Science* (London: UCL Press)

Bird, Alexander 2000 *Thomas Kuhn* (Chesham: Acumen)

Bird, Alexander 2007 'What is scientific progress?' *Noûs* 41, 64–89

Bird, Alexander 2022 *Knowing Science* (Oxford: Oxford University Press)

Brock, William H. 1992 *The Fontana History of Chemistry* (London: Fontana Press)

Brock, William H. and David M. Knight 1967 'The atomic debates' in W. H. Brock (ed.) *The Atomic Debates: Brodie and the Rejection of Atomic Theory: Three Studies* (Leicester: Leicester University Press), 1–30

Brush, Stephen G. with Ariel Segal 2015 *Making 20th Century Science: How Theories Became Knowledge* (Oxford: Oxford University Press)

Chalmers, Alan 2009 *The Scientist's Atom and the Philosopher's Stone: How Science Succeeded and Philosophy Failed to Gain Knowledge of Atoms* (Dordrecht: Springer)

Chang, Hasok 2012 *Is Water H_2O? Evidence, Realism and Pluralism* (Dordrecht: Springer)

Dellsén, Finnur 2016 'Scientific progress: Knowledge versus understanding' *Studies in History and Philosophy of Science* 56, 72–83

Dellsén, Finnur 2018 'Scientific progress: Four accounts' *Philosophy Compass* 13(11), e12525

Glazer, A.M. 2017 *A Journey into Reciprocal Space: A Crystallographer's Perspective* (Bristol: Institute of Physics)

Goodwin, W.M. 2007 'Scientific understanding after the Ingold revolution in organic chemistry' *Philosophy of Science* 74, 386–408

Hendry, Robin Findlay 2005 'Lavoisier and Mendeleev on the elements' *Foundations of Chemistry* 7, 31–48

Hendry, Robin Findlay 2008 'Two conceptions of the chemical bond' *Philosophy of Science* 75, 909–920

Hendry, Robin Findlay 2010 'The elements and conceptual change' in Helen Beebee and Nigel Sabbarton-Leary (eds.) *The Semantics and Metaphysics of Natural Kinds* (London: Routledge), 137–158

Hendry, Robin Findlay 2018 'Scientific realism and the history of chemistry' *Spontaneous Generations* 9, 108–117

Hendry, Robin Findlay 2020 'Trusting atoms' in Ugo Zilioli (ed.) *Atomism in Philosophy: A History from Antiquity to the Present* (London: Bloomsbury), 470–488

Hendry, Robin Findlay 2021 'Elements and (first) principles in chemistry' *Synthese* 198, 3391–3411

Hendry, Robin Findlay 2022 'Quantum mechanics and molecular structure' in Olimpia Lombardi et al. (eds.) *Philosophical Perspectives on Quantum Chemistry* (Cham: Springer), 147–172

Hodgkin, D. C. 1964 'X-ray analysis of complicated molecules' in *Nobel Lectures, Chemistry 1963–1970* (Amsterdam: Elsevier), 71–91

Kitcher, Philip 1993 *The Advancement of Science: Science without Legend, Objectivity without Illusions* (New York: Oxford University Press)

Klein, Ursula 1994 'Origin of the concept of chemical compound' *Science in Context* 7, 163–204

Klein, Ursula 2015 'A revolution that never happened' *Studies in History and Philosophy of Science* 49, 80–90

Kuhn, Thomas 1970 *The Structure of Scientific Revolutions*, Second Edition (Chicago: University of Chicago Press)

Kuhn, T. S. 1977 'Objectivity, value judgment, and theory choice' in *The Essential Tension: Selected Studies in Scientific Tradition and Change* (Chicago: University of Chicago Press), 320–339

Lavoisier, Antoine 1790 *The Elements of Chemistry* (Edinburgh: William Creech), translation by Robert Kerr of *Traité Élémentaire de Chimie* (Paris, 1789)

Lonsdale, Kathleen 1928 'The structure of the benzene ring' *Nature* 122, 810

Lonsdale, Kathleen 1929 'The structure of the benzene ring in $C_6(CH_3)_6$' *Proceedings of the Royal Society of London* A123, 494–515

Maxwell, G. 1962 'The ontological status of theoretical entities' in H. Feigl and G. Maxwell (eds.) *Scientific Explanation, Space and Time* (Minneapolis: University of Minnesota Press), 3–26

Meinel, Christoph 2004 'Molecules and croquet balls' in Soraya de Chadarevian and Nick Hopwood (eds.) *Models: The Third Dimension of Science* (Stanford: Stanford University Press), 242–275

Mendeleev, D. I. 1891 *The Principles of Chemistry*, First English Edition, 2 vols, trans. G. Kamensky (London: Longmans, Green)

Morris Peter J. T. and Anthony S. Travis 2002 'The rôle of physical instrumentation in structural organic chemistry in the twentieth century' in Peter J. T. Morris (ed.) *From Classical to Modern Chemistry: The Instrumental Revolution* (London: Royal Society of Chemistry), 57–86

Needham, Paul 2004a 'When did atoms begin to do any explanatory work in chemistry?' *International Studies in the Philosophy of Science* 18, 199–219

Needham, Paul 2004b 'Has Daltonian atomism provided chemistry with any explanations?' *Philosophy of Science* 71, 1038–1047

Needham, Paul 2020 'Atomism and chemistry: Not a success story' in Ugo Zilioli (ed.) *Atomism in Philosophy: A History from Antiquity to the Present* (London: Bloomsbury), 457–469

Niiniluoto, Ilkka 2014 'Scientific progress as increasing verisimilitude.' *Studies in History and Philosophy of Science Part A* 46, 73–77

Pulkkinen, Karoliina 2019 'The value of completeness: How Mendeleev used his periodic system to make predictions' *Philosophy of Science* 86, 1318–1329

Ramberg, Peter 2003 *Chemical Structure, Spatial Arrangement: The Early History of Stereochemistry, 1874–1914* (Aldershot: Ashgate)

Rocke, Alan J. 1984 *Chemical Atomism in the Nineteenth Century: From Dalton to Cannizzaro* (Columbus, OH: Ohio State University Press)

Rocke, Alan J. 1985 'Hypothesis and experiment in the early development of Kekulé's Benzene theory' *Annals of Science* 42, 355–381

Rocke, Alan J. 2005 'John Dalton and the origins of the atomic theory' *Social Research* 72, 125–158

Rocke, Alan J. 2010 *Image and Reality: Kekulé, Kopp and the Scientific Imagination* (Chicago: Chicago University Press)

Sankey, Howard 1993 'Kuhn's changing concept of incommensurability' *British Journal for the Philosophy of Science* 44, 759–774

Scerri, Eric 2007 *The Periodic Table: Its Story and Its Significance* (Oxford: Oxford University Press)

Schummer, Joachim 1997 'Scientometric studies on chemistry I: The exponential growth of chemical substances, 1800–1995' *Scientometrics* 39, 107–123

Shan, Yafeng 2019 'A new functional approach to scientific progress' *Philosophy of Science* 86, 739–758

Shechtman, Dan 2019 'Quasi-periodic materials: A paradigm shift in crystallography' *Nova Acta Leopoldina Neue Folge Supplementum* 34, 11–36

Sidgwick, N. V. 1936 'Structural chemistry' *Journal of the Chemical Society* 149, 533–538.

Suárez, Mauricio 2004 'An inferential conception of scientific representation' *Philosophy of Science* 71, 767–779

Thagard, Paul 1990 'The conceptual structure of the chemical revolution' *Philosophy of Science* 57, 183–209

Wray, K. Brad 2018 'The atomic number revolution in chemistry: A Kuhnian analysis' *Foundations of Chemistry* 20, 209–217

Zwier, Karen R. 2011 'Dalton's chemical atoms vs. Duhem's chemical equivalents' *Philosophy of Science* 78, 842–853

7 Progress in Chemistry
Themes at the Macroscopic and Microscopic Levels

Paul Needham

1. Introduction

From ancient times, chemistry has always been 'concerned', as a prominent organic chemist put it, 'with substances and with their transformations into other substances' (Benfey 1963, p. 574). It was left aside in the development of mechanics in the two centuries following Newton's *Principia*, his interest in alchemy notwithstanding, because mechanics made no distinction between the different kinds of matter that were of interest to chemists. Aristotle was on the right track with his notion of a single substance as a homogeneous body, and his theory of mixts and chemical change based on four elements which could exist as such in isolation. Chemistry progressed during the ensuing two and a half millennia with the discovery, by new methods of synthesis and separation, of new substances unknown to Aristotle and distinguished with increasing clarity as criteria of individuation developed with advances in experiment and theory. Aristotle's notion of a mixt gave way to a twofold distinction between compounds and solutions, and an elaborate theory of mixtures has been forthcoming on the basis of the distinction between phase and substance, recognising the possibilities of single substances exhibiting several phases, single phases containing several substances, and complex mixtures of several substances distributed over several phases. Gaining a proper understanding[1] of the character and interrelations of these fundamental macroscopic concepts in the nineteenth century, followed by a sophisticated understanding of microstructure since the beginning of the twentieth century, has brought the subject ever closer to physics and marks a progression in the direction of unity.

These developments have been achieved as a result of painstaking experiment and observation, necessitating the modification, and sometimes the overthrow, of older ideas together with the introduction of new theories along with the associated conceptual apparatus. The enormity of the mass of factual material concerning the variety, chemical reactivity, and physical properties of chemical substances and their systematic categorisation, filling weighty tomes of organic and inorganic chemistry

DOI: 10.4324/9781003165859-10

and standard reference works, is seldom conveyed in philosophical texts where surprising and exotic theoretical revelations furnish more lively subjects. But it represents a major and continuing achievement, forming the backbone of analytic chemistry and the methodology of synthesis, without which much of forensic science, food control and molecular biology, for example, would not be where they are today. The progress represented by this accumulation of knowledge is not understood on the basis of a simplistic notion of observation of the kind promulgated by traditional empiricism, logical positivism and lately by constructive empiricism. The electron as the paradigm of an unobservable entity is particularly useless and misleading as a lead in understanding the empirical basis of chemistry. Allusions to the micro-realm are irrelevant for understanding that it has never been possible to perceive macroscopic features such as distinct substances in a solution, let alone the energy and entropy changes in chemical reactions or even the distinct phases in the solid white mass visible in a closed vessel containing calcium carbonate raised to higher temperatures. Duhem carefully explained the sort of commonplace, not necessarily exotic, subject-wide or theoretical intricacies involved in rendering scientifically interesting phenomena amenable to the useful analysis and reporting of experimental results – intricacies that are lost on the amateur without necessarily being about microscopic entities, which just involves more of the same.

The subject is vast, with chemists outnumbering physicists or biologists and breaking ever new ground. Covering the breadth of the subject would be out of the question in a book, let alone a short article. I will focus on how the principal macroscopic concepts were clarified during the nineteenth century and an understanding of microscopic phenomena was developed in the first decades of the twentieth century, mentioning some controversial issues arising in its wake. This bears more on the acquisition of knowledge at the theoretical level. The massive accumulation of factual knowledge doesn't lend itself to a brief overview. But it contributes to the progress of chemistry along with conceptual developments and more theoretical knowledge to which it lends support and in terms of which it is formulated. This illustrates some aspects of progress in chemistry and some comments in the final section suggest that the epistemological approach seems to give the best account of it amongst the current philosophical views of progress.

2. The Development of the Basic Macroscopic Distinctions

The major advance in Dalton's time was the establishment of the law of definite or constant proportions. Although exceptions were recognised a century later with the discovery of non-stoichiometric compounds (Berthollides), it was considerably more robust than Dalton's law of multiple proportions, which was soon to see numerous exceptions with the recognition of homologous series of substances in organic chemistry. Definite proportions of elements in compounds were implicitly recognised by Lavoisier in his earlier study of the composition of water, but widely acknowledged shortly after the turn of the century with Proust's refutation of Berthollet's view that the strength of chemical affinity between given elements varies according to circumstances. Berthollet compared compounds with what came to be distinguished, after acceptance of the law, as saturated solutions, whose composition is fixed by the circumstance of the solvent being unable to take up more solute. We think, for example, of zinc forming a compound by reaction with sulphuric acid, whereas adding sugar to water forms a solution. But the process of adding zinc to sulphuric acid is like that of adding sugar to water. At first, the sugar, like the zinc, disappears. But beyond a certain point, no more zinc is taken up into the liquid and it remains as a solid, just as continually adding more sugar to the water eventually leads to two phases after the power of the liquid to take up the solid has been saturated. By Berthollet's lights, the combination proceeds in both cases with the production of a homogeneous product until the affinities are exhausted. Careful determination of the elemental proportions of a number of compounds led Proust to conclude that the proportions are the same, whatever the natural source or route of artificial synthesis, and remained constant as temperature and pressure varied. Although he investigated only a small number of compounds, further confirmation soon accrued as Berzelius and others analysed many more compounds and confirmed that they also exhibited constant proportions. A solution like brine, on the other hand, may be anything from a very dilute to a concentrated solution of salt in water, and the fixed proportions of saturated brine are only fixed at a specific temperature and pressure.

This decided an issue about the nature of air between Davy, who thought it to be a compound, and Dalton, who took it to be a homogeneous mixture with variable composition. It was supposed that whilst the component elements of compounds were chemically combined, the component substances of homogeneous mixtures (solutions) were merely mechanically mixed. In both cases, the properties of the original components are changed – the compound sodium chloride is neither a soft grey metal like sodium nor a greenish-yellow gas like chlorine (at room temperature and pressure), and brine is neither a white solid like sodium chloride nor a tasteless liquid like water. The law of constant proportions was thus taken to provide a criterion on the basis of which a definition of a compound could be formulated. But it only gave a criterion and provided no explanation of chemical combination. Later in the century, when solutions were studied more thoroughly, the idea of a purely mechanical mixture was understood to be an ideal, characterised by behaviour which very few solutions even approximated. Mixing two quantities of distinct paraffins in the liquid state, for example, yields a liquid with a volume equal to the sum of the volumes of the two paraffins (taken at the same temperature and pressure). But in general, when the result of mixing two quantities of different substances is homogeneous matter, the volume of the latter is not equal to the sum of the volumes of the separate quantities. This indicates some sort of interaction between the substances, suggesting that the distinction between compounds and solutions doesn't mark a clear-cut distinction between the occurrence of chemical combination (exhibition of chemical affinity) and its absence. The discovery of non-stoichiometric compounds whose elemental proportions do vary with temperature and pressure was a further nail in the coffin of any such simple correspondence.

The focus on homogeneous quantities of matter at the beginning of the nineteenth century is a vestige of Aristotle's understanding of mixts from which Lavoisier hadn't entirely extricated himself. He understood the transition from solid to liquid to involve a chemical change due to combination with caloric – a substance featuring in his list of elements – and the transition from liquid to gas likewise to be a chemical change involving combination with more caloric. (He distinguished the circumstance of caloric fixed in matter by such combination from that of free caloric entering a body and constituting an increase in temperature.) Thus, what came to be called the phase exhibited by a quantity of matter was at this time still held, as it was for Aristotle, to be characteristic of a substance. Indeed, this understanding lingers on in ordinary usage, as when the *Shorter Oxford English Dictionary* defines water as 'the liquid . . . which forms the main constituent of seas, lakes, rivers and rain . . .; it is a compound of hydrogen and oxygen (formula H_2O), and was formerly regarded as one of the four elements'. However, the compound of hydrogen and oxygen with formula H_2O is not necessarily liquid; ice and steam are water, as are heterogeneous mixtures of two or three different phases (solid, liquid, and gas) under conditions of equilibrium consistent with thermodynamics (specifically, the phase rule). There may be a phase-restricted usage of certain common substance names, as when 'water' is understood to exclude ice and steam; but such everyday usage diverges from scientific usage. In chemistry, a substance may in general occur in several phases (although there may be exceptions due to special circumstances, for example, heating a solid protein may well decompose it before it can attain the liquid or gas phase). The expression 'H_2O' is a compositional formula deriving from a gravimetric ratio of hydrogen to oxygen of 1:8, which is independent of the phase (and thus the temperature and pressure) of these elements and independent of the phase (and thus the temperature and pressure) of water.

We thus have the general conception of a mixture, which is a quantity of matter exhibiting several phases over which a number of substances are distributed. Quantities of matter typically encountered on the earth, such as the matter lying in the earth's crust yesterday, the matter constituting a particular person at any time during a specific interval, a stone, a flake of sawdust, and so on, are mixtures. A special case is a solution, which is a single phase (not necessarily liquid, but possibly solid or gas) containing several substances. An isolated quantity may consist of a single substance, that is be a pure sample of the substance, but such a thing is rare in nature. (Chemists typically speak of the degree of purity; distilled water, for example, may not be sufficiently pure for certain purposes.) Sometimes the heterogeneous character of a quantity of matter may not be apparent, as in the case of colloids consisting of small particles (smoke) or droplets (mist) of one phase distributed throughout another, or the calcium carbonate example referred to earlier. But numbers of substances and of phases are not arbitrarily related; the equilibrium between substances and phases in a mixture is governed by the laws of thermodynamics, and in particular, the phase rule.

3.   Thermodynamics: Theory at the Macroscopic Level

The caloric theory of warmth and heating finally proved untenable and was abandoned with the formulation of the principle of the conservation of energy or first law of thermodynamics. Oxygen and hydrogen gas could then be seen as genuine elements rather than compounds of base of oxygen and base of hydrogen with caloric as Lavoisier had proposed. But thermodynamics has done much more for chemistry than simply removed an obstacle to recognising that Lavoisier actually succeeded in isolating elemental hydrogen in his experimental work. Although the theory originated by coming to grips with the theoretical principles on which steam engines operated, the concepts of energy and entropy were harnessed, principally by Gibbs in his ground breaking work 'On the Equilibrium of Heterogeneous Substances' published by in two parts in 1876 and 1878, to provide what might be regarded as a pinnacle in the development of the notion of a chemical substance, and to offer the first workable theory of chemical combination. Despite some controversy and complementary developments in other areas this work is, in a sense, final: 'Gibbs's work . . . has required no correction since it was published, and remains to this day the foundation for the study of phase separation. The underlying principles are few, and rigorous' (Sengers 2002, p. 43). Chemists were on the whole slow to recognise the relevance of his work, despite Gibbs' efforts distributing offprints to many European scientists, and several important laws stated for the first time in Gibb's paper were subsequently discovered by others, usually by experiment, in ignorance of Gibbs' priority. The abstract nature of the theory and Gibbs' terse style of presentation combined with chemists' poor training in mathematics to make the theory inaccessible. But physical chemistry did emerge as a recognised branch of the subject at the end of the nineteenth century, with thermodynamics as one of its cornerstones.

One of the most important contributions to physical chemistry from this period was Gibbs' phase rule, presented in characteristically terse style by Gibbs himself (Gibbs 1948, pp. 96–97; for a clear exposition, see Denbigh 1981, pp. 182–195) and applied with great success in the early days in the Netherlands (Daub 1976). This law systematises a general conception of mixture. Like Aristotle, Gibbs recognised that a quantity of matter may comprise different homogeneous bodies, formed from different component substances, constituting a heterogeneous mixture. But whereas Aristotle thought that distinct homogeneous bodies comprise different substances, Gibbs allows that they may themselves be mixtures composed of some or all of the substances present in the quantity as a whole. Gibbs distinguished between the substances making up the total mass of the quantity of matter and the different homogeneous parts of a heterogeneous quantity. He calls the different homogenous bodies which 'differ in composition or state[,] different *phases* of the matter considered' (Gibbs 1948, p. 96), adding in an abstract 'all bodies which differ only in size and form [shape] being regarded as different examples of the same phase' (Gibbs 1948, p. 359). In general, a mixture comprises several coexisting phases over which several substances are distributed.

The concepts of phase and substance can be applied in simple situations without problem. At 0°C under normal atmospheric pressure, water forms a heterogeneous mixture of the solid and liquid – a mixture comprising one substance and two phases. Taste may reveal that a colourless single-phase liquid mixture comprises salt dissolved in water. Such a solution would have to be taken to a temperature below 0°C before it separates into a two-phase system, when the solid would comprise just water. But the interpretation of the constitution of heterogeneous systems and their behaviour with changing conditions can be considerably more complicated than these simple examples would suggest, and distinguishing phases and substances can be a ticklish problem. A systematic law was needed to resolve the many such issues coming to light at the end of the nineteenth century, and Gibbs' phase rule fitted the bill.

The rule prescribes a general pattern of behaviour characteristic of a quantity of matter comprising a single substance, thus serving as a criterion of purity. Specific quantitative details, such as temperatures at varying pressures of phase transitions, are characteristic of particular substances. For example, water has a triple point (with solid, liquid, and gas phases at equilibrium) at 0.01°C and 4.58 mmHg, distinguishing it from all other substances. A fair indication is given by water's freezing at 0°C and boiling at 100°C under normal atmospheric pressure. This would provide scientists visiting a new world with simple criteria (among others) for determining whether the material 'superficially' (as Putnam says in his twin-earth fantasy) similar to water is in fact the same substance

(Needham 2017). This general conception of mixture provides a criterion for the actual presence of several substances in a mixture, and lends itself to a mereological interpretation.

Thermodynamics bears on chemical combination by providing an account of how a system transforms into a more stable state, for example, by bringing different chemical substances into contact and allowing the process to proceed to equilibrium. Gibbs opened 'On the Equilibrium of Heterogeneous Substances' by considering the mechanical criterion of equilibrium of minimum energy. This is not the universal feature it was once taken to be, but is confined by a restriction which cannot be expressed within the purview of mechanics, namely, that the entropy is constant. Thermodynamics opens up new dimensions of change in a landscape invisible from the vista of mechanics, complementing the mechanical property of volume by dividing the mechanical property of mass into the masses of the distinct substances and adding the entropy. Within the new dimensions of change, Gibbs demonstrates that the criterion of minimum energy *along a line of constant entropy* is equivalent to maximum entropy *along a line of constant energy*. But these criteria are often not easily applied because conditions of constant energy or constant entropy are not always the natural controlling factors. Frequently, it is more practical to maintain constant volume and pressure, or even more usually when dealing with chemical reactions, to maintain constant temperature and pressure.

More practically applicable criteria of equilibrium would be expressed in terms of more practically accessible controlling factors which include the intensive properties of temperature, T, pressure, P, and the chemical potential, μ_k, of the kth substance in the quantity of matter to which the variables apply. For this purpose, instead of an extensive magnitude like the energy, which is a function, $U(S, V, N_1, \ldots, N_k, \ldots)$, of the other extensive variables, entropy, volume, and amounts (masses) of the 1st, . . ., kth . . . substance, this has to be transposed into a function of a set of independent variables in which one or more of the extensive variables has been replaced by intensive variables. These functions are called thermodynamic potentials. The intensive variables are defined in terms of the extensive variables as derivatives of one with respect to the other, thus $T = \partial U/\partial S$, $P = -\partial U/\partial V$ and $\mu_k = \partial U/\partial N_k$, and the desired functions are obtained by a general procedure called Legendre transformation, or rather as partial Legendre transformations where some but not all the extensive variables are replaced by intensive variables (Callen 1985, pp. 131–152). The function so obtained of the variables S, P, N_1, . . ., N_k, . . ., where the volume has been replaced by the pressure, is the *enthalpy*, usually symbolised by H, which is related to the energy by $H = U + PV$. At constant pressure, a change, ΔH, in the enthalpy is given by $\Delta U + P\Delta V$, and so the enthalpy change captures the energy change except any energy gained or lost in doing expansion or contraction work against the constant pressure. Given that the pressure is constant, it measures the available energy.

Perhaps the most familiar thermodynamic potential for chemists is the *Gibbs free energy*, usually symbolised by G. This is a function of the variables T, P, N_1, . . ., N_k, . . ., where both the entropy and the energy have been replaced by the temperature and pressure. It is related to the energy and entropy by $G = U + PV - TS$, or $G = H - TS$. The criterion of equilibrium for a system at constant temperature and pressure, which are the normal control factors where chemical reactions are under consideration, is that the Gibbs potential is a minimum. In other words, a change in the Gibbs potential, ΔG, must be negative if it is to lead to a minimum, and at constant temperature and pressure,

$$\Delta G = \Delta H - T\Delta S = \Delta U + P\Delta V - T\Delta S.$$

We can see from this that a reduction in energy, for which ΔU is negative, tends to favour a reduction in the Gibbs potential and a more stable system, whereas an increase tends to have to opposite effect. But the change in energy is not the only factor, and what would be a destabilising increase in energy may well be counterbalanced by a stabilising increase in entropy, ΔS. Of course, both energy and entropy changes may operate in the same direction, but cases where there is counterbalancing have been particularly interesting. It used to be thought that chemical reactions will only occur spontaneously, at a given temperature and pressure, on mixing the reactants, if the process is exothermic, that is evolves heat. On the basis of the first law, this was taken to indicate that reactions occur spontaneously only when there is a net reduction in energy, and the idea was formulated in the principle of maximum work by the Danish chemist Thomsen in the middle of the nineteenth century, and famously upheld to Duhem's cost by Marcelin Berthelot in the latter half of the century. When Duhem (1886) mobilised the thermodynamic potentials to show that Berthelot was wrong and endothermic reactions (involving heating by the surroundings) are possible because the entropy gain can offset the energy gain, Berthelot arranged that Duhem's thesis was failed.

Another interesting application is in understanding the breakdown of the law of constant proportions with the so-called berthollides or non-stoichiometric compounds discovered at the beginning of the twentieth century (Kurnakow 1914). For example, ferrous oxide rarely occurs as a daltonide obeying the law of constant proportions with a stoichiometric formula FeO. More commonly, the proportions cannot be expressed as a ratio of small integers and vary with conditions of temperature and pressure within the range $Fe_{0.84}O$ to $Fe_{0.95}O$, where the proportion of iron never attains the level of the ideal stoichiometric formula. The deficiency of ferrous ions in the solid crystal gives rise to holes or defects randomly distributed about the lattice, with a greater entropy than the ideal stoichiometric structure. There is an energy cost to this entropy gain because electrical neutrality must be maintained, requiring the transformation of some ferrous ions, Fe^{2+}, into ferric ions, Fe^{3+}, which calls for energy to remove a third electron. The overall stability is governed by the thermodynamic potentials, which come down in favour of a non-stoichiometric compound.

This reaffirms the close tie between compounds and solutions, whose stability is also subject to the same thermodynamic considerations. The idea of a purely mechanical mixture involving no intermolecular forces could be theoretically understood as involving a Gibbs free energy change on mixing the ingredients due entirely to the entropy contribution – the so-called entropy of mixing. Ideal solutions are defined in this way, and as mentioned earlier, are approximately realised by very few real-life solutions.

For a chemical change (reaction or dissolution) to occur at constant temperature and pressure, then, the overall Gibbs potential must decrease. But this says nothing about the rate at which the reaction takes place. Classical thermodynamics deals with equilibrium states, and has no time variable in its equations indicating how fast thermodynamically possible processes actually proceed. Venerable textbooks are pleased to say that reactions in which the Gibbs potential decreases occur 'spontaneously'. But as in the case of a mixture of two moles of hydrogen to one of oxygen, this could take eons in the absence of a catalyst or a spark to initiate the chain reaction. Factors governing the rate at which reactions proceed, and the molecular mechanisms by which the overall conversion of the reactants to the products takes place, which may involve intermediate species generated and subsequently consumed, are studied in chemical kinetics. This is very important in biochemistry, where enzymes function as catalysts controlling the rate at which reactions occur in living organisms. The overriding principle, however, is that the rate of a reaction can be controlled by circumstances such as the presence of a catalyst only provided the reaction is thermodynamically feasible. Unless the appropriate thermodynamic potential for the conditions at hand allows it, the reaction cannot proceed. No kinetic considerations can bring about what is thermodynamically impossible.

Thermodynamics itself has been extended with the introduction of a time variable into a theory of irreversible thermodynamics, based on the principle that thermodynamic equilibrium holds at points for instants and functions of state such as temperature vary continuously over space and time. Beyond the limits of applicability of this theory, chaos sets it and the study of chaotic systems has developed from early ideas of (Alan) Turing structures in the 1950s into a thriving field of chemical research (Kondepudi and Prigogine 1998).

4.   Other Aspects of Progress in the Nineteenth Century

Nineteenth century chemistry made great strides in understanding how to characterise and synthesise enormous numbers of substances, inorganic and organic, both naturally occurring and new to the world. Explicit generalisation led to a sharpening of the concept of a substance and associated concepts.

Aristotelian doctrine finally gave way to the phase-independent notion of substance that came into play after the broad acceptance of the law of constant proportions. Aristotle had gotten it wrong, but he did at least have notions of substance and a thesis about its relation to phase, which is more than can be said for ancient atomism. With the law of definite proportions, compounds are distinguished from mere mixtures. There is some talk of proportions in Aristotle, but it is unclear what the underlying measure of amount is since there seems to be no notion of mass in ancient science and certainly no account of proportions by weight or mass in Aristotle. Extant texts give no hint of a corresponding notion deriving from the ancient atomists. But at the beginning of the nineteenth century, Dalton proposed a version of atomism distinguished from its predecessors by construing atoms as minimal amounts of the elements.

Dalton understood macroscopic proportions to derive from the *proportion by mass* in which atoms of different elements stand to one another and the fixed ratio of the *numbers* of atoms of different elements in a given compound. He wanted to determine relative atomic weights, that is given, say, hydrogen as standard, how much heavier were other, say oxygen, atoms. But the only empirical data he had to go on was the gravimetric proportions of elements in compounds. In order to derive relative atomic weights from this, he also needed to know the relative numbers of, say, hydrogen and oxygen atoms in a quantity of water. (He had no conception of molecules; compounds comprise an array of atoms held in place by caloric repulsion and a minimal amount of water comprises an oxygen atom paired with *any* one of the nearest hydrogen atoms.) Dalton assumed there were as many atoms of hydrogen as of oxygen in water, in effect representing water as HO, in accordance with a 'rule of greatest simplicity'. He tried to justify this general assumption with an argument about the stability of an arrangement of closely packed atoms on the basis of further assumptions of a mutual repulsion between atoms of the same but not of different kinds (in virtue of how the caloric is arranged around the massive portion of the atom), for which there was no independent empirical evidence.[2]

Many of Dalton's contemporaries were sceptical and atomism remained controversial in chemistry throughout the nineteenth century, presumably because macroscopic proportions are not independently explained by atomic masses standing in the same proportion. After Wöhler's laboratory synthesis of urea in 1828 by heating ammonium cyanate, demonstrating that substances previously known only as products of living processes could have a wholly non-biological source, tremendous strides were taken in organic chemistry. Although organic chemistry supplied innumerable cases confirming the law of constant proportions, it also provided many counterexamples to Dalton's 'law' of multiple proportions with the discovery of systematic series of homologous compounds such as the paraffins (alkanes), with compositional formulas of the form C_nH_{2n+2}, olefins (alkenes) C_nH_{2n}, alcohols $C_nH_{2n+2}O$, aldehydes $C_nH_{2n}O$, etc. It also became apparent in organic chemistry (and only much later in inorganic chemistry) that composition doesn't necessarily characterise substances.

Wöhler's analysis of cyanic acid in 1825 showed it to have precisely the same composition as fulminic acid previously analysed by Liebig in 1824, and so with compositional formula HCNO in common, although they were clearly distinct substances by the macroscopic criteria known at the time. Organic chemistry soon provided many other cases establishing that difference in properties is not simply due to difference in composition. Berzelius coined the term 'isomerism' (after the Greek for being composed of equal parts) for this phenomenon, calling distinct substances with the same composition isomers. Compositional formulas represent the gravimetric proportions of constituent elements with a change of scale relative to some standard, such as hydrogen. Structural formulas were developed from this basis to provide distinct representations of isomers, sometimes with parts representing the functional group common to members of a homologous series. Alcohols, for example, have in common the functional group OH, in terms of which similarities in the chemistry of homologous compounds are naturally represented.

Structural formulas have in modern times been taken to provide evidence supporting atomism in the nineteenth century. But there was by no means universal agreement on this at the time, with sceptics and agnostics claiming that the virtues of atomism were merely pedagogical, facilitating the visualisation of the representations. The vindication of the agnostic position was provided in retrospect by Duhem at the turn of the century. Although other evidence was by then beginning to appear which put a different complexion on the issue and which Duhem ignored, he was the man with the logical acumen to provide a concise explication of the theory of chemical structure underlying structural formulas accommodating all and only the evidence available throughout most of the nineteenth century. He built up the topologically ordered structures representing distinct isomers beginning with the facts of constant proportions. This justified the assignment of what he called a proportional number to each element, represented by the familiar letters H for hydrogen, O for oxygen, C for carbon, etc., and all compounds could be represented by a compositional formula with their constituent elements standing to one another in integral multiples of these proportional numbers. The possibilities of deriving substances by substitution from archetypal substances and organising substances into chemical types led to the full-blown structural formulas, distinguishing for example, the isomers dimethyl ether and ethyl alcohol, with compositional formula C_2H_6O in common, with structural formulas $(CH_3)_2O$ and C_2H_5OH, respectively, or in more detail:

```
    H   H                      H  H
    |   |                      |  |
  H—C—O—C—H                  H—C—C—O—H
    |   |                      |  |
    H   H                      H  H
```

Stereoisomerism was accommodated by saying that the topological ordering of structural formulas is complemented with full geometry, allowing
the representation of stereoisomers by mirror images of one another.[3]
Duhem thus provides for an incomplete theory of chemical structure,
neutral on the question of what might have been articulated under the
term 'atomism' or any alternative to it, open to further explication when
empirical and theoretical advances warrant it.

5. Atomic and Molecular Structure

Some say Kant's view of chemistry as inferior to physics because entirely
qualitative was reputed by Lavoisier's estimation of the relative gravimetric proportions of elements in water and the ensuing confirmation of the
law of definite proportions. Others have claimed Boyle took a major stride
towards modern chemistry with his corpuscular speculations. Lavoisier's
was the crucial experimentally confirmed advance, and he notoriously
rejected atomism:

> if, by the term *elements*, we mean to express those simple and indivis-
> ible atoms of which matter is composed, it is extremely probable we
> know nothing at all about them; but if we apply the term *elements*,
> or *principles of bodies*, to express our idea of the last point which
> analysis is capable of reaching, we must admit, as elements, all the
> substances into which we are capable, by any means, to reduce bodies
> by decomposition.
>
> (Lavoisier 1789 [1965], p. xxiv)

where 'the last point [of] analysis' is a substance that cannot be decom-
posed into other substances, as opposed to the mechanical idea of an
indivisible atom. A century later, Duhem's appeal to Gibbs' thermodynamics to refute the last vestiges of the delimitation between physical
and chemical phenomena[4] was a sign of the first real rapprochement
between physics and chemistry (Duhem 1886). This band was to deepen
with revolutionary experimental discoveries giving the microstructure of
matter significant empirical support at the turn of the century and the
emergence of the old quantum theory. The development of spectroscopy
in the nineteenth century gave little hint of atomic structure (McGucken
1969). But the empirical equations formulated by Rydberg famously
yielded to an atomic interpretation at the hand of Bohr in 1913. Shortly

thereafter, Bjerrum deployed the old quantum theory to infrared spectra, solving the decades-old problem for atomism of the anomalous specific heat ratio of diatomic molecules because of an unsuspected temperature variation.

Despite Bohr's promising explanation of electronic spectra, these new ideas about atomic structure were not straightforwardly applicable to chemistry. Gilbert Lewis suggested that molecules are built by linking atoms with chemical bonds – either ionic, when one or more electrons are transferred from one atom to another, or covalent, comprising a pair of electrons, one donated from each atom. But if the atomic interpretation of structural formulas is to be believed – particularly of stereoisomers – then these covalent bonds are directed along specific directions in space, and it is a mystery how this could result from the spherically symmetrical Bohr atom. Lewis proposed instead a cubic atom, with stationary valence (outer-shell) electrons occupying the corners of the cube. This was easy prey for Bohr (1965, p. 35), who pointed out in his 1922 Nobel acceptance speech that stationary charges conflicted with Earnshaw's theorem of classical electromagnetism. But he had nothing to say about how the spatial structure of molecules is related to atomic structure. Lewis came to see the possibility of a reconciliation when Bohr abandoned the idea of electrons revolving in rings in favour of electrons in shells, but remained clear about the status of the chemical evidence:

> No generalization of science, even if we include those capable of exact mathematical statement, has ever achieved a greater success in assembling in simple form a multitude of heterogeneous observations than this group of ideas which we call structural theory.
>
> (Lewis 1923, pp. 20–21)

This called for a notion of a bond which could certainly be more deeply understood, but whose status he took to be established:

> The valence of an atom in an organic molecule represents the fixed number of bonds which tie this atom to other atoms. Moreover in the mind of the organic chemist the chemical bond is no mere abstraction; it is a definite physical reality, a something which binds atom to atom. Although the nature of the tie remained mysterious, yet the hypothesis of the bond was amply justified by the signal adequacy of the simple theory of molecular structure to which it gave rise.
>
> (Lewis 1923, p. 67)

This is not to say that the claims of structural theory didn't go unchallenged, even within organic chemistry. In a landmark 1920 paper Hermann Staudinger advocated abandoning nineteenth century concepts of purity in order that materials such as rubber, starch, and cellulose could be recognised as single substances despite consisting of polymers with a spread of chain lengths.

> under Staudinger's interpretation, it was not anymore possible to characterize pure substances by one structural formula. Also, something like *the* molecular mass of a pure polymer did not exist anymore. Finally, such pure polymers did not possess sharp values for physical properties like melting and boiling point. All this meant that much of the canonical method of classical organic chemistry became useless for the study of pure substances in the sense of Staudinger.
>
> (Zandvoort 1988, p. 501)

Staudinger's innovations were resisted by the chemical community for years, but he was finally awarded the Nobel Prize in 1953 for his introduction of the concept of macromolecular substances. This work paved the way to the detailed study of polymers and development of protein chemistry and DNA synthesis.

Lewis's ideas about electrons in atoms in molecules soon inspired Langmuir, Robinson, and others to develop fruitful theories of chemical reaction mechanisms, even if they couldn't at first be related to radioactivity, spectroscopy, and scattering experiments that provided physicists with the first evidence of atoms. But applications of the new quantum mechanics from 1927 to some simple tractable cases pointed more firmly to links with physics, from which two schools of quantum chemists emerged.

The result of the first application of quantum mechanics to the calculation of the interatomic distance and binding energy of the diatomic hydrogen molecule by Heitler and London (1927) was 'only fair' (Levine 2000, p. 413). Taken in isolation, it would hardly be considered as confirming this treatment of the hydrogen bond, perhaps even as falsifying it. But other considerations bore on the case. In particular, it suggested a resolution of the apparent paradox facing Lewis's conception of a covalent bond comprising shared electrons, which would be unstable on classical theory because of mutual repulsion. Despite the quantitative inadequacy, the treatment gave some definite indication of distinctly quantum mechanical sources of bond stability: the indistinguishability of microentities of the same kind and electron spin, jointly requiring a so-called antisymmetric solution to the wave equation. Accordingly, since Heitler and London's treatment was not an exact mathematical solution, this prompted a succession of attempts to improve one or other aspect of the approximation procedure, leading to James and Coolidge's (1933) very good estimate of the bonding energy. Around the same time, Burrau (1927) published an exact solution of the single electron orbital (given the clamped nuclei assumption of the so-called Born-Oppenheimer approximation) in the hydrogen molecule ion, H_2^+, first discovered in cathode ray tubes by J. J. Thomson, demonstrating that bonding doesn't require electron pairs.

There is an exchange integral arising from the indistinguishability of electrons in this case too, corresponding to the single electron being associated either with the one atom or the other, and this leads to all the calculated binding energy of H_2^+.

These are the two simplest cases of bonding. Extending these results to provide a general theory of bonding was no mean feat. Dirac famously claimed that with the advent of quantum mechanics

> The underlying physical laws necessary for the mathematical theory of a large part of physics and the whole of chemistry are thus completely known, and the difficulty is only that the exact application of these laws leads to equations much too complicated to be solvable. It therefore becomes desirable that approximate practical methods of applying quantum mechanics should be developed, which can lead to an explanation of the main features of complex atomic systems without too much computation.
>
> (Dirac 1929, p. 714)

But how much chemistry thereby came to be known is a moot point. Whatever comfort was afforded by postulating that all of chemistry is *in principle* derivable (which can't be justified until the presumed reduction is actually carried through), it doesn't yield any actual knowledge about the subject matter of interest to chemists without successful application calling for approximate methods and corresponding justifications of specified degrees of accuracy and precision.

Building on Heitler and London's work, Pauling promoted his Valence Bond (VB) approach in which molecules were construed as built up from individual atoms, facilitating an interpretation in terms of ideas from the classical structural theory of organic chemistry with which chemists were familiar and Lewis had sought to retain. But Coulson famously declared 'a chemical bond is not a real thing: it does not exist: no one has ever seen it, no one ever can. It is a figment of our own imagination' (Coulson 1955, p. 2084). By 1966 the awarding of the Nobel Prize in chemistry to Robert Mulliken apparently acknowledged the superiority of the alternative, quantum mechanically more purist, Molecular Orbital (MO) approach he and Friedrich Hund had developed. This approach threatened the classical idea of a bond as a localised material connecting link between specific pairs of atoms in a molecule that Lewis had singled out and Pauling had retained. MOs were delocalised, defined over the whole molecule in which the individual atoms were no longer seen as the integral parts that they were on the classical theory. 'It is really not necessary', according to Mulliken, 'to think of valence bonds as existing in the molecule' (1931, p. 369). After 1960, as more tractable calculations become feasible for ever more complex systems when formulated in MO terms, this approach gained the upper hand.[5] But it meant that the world of quantum chemistry became ever more foreign to the ideas which chemists had found so useful. This only reinforced the view of many chemists, who for many decades remained sceptical of the value of all this number crunching that had produced little in the way of solid chemical predictions.[6] In that respect the tide has turned, however, and quantum chemistry has over recent decades been earning its keep, although not by reinstating the structuralist conception of molecules as atoms linked by single, double, and occasionally triple bonds. Theory always provides a delocalised description. Sometimes equivalent localised descriptions corresponding more nearly to classical bonds are available, namely, when the number of bound neighbours, the number of available valence electrons and the number of valence atomic orbitals participating in the bonding are all equal (known as Hund's criteria; Kutzelnigg 1996, pp. 578–579). In BeH_2, for example, these numbers are 2 for Be and 1 for each H atom. Even here, however, the classical picture cannot lay claim to being uniquely true, but is equivalent to a delocalised one.

Bonds, according to classical structure theory, bind nuclei in a stable molecular structure. But even the nuclear framework underlying the traditional notion of a molecule has, according to Woolley, Sutcliffe and Primas, a questionable basis in quantum mechanics. Nuclei of the same kind should, like electrons, be indistinguishable, which should be reflected in a permutation symmetry of their quantum mechanical description. Quantum mechanics, they argue, suggests that traditional structure is not an intrinsic feature of molecules but acquired as a result of interaction with the environment.[7] Not all the cognoscenti agree, but Claverie and Diner (1980, p. 79) say

> no convincing derivation of . . . classical [molecular] structure, based upon a strict quantum treatment, exists at the present time, essentially because this connection is simply a special case to the general and profound problem of the explanation of classical physics, starting from a strict quantum theoretical basis. . . . [A]s Bohm [1951, p. 625] [puts it]: "quantum theory . . . does not deduce classical concepts as limiting cases of quantum concepts (as, for example, one deduces Newtonian mechanics as a limiting case of special relativity)". This situation manifests itself very clearly in the problem of the "spreading of the wave-packet" . . . [E]ven initially localized wave packets do not enable us to consider molecules as well-localized objects . . . Spreading over a length of macroscopic order . . . occurs over times of the order of one minute.

A lifetime of investigating processes of transformations of substances provided another source of objections to Dirac's programme. Whereas quantum mechanics gives 'a deterministic time-reversible description'

Chemical reactions correspond to irreversible processes creating entropy. That is, of course, a very basic aspect of chemistry, which shows that it is not reducible to classical dynamics or quantum mechanics. Chemical reactions belong to the same category as transport processes, viscosity, and thermal conductivity, which are all related to irreversible processes.

(Prigogine 2003, p. 128)

Following the course of processes over time falls within chemical kinetics, a discipline first appearing towards the end of the nineteenth century that played a central role in understanding the mechanisms of chemical reactions throughout the twentieth and beyond. With modern techniques, it is possible to follow the course of extremely fast processes, of the order of femtoseconds. Space hasn't allowed following up progress in understanding chemical processes in any detail, however.

6. Final Comments

Foregoing sections relate something of the ups and downs in the progress of chemistry in pursuit of its goal of investigating the variety of substances and how and why they interact by developing, adjusting, and clarifying the conceptual apparatus in conjunction with the development of theory. The concept of a substance has been finessed to accommodate a systematic relation to phase and the notion of a mixture has been properly understood. This understanding has been deepened by relating macroscopic features to microscopic phenomena. Reduction would be progress in the direction of deepening the unity of science, but has yet to be established. As matters stand, (actual) progress (as opposed to what would, or possibly will, be progressive) towards a more modest goal has been achieved by overcoming obstacles to construing macroscopic and microscopic facts and theory as mutually consistent and complementary. Articulating the actual steps poses a challenge, however. Whether Dalton made any advance on Lavoisier's discovery of the composition of water and if so, what, is controversial. Gravimetric proportions were more accurately determined. But I've suggested that only the less committed claims captured by Duhem's interpretation of structural formulas were justified, not Daltonian atomism. (This doesn't detract from its value as qualified speculation and interpretation for the promotion of scientific progress.) Positive evidence for atomic structure began to accrue at the very end of the nineteenth century. But although it took some time for a basic theoretical conception of molecular structure adequate for chemistry to emerge meeting Lewis's objections, and the full-blown Schrödinger equation for a molecule is still spherically symmetric, some notion of microstructure has served chemistry well, facilitating for example, a rich theory of organic reaction mechanisms.

Bird (2007) points out that problem-solving and understanding, sometimes taken to be criteria of scientific progress, involve knowledge – of solutions to puzzles or theories that correctly explain. As well as guiding and facilitating the definition of useful problems, knowledge makes a claim to truth, which is obscured in Dellsén's understanding of understanding. Explanation in terms of minimal idealisations encapsulating a simplified context, overemphasising factors which would otherwise make models complex and intractable and overlooking 'explanatorily irrelevant factors' by neglecting air resistance, assuming infinite populations, neglecting intermolecular collisions, and so on, render models, in Dellsén's view, 'not true' (2016, p. 81). But this is surely a misleading oversimplification, disregarding the estimated degree of precision and acknowledgement of systematic error characteristic of professionally justified empirical claims. Such models are only accepted as holding approximately within specified limits of approximation of the relevant factors and on the explicit understanding that the neglected factors don't come significantly (beyond the specified limits of approximation) into play. The kinetic model of the ideal gas law neglecting intermolecular collisions holds only under low pressure; where greater precision calls, more accurate gas laws are employed. Quantum chemists seeking to understand the source of chemical bonding have debated whether it is essentially electrostatic attraction between nuclei and electron clouds or increased kinetic energy as the electrons move more freely around the nuclei. They are after the truth. Another example of insisting on the correct explanation is the critique of the secondary electrostatic interaction model formulated in 1990 to explain differences in binding strengths between multiple hydrogen bonded arrays. Although successful in predicting experimentally determined binding strengths, Lubbe et al. (2017, 2019) endeavour to replace arbitrarily oversimplified features with a more accurate and sophisticated picture.

As I emphasised in the Introduction, the epistemological approach isn't confined to high-level theoretical progress. The accumulation of facts (as distinct from faked 'discoveries' and spurious observations) is as important to the progress of chemistry as the considerable body of knowledge accumulated by naturalists is to biology, as is knowhow embodied in the trained eye and meticulously applied experimental techniques (cf. Mizrahi 2013). Laboriously acquired knowledge of the properties of nicotine and techniques of separation famously permitted Jean Servais Stas to identify it in the body of a murder victim in 1850. This is the stuff of analytic chemistry, which deploys an encyclopaedic battery of factual knowledge in the identification of substances in mixtures (advancing the understanding of the constitution of matter in various contexts). Organic synthesis similarly calls upon vast tracts of accumulated chemical knowledge, as does detecting signs of disease in blood, controlling food and searching for life in the universe. It is apposite to point out that progress needn't be abrupt, startling or revolutionary. Progress can be steady, gradual, or minute, and if less significant than other advances, is progress neverthe-less. Such is the determination of melting and boiling points and other physical properties characterising yet another alkane, which would hardly further understanding of alkanes, although it was a great help in discerning the contents of crude oil (Edgar et al. 1929).

Notes

1. Rather than following a particular stipulative definition of 'understanding', the strategy will be to illustrate cases which any adequate analysis would have to accommodate.
2. The compositional formula H_2O now accepted for water was determined without resort to atomic hypotheses on purely chemical grounds concerning the reactions of alcohols and ethers by Williamson in 1851 (Duhem 1902, pp. 66–67).
3. The details of this account of chemical formulas are meticulously spelt out in Duhem (1902). The essential points are summarised in Needham (1996, 2008).
4. Throughout the nineteenth century, it was common for chemists to distinguish between 'physical properties' of substances such as melting and boiling points, latent heats etc., and 'chemical properties' concerning their reactivity with other substances.
5. For a more general comparison between the VB and MO approaches, see Kutzelnigg (1996, pp. 576–578).
6. A fair indication of this attitude is given by the science historian Stephen Brush, himself 'a former Coulson student who abandoned quantum chemistry in the late 1950s because of dissatisfaction with the scientific value of the results obtained by MO calculations' (1999, p. 279).
7. See Weininger (1984) and Woolley (1988) for accessible discussions.

References

Benfey, O. T. (1963), "Concepts of Time in Chemistry", *Journal of Chemical Education*, 40, 574–577.

Bird, Alexander (2007), "What is Scientific Progress?", *Noûs*, 41, 64–89.

Bohm, D. (1951), *Quantum Theory*, Prentice-Hall, New York.

Bohr, Niels (1965), "The Structure of the Atom", in *Nobel Lectures, Physics 1922–1941*, Elsevier Publishing Company, Amsterdam, 7–43.

Brush, Stephen (1999), "Dynamics of Theory Change in Chemistry: Benzene and Molecular Orbitals, 1945–1980", *Studies in History and Philosophy of Science*, 30, 263–302.

Burrau, Olaf (1927), "Berechnung des Energiewertes des Wasserstoffmolekel-Ions (H_2^+)", *Videnskabernes Selskab Matematisk-Fysiske Meddelelser*, 7, 1–18".

Callen, Herbert B. (1985), *Thermodynamics and an Introduction to Thermostatistics*, Wiley, New York.

Claverie, P. and S. Diner (1980), "The Concept of Molecular Structure in Quantum Theory: Interpretation Problems", *Israel Journal of Chemistry*, 19, 54–81.

Coulson, C. A. (1955), "The Contributions of Wave Mechanics to Chemistry", *Journal of the Chemical Society*, 2069–2084.

Daub, Edward E. (1976), "Gibbs' Phase Rule: A Centenary Retrospect", *Journal of Chemical Education*, 53 (December), 747–751.

Dellsén, Finnur (2016), "Scientific Progress: Knowledge Versus Understanding", *Studies in History and Philosophy of Science*, 56, 72–83.

Denbigh, Kenneth (1981), *The Principles of Chemical Equilibrium*, 4th. ed., Cambridge University Press, Cambridge.

Dirac, P. A. M. (1929), "The Quantum Mechanics of Many-Electron Systems", *Proceedings of the Royal Society of London*, A123, 714–733.

Duhem, Pierre (1886), *Le potentiel thermodynamique et ses applications à la mécanique chimique et à l'étude des phénomènes électriques*, A. Hermann, Paris.

Duhem, Pierre (1902), *Le mixte et la combinaison chimique: Essai sur l'évolution d'une idée*, C. Naud, Paris; reprinted Fayard, Paris, 1985. Translated in *Mixture and Chemical Combination, and Related Essays*, by Paul Needham, Kluwer, Dordrecht, pp. 1–118.

Edgar, G., G. Calingaert and R. E. Marke (1929), "The Preparation and Properties of the Isomeric Heptanes. Part I. Preparation; II. Physical properties", *Journal of the American Chemical Society*, 51, 1483–1491, 1540–1550.

Gibbs, J. W. (1876–1878), "On the Equilibrium of Heterogeneous Substances", *Transactions of the Connecticut Academy of Arts and Sciences*, 3, 108–248, 343–520; reprinted in *The Collected Works of J. Willard Gibbs*, Vol. I, Yale University Press, New Haven, 1948, pp. 55–353.

Heitler, Walter and Fritz London (1927), "Wechselwirkung neutraler Atome und homöopolare Bindung nach der Quantenmechanik", *Zeitschrift für Physik*, 44, 455–472. Trans. in Hinne Hettema, *Quantum Chemistry: Classic Scientific Papers*, World Scientific, Singapore, 2000, pp. 140–155.

James, Hubert M. and A. S. Coolidge (1933), "The Ground State of the Hydrogen Molecule", *Journal of Chemical Physics*, 1, 825–834.

Kondepudi, Dilip and Ilya Prigogine (1998), *Modern Thermodynamics*, John Wiley & Sons, New York.

Kurnakow, N. S. (1914), "Verbindung und chemisches Individuum", *Zeitschrift für anorganische und allgemeine Chemie*, 88, 109–127.

Kutzelnigg, W. (1996), "Friedrich Hund and Chemistry", *Angewandte Chemie, International Edition*, 35, 573–586.

Lavoisier, Antoine-Laurent (1789 [1965]), *Elements of Chemistry*, Dover Publications, New York. Translated by Robert Kerr.

Levine, Ira (2000), *Quantum Chemistry*, 5th ed., Prentice Hall, Hoboken, NJ.

Lewis, G. N. (1923), *Valence and the Structure of Atoms and Molecules*, Chemical Catalog Co., New York; Dover reprint, New York.

Lubbe, Stephanie C. C. van der and Célia Fonseca Guerra (2017), "Hydrogen-Bond Strength of CC and GG Pairs Determined by Steric Repulsion: Electrostatics and Charge Transfer Overruled", *Chemistry – A European Journal*, 23, 10249–10253.

Lubbe, Stephanie C. C. van der, Francesco Zaccaria, Xiaobo Sun and Célia Fonseca Guerra (2019), "Secondary Electrostatic Interaction Model Revised: Prediction Comes Mainly from Measuring Charge Accumulation in Hydrogen-Bonded Monomers", *Journal of the American Chemical Society*, 141, 4878–85.

McGucken, W. (1969), *Nineteenth-Century Spectroscopy: Development of the Understanding of Spectra, 1802–1897*, Johns Hopkins University Press, Baltimore.

Mizrahi, Moti (2013), "What is Scientific Progress? Lessons From Scientific Practice", *Journal for General Philosophy of Science*, 44 (2), 375–390.

Mulliken, R. S. (1931), "Bonding Power of Electrons and Theory of Valence", *Chemical Reviews*, 9, 347–388.

Needham, Paul (1996), "Substitution: Duhem's Explication of a Chemical Paradigm", *Perspectives on Science*, 4, 408–433.

—— (2008), "Resisting Chemical Atomism: Duhem's Argument", *Philosophy of Science*, 75, 921–931.

—— (2017), "Determining Sameness of Substance", *British Journal for the Philosophy of Science*, 68 (4), 953–979.

Prigogine, Ilja (2003), "Chemical Kinetics and Dynamics", *Annals of the New York Academy of Sciences*, 988, 128–133.

Sengers, Johanna Levelt (2002), *How Fluids Unmix: Discoveries by the School of Van der Waals and Kamerlingh Onnes*, Koninklijke Nederlandse Adakamie van Wetenschappen, Amsterdam.

Weininger, S. J. (1984), "The Molecular Structure Conundrum: Can Classical Chemistry be Reduced to Quantum Chemistry?", *Journal of Chemical Education*, 61 (11), 939–944.

Woolley, R. G. (1988), "Must a Molecule Have a Shape?", *New Scientist*, 120 (22 October), 53–57.

Zandvoort, H. (1988), "Macromolecules, Dogmatism, and Scientific Change: The Prehistory of Polymer Chemistry as Testing Ground for Philosophy of Science", *Studies in History and Philosophy of Science*, 19, 489–515.

8 Epigenetic Inheritance and Progress in Modern Biology

A Developmental System Approach

Eva Jablonka

Progress is a notoriously elusive notion. It implies there is a normative standard relative to which a historical trend can be described. The term is used not only in the context of human affairs that are guided by societal or scientific norms, but is also used and debated within evolutionary biology. The evolutionary notion of progress, which has its roots in enlightenment ideas of social improvement, is measured against standards such as increased adaptability, increased complexity, and increased number of levels of hierarchical organisation (for discussions Nitecki 1988; Ruse 1996). In the last 25 years, evolutionary progress has been studied within the framework of major evolutionary transitions (Maynard Smith and Szathmáry 1995; see Calcott and Sterelny 2011 for extensions and qualifications), where an evolutionary transition involves the addition of new processes and structures for acquiring, storing, processing, and transmitting information, and is associated with the addition of a new level of hierarchical organisation. The study of every *particular* major transition requires ecological, developmental, and genetic considerations, which interact to form the enabling conditions for that evolutionary transition. Hence, an approach that stresses the stabilising or the destabilising interactions among different extrinsic and intrinsic factors, an evolutionary-ecological-developmental system approach, is required for an in-depth study of evolutionary transitions (see e.g. Gilbert 2019).

The standards according to which *scientific* progress is defined partially overlap those suggested for evolutionary progress during evolutionary transitions. Some philosophers suggest that scientific progress entails greater approximation to the Truth, others that it requires an increase in Knowledge (a notion related to information), and still others that it involves an increased ability to define and solve problems, a notion that is related to adaptability (Shan 2019; see Niiniluoto 2019 for a review of scientific progress from these different perspectives). Since scientific research is affected by overlapping and interlinked economic, cultural, political, institutional, technological, empirical, and theoretical factors, scientific progress too has many interlinked aspects.

DOI: 10.4324/9781003165859-11

For sociologically oriented philosophers, typically those who are interested in periods of theory change and problematise the notion of progress, the systemic nature of science, which includes social, cultural, political, and technological aspects and which can explain both periods of system stability and instability, is of the essence (Fleck 1935/1979; Kuhn 1970). This system approach to scientific progress is particularly useful during periods of theory change when the contribution of a particular, potentially destabilising research results to the scientific progress of a particular disciple is denied or debated. Since the developmental system (DS) approach highlights the central role of self-sustaining interactions among different internal and external factors that render the dynamics of the system robust, it is useful for the analysis of scientific progress during periods of theoretical transformation, when these interactions are destabilised.

For a working scientist, 'scientific progress' is a context-dependent concept. On the one hand, progress in science seems obvious. Most scientists tend to see the current state of affair in their discipline as the apex of that discipline's historically accumulated knowledge. On the other hand, most scientists would readily agree that what amounts to progress in *current* research is evaluated differently by some of their colleagues, with certain aspects of current scientific practice appearing to some members of the community to lead to a dead end or even to reverse progress. Hence, for a working scientist, the notion of scientific progress is both self-evident (in an idealised, historically naïve, regulative sense), and yet is far from clear when she engages in the debates that accompany her research.

In this article I am going to use a developmental system (DS) framework for discussing scientific progress in one particular aspect of biological research – epigenetic inheritance (EI). The contribution of EI research to progress in evolutionary biology is contested and requires the examination of the self-sustaining social, methodological, and theoretical interactions that have led to the robustness of the traditional evolutionary view, which disallowed or downplayed EI, as well as the current context in which the progressive nature of EI research is discussed. I shall use the DS framework developed by Conrad Waddington for investigating the processes and the effects of embryological development (Waddington 1957, 1975), and combine it with the systems approach of Ludwik Fleck, one of the twentieth century great microbiologists and sociologists of science (Fleck 1935/1979). I am focusing on Fleck's work rather than on that of related studies such as those of Kuhn, Lakatos, and Laudan because of Fleck's biological expertise and appreciation of the variability inherent in biological systems and the pivotal role of methodological standardisation, as well as his explicit sociological approach, which is sensitive to both intrinsic and extrinsic factors. Fleck's analysis therefore seems particularly suitable for the explication of rapid periods of scientific change in biological sciences, where the stabilisation of new experimental methodologies, as well as institutional and theoretical, is pivotal.

Fleck chose the network metaphor to describe what he called the *thought style* of the scientific community – a concept that is roughly equivalent to what Kuhn called, decades later, 'a paradigm' and 'a disciplinary matrix' (Kuhn 1970). Thinking in terms of the construction of the networks that constitute the thought style, Fleck analysed the generation of scientific 'facts', the relatively stable junctions in the network that act as anchors for subsequent consolidation or transformation of the research program. (Fleck's book is tellingly called *Genesis and Development of a Scientific Fact*.) This approach emphasises the functional aspects of scientific progress (Shan 2019) within a sociological–historical system view during relatively short periods of theory change. When there are mounting challenges to an orthodox scientific view and an alternative theoretical framework is being constructed, the functional significance of challenging data and models is openly debated, and the various factors (both intrinsic and extrinsic) that affect how the new results are interpreted come to the fore and become apparent. Even when there is consensus that the new research direction leads to the definition of new problems and to the solution of both old and new problems, the significance of such changes to theory change may be very differently seen by different schools of thought.

The results of research into EI are seen by some biologists to challenge the dominant neo-Darwinian evolutionary thought style known as the Modern Synthesis (MS) and contribute to a new thought style called the Extended Evolutionary Synthesis (EES). This is contested by adherents of the MS, but a dialogue between the two factions is possible because the EI research program grew from research in developmental genetics and molecular biology, which is fundamental to both thought styles. However, the significance and progressive nature of the results of EI studies are seen very differently by members of the two groups, so this case exposes the interacting processes that construct the thought style and makes sense of the difficulties inherent in the concept of scientific progress during periods of theory change.

The article has three sections. In Section 1, I describe Waddington's system's view of biological development and its application to the dynamic persistence and transformation of social systems. In Section 2, I apply the approach to scientific development and combine it with Fleck's framework, using research into EI as a case study. How and whether the notion of progress may apply to the research programme of EI and what the DS view adds to the discussion is the subject matter of Section 3.

1. A Developmental System's View of Scientific Persistence and Change

Developmental system theory (DST) is an approach to the study of biological systems that takes the unit of analysis to be the self-sustaining network of interactions among multiple resources (genetic, developmental, ecological, social) and activities that lead to the dynamic persistence (and in some condition to deviations from persistence) of the system over time (Oyama et al. 2001). It is assumed that the dynamic, cybernetic architecture of the network of interactions channels developmental trajectories so they converge to produce a typical developmental outcome, accounting for the robust nature of development as well as shedding light on the conditions that lead to departures from typical outcomes, leading to new developmental stable states. This approach was pioneered by Conrad Waddington, one of the founders of systems biology and one of the twentieth-century proponents of evolutionary developmental biology (Waddington 1957, 1975; Baedke 2018; Jablonka and Noble 2019).

In developing his ideas, Waddington, an embryologist and evolutionary biologist, coined the term *epigenetics* to describe 'the branch of biology which studies the causal interactions between genes and their products which bring the phenotype into being' (Waddington 1975, p. 218). He was interested in what he called the *epigenotype*, the flexible network of regulatory interactions that underlies the stable trajectories of embryonic development. He depicted embryonic development as the progression of a ball (the developing entity, starting with the fertilised egg) through a sloping landscape of alternative valleys (representing developmental trajectories) (Waddington 1957). This landscape, which he called the *epigenetic landscape*, is shaped by underlying networks of interacting genes and their products, which respond to the environmental conditions in which the developing embryo is embedded. The dynamic regulatory architecture of the networks of interaction constructs a functional, species-typical end state at which the developing entity comes to (dynamically) 'rest'. The end state, also called the attractor state, is relatively stable or canalised – the regulatory interactions and backup systems embedded in the network flexibly reorganise its dynamics when the internal and external conditions fluctuate, ensuring that the developmental trajectories converge at the attractor state (Figure 8.1A). An attractor state can be described at different levels – the entire organism (e.g. a typical healthy new-born), an organ (e.g. a functioning heart), or a cell type (e.g. a muscle cell), and is typically stable even when the underlying (genetic) or overlying (environmental) conditions change. For example, a change in the environment – a period of maternal malnutrition during pregnancy, for example – may not disturb a heart's function because alternative developmental and behavioural strategies are employed and alter gene-expression patterns which compensate for the change (Figure 8.1B). Similarly, a new mutation, even one that affects the pathways leading to the development of a heart, may have no effects on heart function, because the genetic network reorganises and compensates for the effect (Figure 8.1C). However, large fluctuations (e.g. a mutation in a regulatory region or prolonged starvation) or multiple small changes can lead to a new state (Jablonka and Noble 2019).[1]

1A 1B

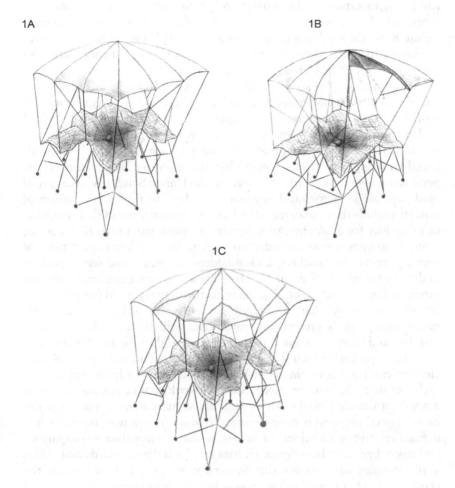

1C

Figure 8.1 The robustness of the epigenetic landscape.

Figure 8.1A shows the part of an epigenetic landscape (middle grey net), which is constructed and constituted by an underlying genetic network and overlying, relatively stable network of external conditions. The circles at the bottom represent genes, and the strings represent the biochemical effects of the genes; the upper umbrella represents an overlying network of environmental conditions external to the focal DS. The depression in the middle of the grey web is the set point or attractor state representing the canalised, robust functional state of the focal system (e.g. a functional heart). Figure 8.1B shows that the landscape can persist even when there is a large change in external condition such as a short period of malnutrition during pregnancy (torn upper umbrella) through the re-organisation of the genetic and environmental-behavioural network, which compensates and buffers the effects of the external change. Figure 8.1C shows that a genetic mutation (larger circle) may also leave the landscape unchanged, since the genetic and environmental networks reorganise to compensate for its effects.

Tavory et al. (2014) adapted Waddington's model to the far more open-ended and flexible case of social systems, suggesting that a regulatory network of interactions among the practices of individuals, groups, medical and nutritional factors, institutional structures, ecological affordances and so on, can lead to the continuous re-construction of the focal cultural-social landscape, as well as explain departures from it. An example is the persistence over time of urban poverty in the United States (the focal social landscape). The factors and processes that lead to the reconstruction of poverty include the developmental effects of malnutrition; the consumption of unhealthy food, alcohol, or other toxins; poor parenting; bad schools; limited job opportunities; residential segregation, the low expectations of peers, parents, and teachers; and outsiders' prejudice and social policies at different levels of government. These factors interact and reinforce each other, sustaining and re-forming the trajectories that lead to poverty, making sense of the difficulty of escaping it. However, although individuals go through partially pre-existing trajectories, which they typically tend to stabilise and 'deepen', their activities may also alter the local features of the landscape with which they interact. A gradual accumulation of such changes can reach a tipping point and lead to a change in the landscape's architecture and a new set point. Such network change can also occur as a result of drastic global changes (e.g. wars, mass immigrations, revolutions). Social progress is not an empty term in this context, but since it is defined relative to moral sets of values, which are not always compatible, and since there may be different entities to which the standards and values apply (human white males, the Western world, the human species, the ecosystem), what counts as progress is inevitably contested.

Like other social-cultural systems, scientific research traditions are complex dynamic entities constructed by the activities of scientists and shaped by intra- and extra-disciplinary theoretical assumptions, experimental practices, and educational and institutional norms. They therefore lend themselves to analysis in DS terms. However, a DS analysis inherits some of the problems inherent in a sociological approach. These include the notion of progress, which is seen as context-relative (Kuhn 1970).

Modern science is based on the time-consuming acquisition of expert skills and is subject to the implicit and explicit conceptual and methodological standards of the research community. All working scientists experience the effortful and often frustrating process of teasing out regularities from the 'blooming, buzzing confusion' (to borrow a phrase from William James) presented by the part of the world they study. Biologists, on whom I focus in this article, confront variability at every level of biological organisation and have to make conscious decisions about what should be construed as noise and what should be considered as relevant parameters that can be intentionally varied under carefully devised, noise-free conditions, which are often extremely sensitive. Decisions about experimental conditions that exclude noise enable the construction of experimental settings that lead to reproducible results and that are assumed to uncover the most important causal, structural, or functional relations underlying the physiological and behavioural development of the object of study. The exclusion of noise is most apparent when scientists develop a new technique that seems to be theory-neutral, but is also evident when theory-laden research questions, such as transgenerational inheritance, are investigated. Changes in what is considered noise and what is considered informative variability may lead to the redefinition of the research problem or of some of the concepts on which the analysis is based, opening up a new framework of research.

Identifying the conditions and methods that reduce noise and lead to the reproducible correlations that we call 'facts' was central to Fleck's analysis of the development of the Wassermann reaction that detected and defined syphilis in the early twentieth century. His study of the ancient and modern history of the concept of syphilis led him to identify four general characteristics of the thought style of a scientific community (which he called the thought collective):

> It [a thought Style] is characterized by common features in the problems of interest to a thought collective, by the judgment which the thought collective considers evident, and by the methods which it applies as a means of cognition. The thought style may also be accompanied by a technical and literary style characteristic of the given system of knowledge.
>
> (Fleck 1979, p. 99)

Fleck expanded on each of these characteristics in his analysis of the changing notion of syphilis. He pointed to the network organisation of the scientific-professional (esoteric) thought style, which is constituted by interactions among beliefs, metaphors, professional jargon, technologies, practices, descriptive and evaluative models and professional institutions, as well as the ways in which the scientific thought style is embedded within broader, non-specialist (exoteric) social-cultural networks. He stressed the importance of scientific education and science popularisation, and discussed the importance of the interchange among scholars from different disciplines in the initial stages of the construction of a new thought style, and its robustness and resistance to challenges when it becomes consolidated:

> What we are faced with here is not so much simple passivity or mistrust of new ideas as an active approach which can be divided into

several stages. (1) A contradiction to the [present] system appears unthinkable. (2) What does not fit into the system remains unseen; (3) alternatively, if it is noticed, either it is kept secret, or (4) laborious efforts are made to explain an exception in terms that do not contradict the system. (5) Despite the legitimate claims of contradictory views, one tends to see, describe, or even illustrate those circumstances which corroborate current views and thereby give them substance.

(Fleck 1935/1979, p. 27)

How does progress fit into this framework? For Fleck, 'progress' is a problematic notion. Although the growth of the network may involve the addition of new research problems, new predictions and new solutions that need to be replicated by different individuals and research groups, a new thought style may employ very different standards for evaluating these solutions compared to the thought style it replaced. Nevertheless, few scientists (including Fleck, I believe) would doubt that the scope and range of testable problems in biology have greatly increased during the last 200 years, and in this sense biological knowledge has also increased or 'progressed'. However, most scientists will also agree that this trend has been neither linear nor uniform and that a close-up perspective reveals fluctuations and even regressions in some research programmes. Research into what is now known as the study of EI is an interesting case study because it shows the combination of internal and external processes and constraints that have led to the re-construction of the scientific-social landscape of research into heredity. Since the significance of the results stemming from the EI research programme is seen differently by scholars adhering to the traditional thought style in evolutionary biology and by those advocating a new emerging thought style (the extended evolutionary synthesis or EES), the contribution of these results to 'progress' is ambivalent.

2.   Epigenetic Inheritance: Growth and Differentiation

Providing a comprehensive Fleckian analysis of EI is beyond the scope of this article. Here I am giving only a sketch of the development of EI based on my work with Marion Lamb on the topic over the last 30 years, which is enough to enable its evaluation in terms of progress (Jablonka and Lamb 1995, 2002, 2011, 2014, 2020 and references therein; see also Baedke 2018; Meloni 2019).

The term 'epigenetic inheritance' was invented and first used during the 1980s. Although it refers to the inheritance of developmentally acquired variations and is therefore conceptually associated with the Lamarckian doctrine of the inheritance of acquired characters, it was at first dissociated from evolutionary biology and applied to cell heredity during development. It grew out of Waddington's more general concept of epigenetics, and initially was used by cell biologists to refer to the mitotic inheritance of determined states in somatic (body) stem cells and cells in culture. Microbiologists used a related, earlier term, 'epigenetic control mechanisms' to refer to the inheritance of environmentally induced variations in microorganisms (Nanney 1958) but this term and the research programme of which it was part remained relatively peripheral until the 1980s. It is important to note that Waddington himself did not use epigenetics-related terms to describe cell heredity or transgenerational non-genetic inheritance. Although he recognised that there is transmission between generations of non-genetic variation, he paid very little attention to such transmission and thought it was unimportant in non-humans (Waddington 1975). And although he was very interested in the processes of cell determination, he did not coin a new (epigenetics based) term to describe the process.

In the 1970s, it was proposed that chemical modifications of DNA, specifically methylation or demethylation of cytosines (the addition or removal of CH_3 groups), affect the activity of genes, and that patterns of DNA methylation can be enzymatically copied following DNA replication, resulting in daughter cells inheriting the same patterns of gene activity as their mother cells. This forged an explicit link between gene regulation and cell heredity: DNA methylation was seen both as part of the regulatory system of the cell and as a cellular memory system. This duality could explain the cell-heredity processes that developmental and cell biologists had been trying to understand for a long time. It could explain cell determination during embryonic development, the inactivation and stable propagation (once established) of one of the X chromosomes during the development of female mammals, and the behaviour of cells in culture and during cancer propagation. Two seminal papers suggested that cell heredity in all these processes involves the establishment of new DNA methylation patterns during development, and their subsequent cellular inheritance (Holliday and Pugh 1975; Riggs 1975). Although the terms 'epigenetic' and 'epigenetic inheritance' were not used in these 1975 papers, these processes were later (in the late 1980s and the 1990s) used as exemplars of intra-organismal forms of EI.

During its short history, the term 'epigenetic inheritance' has gone through various changes in meaning. It was at first used interchangeably with epigenetics. In 1990, Holliday used epigenetics in the Waddingtonian sense, but in 1994 he defined epigenetics as 'The study of the changes in gene expression, which occur in organisms with differentiated cells, and the mitotic inheritance of given patterns of gene expression' (p. 453).

EI at first referred to the inheritance of DNA methylation patterns in somatic cells and cell lines, and occasionally and accidentally through sex cells (for discussions and reviews, see Holliday 1987, 1990). However, quite quickly, EI was broadened to include additional molecular mechanisms of cell heredity, not just the DNA methylation system. Jablonka and Lamb (1989, 1995) characterised several different distinct (though often interacting) cellular processes that lead to cell heredity, calling them Epigenetic Inheritance System (EISs). They assembled evidence showing that cellular epigenetic states can be passed to future generations of cells and organisms through self-sustaining metabolic loops, three-dimensional structural templating, chromatin marks (such as DNA methylation), and RNA-mediated inheritance (Jablonka and Lamb 2005, 2011, 2020). Their synthesis led them to sharpen Waddington's concept of epigenetics:

> *Epigenetics* is the study of developmental processes in prokaryotes and eukaryotes that lead to persistent, self-maintaining changes in the states of organisms, the components of organisms, or lineages of organisms.
>
> (Jablonka and Lamb 2014, p. 393)

This definition does not assume that all heritable phenotypic variation stems from genetic variation. It accommodates the fact that mechanisms of canalisation and plasticity can decouple the genotype and phenotype.[2] In the light of this definition, EI is regarded as a component of epigenetics:

> *Epigenetic inheritance* is a component of epigenetics. It occurs when phenotypic variations that do not stem from variations in DNA base sequence are transmitted to subsequent generations. Variations can be transmitted in mitotically dividing lineages of cells, in lineages of organisms that reproduce asexually, and in lineages of sexually reproducing organisms.
>
> (Jablonka and Lamb 2014, p. 393)

Since the 1990s, epigenetics and the epigenetic mechanisms that underlie EI have been recognised to be fundamental to basic biology and have been integrated into almost every discipline in biology and medicine. Epigenetics has become a buzz word, which, like genetics, is used by both professionals and laypeople and has acquired multiple meanings and connotations. Manuscripts, textbooks, handbooks, manuals and popular books, scientific meetings, popular lectures, Wikipedia entries, and You-Tube videos are growing in number. The titles of books published during the last decade give an indication of the scope and range of epigenetic research: *Handbook of Epigenetics; Epigenetics and Epigenomics; Epigenetics: a Graphic Guide; The Developing Genome: An Introduction to Behavioral Epigenetics; Epigenetics Methods; Epigenetics in Human Disease; Medical Epigenetics; Transgenerational Epigenetics; Epigenetics in Psychiatry; Epigenetics of Ageing; Epigenetic Gene Expression and Regulation; Epigenetics and Regeneration; Plant Epigenetics; Nutrient and Epigenetics; Epigenetics for Drug Discovery*. This is far from being a comprehensive list. Laboratories that investigate these aspects of epigenetics are working at full steam, developing new methods, using new model-organisms, and publishing in a growing number of specialised epigenetics journals. Within this largely development-oriented framework, research into within-organism EI is non-problematic and is seen as progressive, contributing to biological knowledge, opening up new research horizons and leading to the development of new methods, the identification of new research problems, and the solution of old and new puzzles.

However, *transgenerational* EI does not fit well within the traditional evolutionary framework known as the MS of evolution. The MS is a wide-ranging synthesis of the results of laboratory and field studies in genetics, systematics, and palaeontology, which crystallised in the mid-twentieth century and later incorporated molecular biology. The MS thought style was based on the assumption that evolution usually involves a gradual, cumulative change in gene frequencies in populations, brought about by selection acting on the variation among individuals that resulted from random gene (DNA) mutation and recombination, and that such changes can explain both changes occurring during microevolution (within a population of a single species) and macroevolution (speciation events and changes above the species level). The MS excluded Lamarckian evolution through the inheritance of acquires characters, identified hereditary variation with variation in DNA, and assumed that all heritable variations in phenotypes are based on variations in genotypes. From the perspective of the MS, transgenerational EI is a challenge. It introduces a Lamarckian dimension into evolution, suggesting that variations in phenotypes can be developmentally induced and inherited through non-genetic mechanisms, and that the inheritance of acquired characters contributes to evolutionary change (Jablonka and Lamb 1995, 2005; Gissis and Jablonka 2011). The idea that it is important has therefore been belittled and attacked (Charlesworth et al. 2017; see also disparaging discussions in Jerry Coyne's blog 'Why Evolution in True'), and its integration within an evolutionary framework is seen by MS adherents either as regressive or as a greatly inflated and overvalued interpretation of studies that pose no challenge to the MS (see, e.g. Wray et al. 2014).

Though still dominant, the MS is not the only evolutionary framework today. A new competing thought style known as the EES is becoming increasingly influential (Pigliucci and Müller 2010; Laland et al. 2014, 2015; Jablonka and Lamb 2020). The point of departure of the EES is the active phenotype, which constructs its ecological and developmental niche and can bequeath these legacies to descendants. Developmental modifications, which are constructed by the interactions among heritable genetic (DNA), epigenetic (e.g. methylation), behavioural and cultural variations, all contribute to evolutionary change. Within this evolutionary thought style, the EI research program is seen as essential and, of course, progressive.

3.   Progress Is Relative to a Thought Style

The relationship between a particular line of research and a thought style is bidirectional, with the thought style providing the standards for the evaluation of the research results as worthy or unworthy, and the research results contributing to or challenging the thought style. This bi-directional relation is particularly evident when a new thought style is beginning to be constructed because at this point active debates ensure that what has been previously taken for granted is under close scrutiny. The study of EI within the changing thought style of evolutionary biology is a good example.

In the early 1990s, there were only a few examples of transgenerational EI that had been characterised at the molecular level, although numerous cases of heritable developmentally induced phenotypic variations had been described at the behavioural and physiological levels. Accounts of these were scattered in different specialised journals that used different terminologies. There was no theoretical framework that could unify them, and the results were very variable, since epigenetic variations are the outcomes of both developmental noise and inducing environmental conditions, and hence are extremely sensitive to developmental context (at several different scales and levels of organisation). The standardisation of the techniques and experimental protocols that estimate the transmissibility of epigenetic variations were therefore crucial for the consideration of the research results as significant. This was one of the reasons that initially results that challenged the MS view of heredity were treated exactly as Fleck described when discussing the response to challenges of the majority of the scientists who adhere to the dominant thought style: they were overlooked, regarded as irrelevant, 'kept secret', dismissed as noise, or explained away.

There were additional intra-scientific reasons for this attitude. The results did not fit the traditional MS conceptual framework of heredity, which focused on the study of sexually reproducing animals that were amenable to Mendelian analysis and showed early separation between somatic and germline cell lineages. Until the 1990s, reports of transgenerational non-genetic inheritance were seen (when not dismissed as fraud) as exotic exceptions that could safely be ignored (Jablonka and Lamb 1995, 2011). Importantly, there were no known cellular and molecular mechanisms that could explain the results and guide further research. Since the fidelity of transmission of developmentally induced heritable variation was often lower than that of classical mutations, they did not fit the gene-based framework of classical genetic research and were treated as noise. Karl Lindegren, one of the early and rare vocal critics of the MS, wrote as early as 1949:

> The genetical data on which the modern conception of the gene is based are intensively selected data . . . The search for precisely

segregating genes compels the selection of genetical material. In our own work on *Neurospora* we were unable to classify the progeny of over two-thirds of our matings.

(Lindegren 1949, Chapter 20, pp. 6–7)

In addition to intra-scientific reasons, there were powerful extra-scientific political reasons that reinforced the exclusion of the research into transgenerational inheritance in the Western world. In the Soviet Union, Trofim Lysenko published and encouraged fraudulent work on the inheritance of acquired characters and brutally suppressed Mendelian genetics, leading to the wholesale dismissal of Soviet genetic work in the West. However, in spite of Lysenko's corrupting influence, there was a body of honest and careful research in the USSR that showed that the effects of environmental stress can be inherited and that Mendelian heredity cannot provide the full picture of biological inheritance. These studies were largely ignored because of their possible association with Lysenkoism, and because the cold war between the USSR and the West spilt over into science, especially in the USA where, during the McCarthy witch-hunts of the early 1950s, American biologists were very cautious about Soviet-promoted ideas for fear of being labelled communists (for an interesting analysis of the 'Cold War in Biology', see Lindegren 1966).

A less well-known example of a politically influenced effect on the research programme into the inheritance of developmental variations was the 'disappearance' of the results of a wide-ranging corpus of research into transgenerational inheritance conducted in the Institute for Experimental Biology (Biologische Versuchsanstalt, BVA) in Vienna, also known as the Vivarium. In this case it seems that virulent anti-Semitism was to blame. The BVA was established and dominated by Jewish scientists and was taken over by Nazi biologists in the 1930s. The Jewish and anti-Nazi scientists dispersed or were murdered, and the continued supremacy of former Nazi scientists in Austrian academies after the World War II and the deliberate whitewashing of their Nazi past ensured that the legacy of the BVA was not disseminated and was actively forgotten (Müller 2017).

Despite these setbacks, by the early 1990s, there was a large number of studies that showed that all was not well with the view of heredity on which the MS was based. As noted earlier, these studies were studiously ignored (especially in the USA) and had little effect on the MS version of evolutionary theory, which became increasingly dogmatic. Nevertheless, a growing number of uneasy scientists (often people who were also interested in areas that were close, yet external to the immediate interests of most other evolutionary biologists) became aware of these challenges, and there were sporadic suggestions that a new stage in biological thinking was looming. These calls for change were peripheral to the dominant thought style, but as the anomalies grew in number, some of the dissenting scientists who were acquainted with molecular biology started to look for molecular mechanisms that could explain them. These scientists suggested that the few and rather messy cases of transgenerational EI that had been characterised at the molecular level during the late 1980s and early 1990s were the tip of a large iceberg and can mechanistically account for many of the unexplained and overlooked cases (Jablonka and Lamb 1989, 1995). They regarded the experimental results of these few first molecular studies as critical experiments that could transform the way of thinking about inheritance and evolution. These very same experimental results were, of course, regarded as noise – unimportant and irrelevant – by the scientists that held the dominant MS view. Interestingly and typically, during the initial stages of the construction of the alternative EES thought style in the 1990s and early 2000s, there was a lot of exchange of ideas between biologists and scholars from other disciplines, mostly medicine, psychology, ecology, and the philosophy of biology (Jablonka and Lamb 2011; Fisch 2017).

During the last 20 years, a great number of cases of transgenerational EI have been characterised at the molecular level. It has been shown that transgenerational EI is widespread in scope and range, varies with context, shows taxon-dependent stability and involves multiple molecular mechanisms (for reviews, see Jablonka and Raz 2009; Jablonka 2017; Bonduriansky and Day 2018). These studies have spurred the invention of new techniques for detecting and measuring epigenetic variations.[3] New mathematical and simulation models that extended classical models in population and quantitative genetics have been developed to analyse changes in the frequency of epigenetic variants in populations and to estimate the effects of heritable epigenetic variations on heritability. New model organisms, for example, clonal hermaphrodites or self-fertilising organisms (such as *Arabidopsis* in plants, and *Daphnia* in animals), are being used to investigate EI. There are new 'exemplars' – case studies that exemplify EI and are used to enculture students and young researchers into the thought style. There are also institutional changes: new laboratories dedicated to the study of EI have been established, new journals are being published, new university courses are being developed, new interdisciplinary cooperative projects have been set up and the scientists involved in EI research are becoming influential.

The EI research programme contributes to the EES thought style, but despite its growing importance and scope as well as the more general growth of the research programmes in ecology and developmental biology that contribute to the EES, the theoretical conflict has not been resolved. I believe that the reason for this is that what is at stake is far greater than the incorporation of yet another source of heritable variations or even the addition of a Lamarckian dimension to evolutionary theory. The MS was not just a revision and updating of Darwin's theory of evolution – it was intended to be a framework for the unification of biology, providing a new perspective for the entire life sciences (Smocovitis 1996). Similarly, the EES also offers a new unifying framework for biology. It calls for an organism-oriented view with plastic phenotypic modifications paving the way for genetic variations, for an inclusive concept of heredity, for an extension of the notion of individuality, for a rethinking of the relationships between ultimate (evolutionary) and proximate (developmental, mechanistic) causation, for a recognition of the agency of organisms and their role in constructing their ecological and developmental niches, for a process – rather than an object-oriented approach to science. According to Denis Noble, a radical proponent of the EES, these changes in biological outlook are going to lead to changes in whole areas in economics, sociology, political science, and philosophy, which in the past have adopted the gene-centred MS, evolutionary perspective (Noble 2017). How the current unstable situation will be resolved is unclear, but whatever the nature of future changes, the new discoveries in epigenetics can no longer be ignored and will become incorporated in one way or another into the prevailing thought style of the future.

So what about progress? Within an established thought style, the notion of progress is not problematic. The results of the EI research programme are seen as progressive within the consolidated ESS of the 2020s. Not only the number of studies that support and extend the range and scope of the EES but also the quality and significance of such studies is measured according to the explicit and implicit standards imposed by the EES scientific community. However, when the EES thought style began to emerge in the 1990s the situation was far more complex. The few experiments and observations that revealed the molecular mechanisms of transgenerational inheritance and supported the new thought style were evaluated in drastically different ways by the MS and EES thought collectives. There was also much dithering and uncertainty about the significance of the results among many of the scientists involved in this new EES research programme because at the initial stages they were themselves unsure of how important their results were. It was only later, when the EES thought style consolidated, that the progressive nature of the studies was seen as stable and reliable.

Analysing scientific progress in a particular research area from a DSs perspective highlights the context dependence and the ambivalence inherent in the notion of progress when a scientific system is being transformed. I have not attempted to address the question of progress at higher levels, when thought styles are compared – in the present case asking whether the EES is a more progressive thought style than the MS. Answering this question may be possible only when a broader historical perspective and longer time scale is chosen. A broader historical perspective calls for a far more coarse-grained analysis of cultural and social networks. Such an analysis would give greater weight to exoteric, non-scientific aspects of culture, and make the notion of scientific progress more dependent on general social-cultural values and more open to the multiple interpretations that such values permit.

Another way of analysing progress is to borrow the standards that are currently used in biology when analysing evolutionary transitions. During the phylogeny of some lineages, there were increases in complexity and levels of individuality (e.g. during the transition from single cells to cells that combine to form a multicellular individual). These transitions were the result of the evolution of new ways of acquiring, storing, processing, and transmitting information. The increase in complexity involved an increased division of labour and the addition of new hierarchical level/s of control (Maynard Smith and Szathmáry 1995). If we adopt this evolutionary-transition framework for scientific progress, we can regard the differentiation of biology into new sub-disciplines as the equivalent of an increase in differentiation and division of labour, and the embedding of these sub-disciplines within the more generalised biological thought style, which incorporates many of the results and models of the previous thought style as special cases, as the addition of new level/s of hierarchical organisation. Using this transition framework, increased disciplinary differentiation and a broader, more encompassing, hierarchical organisation of scientific theories could serve as rough indicators of progress. However, by adopting this framework we may downplay the logical constraints on scientific thinking as well as the significance of the non-scientific societal values that are part of the matrix of the human scientific endeavour.

A single perspective for defining scientific progress cannot provide a satisfactory notion of progress, because the significance of the factors that contribute to our judgement of progress depends on the historical time-frame chosen, with less contextual characterisations as the time scale is extended and new styles of reasoning and criteria of objectivity become established (Hacking 1992). Switching between perspectives with different standards for estimating progress can lead to new insights, especially when the notions of progress that these different perspectives employ do not fully align.

Acknowledgement

Many thanks to Marion Lamb with whom all these issues were discussed and debated, and who critically read and commented on this manuscript.

Notes

1. The original epigenetic landscape figures by Waddington (Waddington 1957, figures 4 and 5) emphasise canalization and the networks of interactions that compensate for and buffer intrinsic genetic changes. Waddington did not depict external sources of change although he did discuss such inputs in the accompanying text to the figures in his 1957 'Strategy of the Genes' as well as in many other publications. The figure shown in this article depicts the additional,

extrinsic system ('top umbrella') that provides inputs into the developmental system. This addition highlights the environmental/extrinsic inputs that need to be canalised but that can also lead to a change in the attractor, an aspect emphasised by West-Eberhard (2003).

2. Canalization occurs when different genotypes and environments generate the same phenotype. Plasticity refers to the ability of a single genotype to generate different phenotypes when conditions changes.

3. Recent examples are the modification of CRISPR-Cas methods for manipulating methylation patterns, and the invention of methods for detecting methylated and unmethylated cytosines in ancient DNA.

References

Baedke, J. (2018). *Above the Gene Beyond Biology: Towards a Philosophy of Epigenetics*. Pittsburgh, PA: University of Pittsburgh Press.

Bonduriansky, R. and Day, T. (2018). *Extended Heredity: A New Understanding of Inheritance and Evolution*. Princeton, NJ: Princeton University Press.

Calcott, B. and Sterelny, K., eds. (2011). *The Major Transitions in Evolution Revisited*. Cambridge, MA: The MIT Press.

Charlesworth, D., Barton, N. H. and Charlesworth, B. (2017). The sources of adaptive variation. *Proceedings of the Royal Society B* 284:20162864.

Fisch, M. (2017). *Creatively Undecided. Toward a History and Philosophy of Scientific Agency*. Chicago, IL: University of Chicago Press.

Fleck, L. (1935/1979). *Genesis and Development of a Scientific Fact*, trans. F. Bradley and T. J. Trenn. Chicago, IL: Chicago University Press.

Gilbert, S. F. (2019). Evolutionary transitions revisited: Holobiont evo-devo. *Journal of Experimental Zoology Part B Molecular and Developmental Evolution* 332(8):307–314.

Gissis, S. B. and Jablonka, E. (2011). *Transformations of Lamarckism: From Subtle Fluids to Molecular Biology*. Cambridge, MA: MIT Press.

Hacking, I. (1992). 'Style' for historians and philosophers. *Studies in History and Philosophy of Science Part A* 23(1):1–20.

Holliday, R. (1987). The inheritance of epigenetic defects. *Science* 238:163–170.

Holliday, R. (1990). DNA methylation and epigenetic inheritance. *Philosophical Transactions of the Royal Society of London. Series B, Biological Sciences* 326(1235):329–338.

Holliday, R. (1994). Epigenetics: An overview. *Developmental Genetics* 15:453–457.

Holliday, R. and Pugh, J. E. (1975). DNA modification mechanisms and gene activity during development. *Science* 187:226–232.

Jablonka, E. (2017). The evolutionary implications of epigenetic inheritance. *Interface Focus* 7(5):20160135.

Jablonka, E. and Lamb, M. J. (1989). The inheritance of acquired epigenetic variations. *Journal of Theoretical Biology* 139:69–83.

Jablonka, E. and Lamb, M. J. (1995). *Epigenetic Inheritance and Evolution: The Lamarckian Dimension*. Oxford: Oxford University Press.

Jablonka, E. and Lamb, M. J. (2002). The changing concept of epigenetics. *Annals of the New York Academy of Science* 981:82–96.

Jablonka, E. and Lamb, M. J. (2005). *Evolution in Four Dimensions: Genetic, Epigenetic, Behavioral and Symbolic Variations in the History of Life*, 1st edn. Cambridge, MA: MIT Press.

Jablonka, E. and Lamb, M. J. (2011). Changing thought styles: the concept of soft inheritance in the 20th century. In R. Egloff and J. Fehr, eds., *Vérité, Widerstand, Development: At Work with/Arbeiten mit/Travailler avec Ludwik Fleck*. Zürich: Collegium Helveticum, pp. 119–156.

Jablonka, E. and Lamb, M. J. (2014). *Evolution in Four Dimensions: Genetic, Epigenetic, Behavioral and Symbolic variations in the History of Life*, 2nd edn. Cambridge, MA: MIT Press.

Jablonka, E. and Lamb, M. J. (2020). *Inheritance Systems and the Extended Evolutionary Synthesis*. Cambridge: Cambridge University Press.

Jablonka, E. and Noble, D. (2019). Systemic integration of different inheritance systems. *Current Opinions in Systems Biology* 13:52–58.

Jablonka, E. and Raz, G. (2009). Transgenerational epigenetic inheritance: Prevalence, mechanisms, and implications for the study of heredity and evolution. *Quarterly Review of Biology* 84:131–176.

Kuhn, T. S. (1970). *The Structure of Scientific Revolutions*, 2nd edn. Chicago, IL: University of Chicago Press.

Laland, K. N., Uller, T., Feldman, M. W., et al. (2014). Does evolutionary theory need a rethink? *Nature* 514:161–164.

Laland, K. N., Uller, T., Feldman, M. W., et al. (2015). The extended evolutionary synthesis: Its structure, assumptions and predictions. *Proceedings of the Royal Society B* 282:20151019.

Lindegren, C. C. (1949). *The Yeast Cell, Its Genetics and Cytology*. St. Louis: Educational Publishers.

Lindegren, C. C. (1966). *The Cold War in Biology*. Ann Arbor, MI: Planarian Press.

Maynard Smith, J. and E. Szathmáry. (1995). *The Major Transitions in Evolution*. Oxford: Freeman.

Meloni, M. (2019). *Impressionable Biologies: From the Archaeology of Plasticity to the Sociology of Epigenetics*. New York: Routledge.

Müller, G. B., ed. (2017). *Vivarium. Experimental, Quantitative, and Theoretical Biology at Vienna's Biologische Versuchsanstalt*. Cambridge, MA: MIT Press.

Nanney, D. L. (1958). Epigenetic control systems. *Proceedings of the National Academy of Sciences USA* 44:712–717.

Niiniluoto, I. (2019). Scientific progress. In Edward N. Zalta, ed., *The Stanford Encyclopedia of Philosophy*. Winter 2019 edn. https://plato.stanford.edu/archives/win2019/entries/scientific-progress/

Nitecki, M. H., ed. (1988). *Evolutionary Progress*. Chicago: University of Chicago Press.

Noble, D. (2017). *Dance to the Tune of Life: Biological Relativity*. Cambridge: Cambridge University Press.

Oyama, S., Griffiths, P. E. and Gray, R. D. (2001). *Cycles of Contingency*. Cambridge, MA: MIT Press.

Pigliucci, M. and Müller, G. B., eds. (2010). *Evolution – The Extended Synthesis*. Cambridge, MA: MIT Press.

Riggs, A. D. (1975). X inactivation, differentiation, and DNA methylation. *Cytogenetics and Cell Genetics* 14:9–25.

Ruse, M. (1996). *Monad to Man: The Concept of Progress in Evolutionary Biology*. Cambridge, MA: Harvard University Press.

Shan, Y. (2019). A new functional approach to scientific progress. *Philosophy of Science* 86:739–758.

Smocovitis, V. B. (1996). *Unifying Biology. The Evolutionary Synthesis and Evolutionary Biology*. Princeton, NJ: Princeton University Press.

Tavory, I., Ginsburg, S. and Jablonka, E. (2014). The reproduction of the social: A developmental system approach. In L. R. Caporael, J. R. Griesemer and W. C. Wimsatt, eds., *Developing Scaffolds in Evolution, Culture, and Cognition*. Cambridge, MA: MIT Press, pp. 307–325.

Waddington, C. H. (1957). *The Strategy of the Genes*. London: Allen & Unwin.

Waddington, C. H. (1975). *The Evolution of an Evolutionist*. Edinburgh: Edinburgh University Press.

Wray, G. A., Hoekstra, H. E., Futuyma, D. J., et al. (2014). Does evolutionary theory need a rethink? No, all is well. *Nature* 514:161–164.

9   Progress in Seismology
Turning Data Into Evidence About the Earth's Interior

Teru Miyake

1. Introduction

In 1889, a German physicist, Ernst von Rebeur-Paschwitz, recorded anomalous signals on sensitive devices set up for measuring the lunar pull of gravity in Potsdam and Wilhelmshaven. Having chanced upon a report about an earthquake that occurred in Tokyo the day he recorded the signals, he inferred that his devices in Germany had picked up seismic waves that had originated in Japan. This discovery, and its later corroboration by other seismologists through the detection of waves from other distant earthquakes, led to the following questions. Do these waves pass through the earth's deep interior? And if so, can they be harnessed to extract information about the vast unexplored region beneath our feet?

These questions gave rise to a series of investigations by seismologists over the next few decades that successively revealed more and more about the earth's interior. The following half-century would see the identification of all the major features of the earth's interior, including the crust, the mantle, and the inner and outer cores, as well as the development by 1940 of two independently developed models of the earth's interior – the Jeffreys-Bullen model and the Gutenberg-Richter model – that were in close agreement with each other, and would continue to be used in seismological applications throughout the remainder of the twentieth century.

The aim of this chapter is to give an account of how this rapid progress in seismology became possible, but before I give this account, let me briefly explain what makes seismology difficult. Earthquakes occur when a section of a fault suddenly ruptures. These ruptures typically occur between 10 and 100 kilometres beneath the earth's surface – far too deep to be directly accessible. The mechanism by which such ruptures occur are complex, and little was known about it during the period this chapter covers. The rupture creates seismic waves that radiate outwards in all directions, some of which travel deep into the earth's interior before being recorded by seismographs at the earth's surface. As I have already mentioned, the earth's interior contains complex structures, none of which were known to exist in 1889. Seismic wave recordings thus are the downstream effects

DOI: 10.4324/9781003165859-12

of a complex rupture process, modulated by the properties of the complex structure through which the waves have propagated. They are, on the one hand, extremely information-rich, but also extremely complex.[1] The task of extracting the information from the complexity is difficult, and progress in seismology was the result of breakthroughs in methods for doing so.[2]

The view about progress in seismology that I will present in this chapter is closely related to certain views about progress in astronomy. In work on the history of celestial mechanics after Newton, both George Smith (2014) and William Harper (2011) use the phrase 'turning data into evidence' to refer to an important aspect of progress in astronomy. The motions of the planets are extraordinarily complex, depending as they do upon the combined forces from all the massive bodies in the solar system, each of which is itself moving in a complex way. The way in which, according to Smith and Harper, astronomers after Newton dealt with this complexity was by using theory as a means for turning data into evidence – that is, theory is used to *interpret and decompose* observations, and then the interpreted observations are used, in conjunction with theory, as a means for making inferences about the solar system. I will not here go into the details of how this is done in astronomy. I will merely make the claim here that something similar went on in seismology. It turns out that, in seismology too, theory was needed as a means for turning complex data into evidence about the earth's interior.

The story I will tell in this chapter focuses on how seismology developed, in a series of three steps, the methods for doing so. I accordingly divide the rough half-century from 1889 to 1940 into three periods. In the first period, which took place roughly from 1889 to 1906, seismological research focused on the question of whether the waves that are recorded by seismographs are correctly theoretically characterised as elastic waves of a certain type. Although there were difficulties along the way, seismologists came to the conclusion that this is indeed the case, which laid the groundwork for the next step. In the second period, which took place roughly from 1906 to 1926, the research focused on complexities in the seismic wave recordings – anomalies that did not quite fit with the theoretical characterisation. The research problem then became to find what I will refer to as an *interpretation* for every such anomaly. Typically, such interpretations were tied to postulated features within the earth's interior, such as the boundary between the earth and the core. There were, again, significant difficulties, but seismologists eventually gained confidence that the most significant anomalies could be accounted for in this way. This then led to the third period, which took place roughly from 1926 to 1940, during which the research focus was no longer on anomalies, but in making inferences from seismic wave recordings to features of the earth's interior.

This chapter will focus on the first and second periods. For each of these periods, I present a core research problem. As Shan (2019) has emphasised, scientific progress requires two intertwined sets of activities: *problem defining* and *problem-solving* (Shan 2019, p. 745). Accordingly, for each of the two periods, I show how the problem came to be defined precisely, and then how the problem came to be solved. As we will see, it turns out that in seismology, the major obstacles to progress came in the problem-solving phase of research activity. In particular, in the second period, the development of mathematical techniques that were required for solving the problem was of immense importance, a fact that becomes obvious when a comparison is made between British and German seismology during this period. Finally, the focus of this chapter is not on social or political aspects of seismology, but given the nationalities of the protagonists in my story, and the period during which it takes place, political events unavoidably play a major role in it.

2.   First Period (1889–1906)

I place the beginning of the first period in 1889, with Rebeur-Paschwitz's discovery of seismic waves from distant earthquakes. I take the research focus of seismology during this period to be on the question of whether these waves are correctly theoretically characterised as elastic waves of a certain type. The precise formulation of this question requires a theory of elastic waves. Fortunately, such a theory had already been developed decades earlier. Earthquakes had been a topic of interest for British geologists since the early nineteenth century. In 1847, William Hopkins worked out the theory of seismic wave propagation in mathematical detail under the assumption that they are waves in an elastic solid medium. He relied heavily on work from the 1820s and 1830s by Poisson and Cauchy on light waves, which were then conceived of as waves in an elastic solid medium, the ether. According to the theory of Poisson and Cauchy, an isotropic elastic solid medium will support two types of waves – longitudinal waves and transverse waves. Hopkins (1847) gives equations for the respective velocities α and β of these two types of waves, which I have put here in a modern form:

$$\alpha = [(K + 4\mu/3)/\rho]^{1/2} \tag{1}$$
$$\text{and} \quad \beta = (\mu/\rho)^{1/2} \tag{2}$$

Here, ρ is the density of the medium, K is the bulk modulus, and μ is the shear modulus. The two moduli represent elastic properties of the medium.

Throughout this chapter, I will refer to the waves described by Equation (1) as *pressure waves* and the waves described by Equation (2) as *shear waves*.[3] Since the bulk and shear modulus are always positive, pressure waves should travel faster than shear waves. These equations open up a tantalising possibility – if the earth's interior can be approximated as an elastic solid medium, and these wave velocities can be measured, such measurements might yield information about the density and elastic properties of the earth's interior. At the time Hopkins wrote the paper, however, seismic wave observation techniques had not yet been sufficiently developed.

When Rebeur-Paschwitz observed waves from distant earthquakes, then, the theoretical framework was already in place in which to formulate precisely a question regarding the physical nature of these waves: *Are seismic waves elastic waves of the sort described theoretically by Equations (1) and (2) – that is pressure waves and shear waves?* A positive answer to this question would open the door towards extracting information about the earth's interior by means of Equations (1) and (2).

Answering this question is easy in principle. If one hypothesises that seismic waves are elastic waves of the type described by Equations (1) and (2), then two waves – a pressure wave and a shear wave – ought to emanate from any earthquake, and these waves ought to travel at different velocities that depend on the density and elastic properties of the medium through which they are propagating. If the earth's interior is homogeneous and isotropic to a first approximation, then an earthquake would generate two roughly spherical wavefronts that propagate at two different constant velocities. These velocities, being dependent only on the density and elastic properties of the earth's interior, would be the same for all earthquakes. Obviously, one cannot directly track the wavefronts as they travel through the earth's interior, but if a large number of seismographic stations can be established throughout the earth's surface, one can hope to detect the wavefronts as they pass various points on the globe. By recording the times at which each wave passes these various points, the wave velocities can be determined. If it can be shown that there are two distinct velocities at which these waves travel, this would be evidence for a positive answer to the question.

In practice, answering the question turned out to be much more difficult. For starters, it required the establishment of a global network of seismographic stations. Fortunately, this was made possible by rapid developments in observational seismology, led by a group of British scientists in Japan. The most prominent of these early seismologists was John Milne, who had been experimenting with seismographs since arriving at the Imperial College of Technology in Tokyo in 1876 (Wood 1988). Rebeur-Paschwitz's discovery prompted Milne to invent and perfect a seismograph for recording distant earthquakes, and he promoted the idea of having a global network of seismographs in the Report of the 1895 Meeting of the British Association for the Advancement of Science: 'a ring of twelve or twenty-four stations situated round our globe would in a very short time give us valuable information, not simply about its crust, but possibly also about its interior' (Milne 1895, p. 153).[4] The 1897 Report of the newly created Seismological Committee of the British Association contains a letter to be sent to various investigators worldwide, requesting their cooperation, and states that the 'first object in view is to determine the velocity with which motion is propagated round and possibly through our earth' (p. 137). By the time of the 1898 Report of the Seismological Committee, arrangements had been made for seismological instruments to be established at 22 stations all over the globe.

Setting aside the organisational problem of establishing a global network of seismographic stations, the major difficulty of the method I have just described is that it assumes that pressure waves and shear waves can be clearly identified on seismic wave recordings, but this was not yet the case. At the time, there were no standard ways of interpreting the raw data that were recorded on seismographs. Seismic waves were recorded as complex sets of squiggly lines. There were two robust features of these squiggles that seismologists in the 1890s, such as Milne, focused upon. Seismograms typically start off with rumbling motions of a very small amplitude, which are then followed by much more pronounced motions that look like rolling waves. The initial rumbling motions came to be called *preliminary tremors*, while the later motions came to be called *large waves*.[5] Because preliminary tremors always precede the large waves, and since Equations (1) and (2) imply that pressure waves always have a higher velocity than shear waves, one might think that the preliminary tremors are recordings of the pressure wave as it passes through, while the large waves are recordings of the shear waves that arrive a bit later. Complicating matters, however, was the theoretical postulation of a third type of wave by Lord Rayleigh in 1885. This type of wave would propagate along the surface of the earth, and it would travel with a velocity slower than pressure and shear waves. So an alternative is to identify large waves with these surface waves. But what, then, are the preliminary tremors – pressure waves or shear waves?

The way to answer this question is to measure the velocities of preliminary waves and large waves by plotting the time it takes them to travel to various locations on the earth's surface. These plots, called *travel-time curves*, became an important tool of early 20th century seismology. Whenever a large earthquake occurred, preliminary tremors and large waves would be recorded by stations distributed globally. Such recordings could be used to make plots of the time it takes such waves to travel various distances in degrees of arc along the earth's surface. If either preliminary tremors or large waves are surface waves that travel at constant velocity, then one would expect these plots to be linear. If they are, on the other hand, body waves that travel at constant velocity along the chord from the earthquake source to the observing station, these plots would not be linear, but one can easily calculate what these plots would look like, and the wave velocities could be determined from the plots.

The first travel-time curve presented by Milne was in the 1900 Report of the Seismological Committee. His diagram contained three curves – one for the beginning of the preliminary tremor, one for the beginning of the large waves, and one for the difference between them. The curve for the large waves was close to a straight line – which, as I have mentioned, is to be expected if they are surface waves. The Report, however, contains an extended criticism of the identification of large waves with surface waves, by C. G. Knott, a Scottish physicist who had also been stationed in Japan. Knott favoured the view that the large waves are shear waves that have travelled through the earth's interior. The near-linear travel times for the large waves could be accommodated by throwing out the assumption that shear waves travel at constant velocity – a reasonable suggestion, given that shear waves are dependent on the density and elastic properties of the medium, and there is no reason to think that these are constant through-out the earth's interior. The Report ultimately comes to the conclusion that the data are inconclusive.

A major breakthrough came in a 1900 paper published by Richard Oldham, a British geologist who was then Director of the Geological Survey of India. Oldham had, in 1899, published a 500-page report on a major earthquake that had occurred in the Indian state of Assam in 1897. In preparing this report, he carefully studied recordings of seismic waves from this earthquake that had been made on seismographs located in Italy. He found that he could identify two distinct *phases* of the preliminary tremors, which he suggested treating as two different waves. He subsequently examined Italian seismograms for 11 other earthquakes, and identified these two phases on all of them. The 1900 paper contains a plot of the travel times for the two phases of preliminary tremors, showing that distinct travel time curves can be plotted for the two phases. He concludes that the *first phase* can be identified with pressure waves, and the *second phase* can be identified with shear waves.

He makes one other suggestion in the 1900 paper that is of great significance for the history of seismology. Noting the possibility that the earth might have a metallic core, he suggests that an investigation of travel times of the first and second phases of the preliminary tremors, under the assumption that they are respectively pressure and shear waves, might reveal the existence of such a core.

Oldham's distinction between two phases of preliminary tremors appears to have been quietly accepted by Milne. The 1902 Report of the Seismological Committee, written by Milne, contains a plot with three travel time curves: the first phase, the second phase, and the large waves. It also contains (pp. 65–67) an excerpt from a letter from Knott, grudgingly acknowledging that the large waves are likely to be surface waves ('facts are chiels that winna ding' he writes, quoting Robert Burns), and that the data for the first phase of the preliminary tremors appear to be consistent with their interpretation as a wave that travels at constant velocity within the earth's interior, along the chord between two points on the surface of the globe. Knott comments that the data for the second phase show much less agreement. By 1902, then, Oldham appears to have convinced Milne that there is good reason to think that the first phase corresponds to pressure waves and the second phase to shear waves.

By no means, however, was this established beyond all doubt. As Milne, Knott, and Oldham all well knew, the observations at the time were not entirely reliable. The first and second phases of the preliminary tremors were not always clearly distinguishable. The origin times of most earthquakes were not directly known – they were typically inferred – and thus most of the travel times were inferred. The data itself was meagre and biased. The suggestion of Oldham was based on just seven earthquakes, and relied heavily on data from just a handful of seismographic stations, mostly in Europe. By 1905, Knott had become convinced that the first and second phases are respectively pressure and shear waves, but he (Knott 1905, pp. 577–580) also found it worthwhile considering in great detail the views of a group of Japanese seismologists who considered all seismic waves to be transmitted along the surface of the earth. Within a few years, however, seismology turned to a different set of questions, raised in 1906 by Oldham.

3.   Second Period (1906–1926)

By the beginning of the second period, seismologists now had some confidence in the theoretical characterisation of the first and second phases of the preliminary tremors as, respectively, pressure and shear waves. In coming to this conclusion, however, most of the immensely complex structure in actual seismic wave recordings was ignored. The key to unlocking more information from the seismic wave recordings lay in now *assuming that the theoretical characterization of seismic waves is correct, and using theory as a tool for making sense of these complexities*. More specifically, theory was used to make predictions about what the seismic wave recordings ought to be like, and research then focused on anomalies[6] that stood out against such predictions. As seismologists at the time well knew, such anomalies could be the effect of as-yet unknown structures within the earth, so they held the promise of unlocking new discoveries.

The greatest obstacle to progress during this period was that in order to make predictions about seismic wave recordings, one needed mathematical techniques for calculating wave paths in the earth's interior, and these had to be developed. As we will see, the lack of such techniques stifled progress in British seismology during this period. German seismologists were able to develop such techniques, but were unable to make much progress for an entirely different reason, as we will see.

I date the start of the second period in 1906, with the presentation by Oldham of a paper titled 'The Interior Constitution of the Earth as Revealed by Earthquakes' before the Geological Society of London. There are three significant features of this chapter. First, Oldham explicitly states that, although the identification of the first and second phase of the pre-liminary tremors is not established with certainty, he will proceed under the assumption that it is correct:

> I shall take it that the first-phase waves are condensational [i.e. pressure waves] – this being generally acknowledged – and that the second-phase waves are distortional [i.e. shear waves], an assumption which I regard as more probable, and on these assumptions it is pos-sible to estimate the proportion which the modulus of rigidity bears to the bulk-modulus, or resistance to compression [within the earth].
> (Oldham 1906, p. 466)

Second, Oldham (1906, p. 462) presents new travel-time curves for the first and second phases for 14 earthquakes, but notes a major anomaly in the travel time curve for the second phase – a sudden jump in the travel time between 120 and 130 degrees of arc. Third, in order to account for the anomaly in the second phase, Oldham (1906, p. 468) postulates the existence of a central core in which the velocity of shear waves is signifi-cantly lower than in the outer layer. On the basis of the fact that the jump occurs at around 120 degrees, and assuming for simplicity that the path taken by the second phase in the earth's interior is a chord, he proposes that the earth has a central core with a radius 0.4 times that of the earth.

Oldham (1906) displays what I take to be the key features of research during this period – namely, it makes the assumption that the first and second phases are correctly theoretically characterised as pressure and shear waves, it focuses on an anomaly (more specifically: a deviation between what one would expect seismic wave recordings to be like based on the assumption, and what the actually recorded seismic wave data looks like), and it postulates a feature of the earth's interior to account for that anomaly. The anomaly that Oldham points out is just one of many anomalies in the seismic wave data that were identified by seismologists during this period.

I take progress during this period to be the result of the pursuit of the following research problem: *For each robust anomaly in the seismic wave recordings, find an independently confirmable interpretation for that anomaly.* By *robust*, I mean that steps must be taken to ensure that the purported anomaly is not a mere artefact – always a possibility at the time given the very small number of observations, the limited distribu-tion of seismographic stations, and the instrumentation then available. By *interpretation*, I mean an account of how that anomaly arises that is consistent with the theory of elastic waves. In most cases, it involves the postulation of some feature of the earth's interior. In Oldham's case, the anomaly was interpreted as being the result of the retardation of the shear wave as it travels through a central core. The requirement that the interpretation be *independently confirmable* is to prevent the possibility of simply accommodating the data. Once a feature such as a central core was postulated, seismologists were then tasked with finding ways of independently confirming whether it exists. I claim that it is through the pursuit of this research problem that seismologists gradually came to grips with the complexity of seismic wave recordings.

As I have already mentioned, the greatest obstacle to progress during this period was that in order to characterise the anomalies precisely, one must have a means for calculating the expected seismic wave recordings precisely, and for that one needs mathematical techniques for calculating wave paths in the earth's interior. Oldham and many of the British seismologists active at the time (with the exception of Knott) did not have the requisite mathematical training for developing such techniques, and it was German seismologists who were doing the research that turned out, in hindsight, to be the cutting edge.

Particularly important was a group at the University of Göttingen including the physicist Emil Wiechert and his students, among whom are a number of major figures of twentieth century seismology, such as Karl Zoeppritz, Ludwig Geiger, and Beno Gutenberg. In 1907, Wiechert and Zoeppritz published the first in a series of papers with the title 'Über Erdbebenwellen' ('On Earthquake Waves'). Their aim was the same as Oldham's – to use seismic wave observations as a tool for making inferences about the earth's interior. The paper is highly theoretical in its approach, covering in detail the theory of elastic waves, including derivations of Equations (1) and (2), treatments of the reflection and refraction of elastic waves at boundaries within the earth, and possible wave paths in an earth with varying density and elastic constants.

A section of the paper, written by Wiechert, discusses Oldham (1906), and is dismissive of his inference that there is a central core. Wiechert's criticism involves the realisation that, in addition to pressure and shear waves that travel directly from the earthquake origin to the observing station, there are also *surface reflections* – pressure and shear waves that travel from the origin to a point on the earth's surface approximately halfway between the origin and point of observation, reflect off the surface, and then travel to the point of observation.

Here, in order to simplify the exposition, I shall use modern terminology for the first and second phases.[7] According to modern terminology, the pressure wave that travels directly from the origin to the point of observation is called the P phase. The shear wave that similarly travels in a direct route is called the S phase. The pressure wave that reflects once off the surface is called the PP phase, while the once-reflected shear wave is called the SS phase. Waves that undergo two or more reflections at the surface are represented by increasing the number of letters – for example the PPP phase is a twice-reflected P wave and the SSS phase is a twice-reflected S wave. Importantly, when an elastic wave (whether pressure or shear) is reflected, it produces both pressure and shear waves. Thus, the PS phase is a wave that is a pressure wave at the origin, and is converted into a shear wave upon reflection. The SP phase is the reverse of this. One can also have, for example the PSP phase, the SPS phase, and the PSS phase.

Wiechert and Zoeppritz (1907, p. 518) claim that Oldham's anomaly can be explained as a case of misinterpretation. What Oldham had taken to be the S phase for observations beyond 120 degrees was, in actuality, the SS phase. The modern view is that Wiechert's criticism of Oldham is correct (see Brush 1980, p. 711). More importantly, Wiechert had made clear that in order to correctly interpret anomalies, one must be able to correctly identify the P and S phases on seismograms, but this requires being able to pick them out from a tangle of other phases for which they might be mistaken.

The way to do this is to come up with estimates of their travel times, as well as those of the phases for which they might be mistaken, such as SS or PP. This is complicated, however, since by then it was clear that the velocities of both pressure and shear waves vary as a function of depth within the earth, and this would result in curved wave paths. If that function could be determined at least approximately, then the wave paths could be determined, and the expected travel times for various phases calculated. The Göttingen seismologists thus concentrated their efforts on solving the following mathematical problem: *determine, from travel times of pressure and shear waves, their velocities as a function of depth within the earth.* This problem was solved through a method developed in 1907 by the Göttingen mathematician Gustav Herglotz (Schweitzer 2003). Wiechert simplified the method and applied it to seismology, and it is now known as the Herglotz-Wiechert method. Using this method, Wiechert claims, 'it was easy to find reflections in the tangle of earthquake waves'.[8]

There is one further historically important paper by a member of the Göttingen group. Beno Gutenberg (1914) used techniques he developed with Ludwig Geiger for determining seismic velocities in order to make inferences about the earth's interior. The paper argues that the earth has a core with a distinct boundary, and calculates travel times for various *core phases*, which are waves that either travel through the core or reflect off of it. For example, the phase he calls P_cP is a phase that starts off as a pressure wave, is reflected as a pressure wave off the core, and returns to the surface as a pressure wave. An example of a phase that travels through the core is the S_cP_cS phase, which starts off as a shear wave, is a pressure wave within the core, is converted back to a shear wave upon emergence from the core, and is observed at the earth's surface as a shear wave.

Gutenberg calculated theoretical travel times for core phases, used these travel times to identify core phases in seismographic data, and then used these core phases to calculate details about the core, such as its size. In this way, Gutenberg was able to identify a large number of phases, determine their travel times, and make inferences about the earth's interior. Gutenberg (1914) represents in many respects the culmination of the style of research being done in the second period of using sophisticated theoretical techniques to make sense of complexities in seismic wave recordings and thence to make inferences about the earth's interior.

Unfortunately, this work had virtually no impact in seismology for more than a decade. The year Gutenberg's paper was published, Europe was plunged into a devastating war, which made it difficult for the members of the Göttingen group to continue doing seismology (Schweitzer 2003, pp. 17–18). Geiger lost his job as a seismologist (among a number of unfortunate events, including being taken as a prisoner of war and being accused of being a spy) and eventually became a successful businessman. Gutenberg also lost his job as a seismologist and spent the next decade as the manager of a soap factory while working on geophysics at night. He remained remarkably productive given the circumstances, but he would not land a permanent academic position until Caltech hired him as chair of geophysics in 1930. Another effect of the war was that it disrupted communication between British and German seismologists until the mid-1920s. This would have a significant impact on the progress of seismology.

The importance of the mathematical techniques developed by the Germans for progress in seismology becomes particularly clear when one examines British seismology during the second period. While I have given credit to Oldham for first developing the approach of the second period, he failed to convince any of his British colleagues of the existence of a core. Knott (1908, p. 232) took the Oldham anomaly to be an effect of the steep angle at which seismic waves emerge at longer arcual distances, and states that in any case 'we can hardly regard Oldham's hypothesis as at all convincing until a great many more observations are to hand'. G. W. Walker's (1913) *Modern Seismology*, strongly influenced by continental approaches, contains a discussion of both SS waves (p. 42) and PS waves (p. 41), arguing that these phases tend to be confused with the S phase, and as a result it makes the detection of S difficult at certain distances. Without naming Oldham, Walker (1913, p. 45) writes: 'It has indeed sometimes been asserted that S never reaches beyond a certain distance, and to explain this an impenetrable core of the earth has been assumed. We see that no such hypothesis is at all necessary to explain the observations'.

The Oldham core seems to have fallen by the wayside by the mid- to late 1910s. The only significant British writing I can find from this period that mentions it favourably is Knott (1919), his view about it having changed by then.[9] There is no mention of even the possibility of a core existing in the Reports of the Seismological Committee of the BAAS in the 1910s.

Meanwhile, the Seismological Committee concentrated its efforts on the collection and reduction of seismological data flowing in from the global network of seismographs. For the purpose of reducing data, the Seismological Committee adopted a table of travel times for the P and S phase created by Walker (1913) by interpolation from the travel times for a set of well-determined earthquakes given in Wiechert and Zoeppritz (1907). H. H. Turner then set about comparing these travel times with the observed travel times for newly recorded earthquakes in order to make corrections to the tables.

In the late 1910s and into the 1920s, Turner inspected the residuals – that is, the difference between the expected travel times for the P and S phase according to the travel time tables and the actually observed travel times – and noticed several significant anomalies in these residuals. For example, in the 1915 Report of the Seismological Committee of the British Association for the Advancement of Science, Turner reported on a set of deviations of the S phase from the standard tables. These seemed to be systematic, and it was suspected that there was some other phenomenon that was being mistaken for the S phase. Various suggestions were made, including the idea that it was a PS phase, or that it was a pressure wave that had undergone multiple surface reflections, but no conclusive identification could be made. Turner labelled it the Y-*phenomenon*, and it was the subject of much discussion in the BAAS Seismological Reports of the late 1910s. The Y-phenomenon is just one of several other anomalies that are mentioned in the Seismological Reports of this period.

In hindsight, very little progress appears to have been made in British seismology in the 1910s and well into the 1920s. Of course, some of this lack of progress can be attributed to the effects of the war, but another major factor was the lack of a well-developed set of mathematical techniques for calculating travel times, something that is needed for properly interpreting seismic wave recordings. Nothing illustrates this better than the ultimate resolution of the question of the Y-phenomenon, which remained unresolved until 1926.

The BAAS Seismological Report for that year, written mainly by Turner, recounts that he had been examining the residuals for the S phase for an earthquake that had occurred on 11 October 1922, and had noted that the readings from between 80 degrees and 110 degrees probably were the result of a phase mis-identification. The question was: which phase? Turner had just tabulated these results when

> a letter was received from Dr. Harold Jeffreys calling attention, in enthusiastic terms, to [Gutenberg (1914)], and it was at once seen that the readings tabulated as S refer to Gutenberg's ray S_cP_cS; that is, a ray which travels as S until it reaches the liquid core of the earth, is then transformed into P, and finally emerges as S.
>
> (Turner 1926, p. 271)

It was found that the S_cP_cS phase would arrive slightly *before* the S phase at great distances, and is approximately the same amplitude. Because stations usually would record, as the S phase, the earliest wave arriving at around the time S was expected, the S_cP_cS phase was mis-identified as S, and in fact Jeffreys, looking back on this period, states that at angular distances of over 83 degrees, 'there were hardly any genuine readings of S' (Jeffreys 1952, p. 86).

The British seismologists at the time well knew of the work of Wiechert and Zoeppritz, but they apparently were not aware of the important later papers of Geiger and Gutenberg from just prior to the start of World War I. The war is an obvious factor – not only did Geiger and Gutenberg lose their jobs, but also German seismologists were prevented for political reasons from participating in international seismological conferences throughout the 1920s (Schweitzer 2003, p. 12). The British were thus unaware of the theoretical work on core phases and its significance for resolving some of the problems with travel time tables until Jeffreys pointed it out in 1926. Turner, who was revising the tables, and Jeffreys, who took on the task after Turner's death in 1930, immediately saw the importance of the work of Gutenberg, including not just the core phases but also the method that Geiger and Gutenberg had developed for determining seismic velocities from phase amplitudes.

The re-discovery of Gutenberg changed the nature of the question being pursued by seismologists once again. The new central question became: *Assuming that the observations have been properly interpreted, where are the major discontinuities within the earth, what phases do they give rise to, and what are the travel times for these phases?* The next significant period of seismology, in which this question was at the forefront, covers the period 1927 to 1940.

Harold Jeffreys, a particularly adept applied mathematician, was equipped with the right set of tools to become one of the two leading seismologists of this period, along with Beno Gutenberg, newly hired by Caltech. During this period, Jeffreys began revising the travel time tables, adding a large number of phases including surface reflections and core phases. Jeffreys was joined by his student Keith Bullen in the 1930s, and their work culminated in the publication of the Jeffreys-Bullen travel time tables of 1940 (Jeffreys and Bullen 1940). This period is worth examining in its own right, but I shall end my account of the progress of seismology in 1926.

4.   Turning Data Into Evidence in Seismology

Let me recap my account of progress in seismology. Understanding how progress was made requires understanding exactly what makes seismology difficult. At the beginning of this chapter, I stated that seismic wave recordings are, on the one hand, extremely information-rich, but also extremely complex. One must find a way to make sense of the complexity and unlock the information within – to turn data into evidence. The way this was accomplished in seismology is in a series of three steps.

In the first step, most of the complexity was ignored, and the aim was to determine whether a pre-existing theoretical characterisation of seismic waves could plausibly account for major features in the seismic wave recordings. The major breakthrough in this period was the recognition by Oldham that the preliminary tremors contain finer-grained structures – the first phase and the second phase – which support the theoretical characterisation when interpreted respectively as pressure waves and shear waves.

In the second step, seismologists, now confident in the theoretical characterisation of seismic waves, focused on the complexities in the seismic wave data. In particular, for each significant anomaly, seismologists used theoretical tools to seek out an interpretation that could properly account for the anomaly. Because seismologists were by now examining finer grained details in the seismic wave data, the development of mathematical tools for determining wave paths and travel times within the earth was crucial. The major breakthrough during this period was the development of these tools by German seismologists, but the tumultuous political events of the period led to a delay in recognising the importance of this work, especially the important pre-war work of Gutenberg, until its re-discovery by Jeffreys. This opened the door to the third step, in which the focus was no longer on anomalies, but on making inferences from interpreted seismic wave data to features of the earth's interior.

What does all this show, more generally, about scientific progress? My view is that for two sciences at least – astronomy and seismology – an understanding of how progress was made requires coming to grips with the complex relation between theory and observation, especially the role of theory in interpreting data. The history of development of methods for extracting information from complex data, among which I count methods of idealisation and approximation in astronomy and the earth sciences, is underappreciated by philosophers. A better understanding of how such methods were developed in individual sciences will give us a more complete picture of scientific progress.

Acknowledgements

The research for this chapter was supported by the Ministry of Education, Singapore, under its Academic Research Fund Tier 1 Grant, No. RG156/18-(NS). I would like to thank my assistant Ivan Ho, hired under the Singaporean SGUnited Traineeship Programme, for compiling some of the materials, and Yafeng Shan for inviting me to write this chapter.

Notes

1. On top of this complexity is the standard noise that occurs along with any measurement process, but I will not, in this chapter, get into the problem of dealing with noise.
2. This chapter will focus on the problem of extracting earth structure from seismic wave observations. See Miyake (2017) for an account of the history of the parallel problem of extracting information about seismic sources from seismic wave observations.
3. What are now called pressure waves and shear waves often were referred to respectively as 'condensational waves' and 'distortional waves' in the nineteenth and early twentieth century. In order to avoid confusion, I am sticking with modern terminology throughout.
4. Rebeur-Paschwitz also had the same idea at around the same time (see Schweitzer 2003).
5. Sometimes called 'long waves'.
6. I am aware of the Kuhnian overtones of my use of the term 'anomaly'. I am broadly sympathetic to a Kuhnian (or a functional) way of understanding how progress was made in this case, but I will lay aside the obvious question about whether this marks the rise of a seismology paradigm because I think it is beside the point I want to make in this chapter.
7. The word 'phase' when used in modern seismology in terms like 'P phase' and 'S phase' always means an elastic wave that follows a particular wave path in the earth's interior. In the first decade of the twentieth century, however, the word 'phase' just meant a portion of the seismogram (as in 'first phase of the preliminary tremor'). Sometime after 1910, the word appears to have shifted its meaning towards the modern usage.
8. This is my translation of the original German: 'Da war es denn nach alledem leicht, die Reflexionen in dem Gewirre der Erdbebenwellen aufzufinden, zu identifizieren.' (p. 519)
9. Knott (1919) is a very interesting paper which not only revives the Oldham core, but argues that it is liquid – anticipating the ideas of Harold Jeffreys that were to appear in the mid-1920s.

References

Brush, S. (1980) Discovery of the Earth's Core. *American Journal of Physics*. 48, 705–724.

Gutenberg, B. (1914) Ueber Erdbebenwellen. VII A. *Nachrichten von der Gesellschaft der Wissenschaften zu Göttingen, Mathematisch-Physikalische Klasse 1914*, 125–176.

Harper, W. (2011) *Isaac Newton's Scientific Method: Turning Theory into Evidence about Gravity and Cosmology*. Cambridge: Cambridge University Press.

Hopkins, W. (1847) Report on the Geological Theories of Elevation and Earthquakes. *Report of the British Association for the Advancement of Science*. 17, 33–92.

Jeffreys, H. (1952) *The Earth: Its Origin, History and Physical Constitution*, Third Edition. Cambridge: Cambridge University Press.

Jeffreys, H. and Bullen, K. E. (1940) *Seismological Tables*. London: Office of the British Association.

Knott, C. G. (1905) Seismological Studies. *Scottish Geographical Magazine*. 21:11, 569–582.

Knott, C. G. (1908) *The Physics of Earthquake Phenomena*. Oxford: Clarendon Press.

Knott, C. G. (1919) The Propagation of Earthquake Waves through the Earth and Connected Problems. *Proceedings of the Royal Society of Edinburgh*. 39, 157–208.

Milne, J. (1895) Investigation of the Earthquake and Volcanic Phenomena of Japan. Fifteenth Report of the Committee. *Report of the British Association for the Advancement of Science*. 65, 113–185.

Milne, J. (1897) Seismological Investigations – Second Report of the Committee. *Report of the British Association for the Advancement of Science*. 67, 129–205.

Milne, J. (1898) Seismological Investigations – Third Report of the Committee. *Report of the British Association for the Advancement of Science*. 68, 179–276.

Milne, J. (1900) Seismological Investigations – Fifth Report of the Committee. *Report of the British Association for the Advancement of Science*. 70, 59–120.

Milne, J. (1902) Seismological Investigations – Seventh Report of the Committee. *Report of the British Association for the Advancement of Science*. 72, 59–75.

Miyake, T. (2017) Magnitude, Moment, and Measurement: The Seismic Mechanism Controversy and Its Resolution. *Studies in History and Philosophy of Science*. 65–66, 112–120.

Oldham, R. D. (1899) Report on the Great Earthquake of 12th June 1897. *Memoirs of the Geological Survey of India*. 29, v–379.

Oldham, R. D. (1900) On the Propagation of Earthquake Motion to Great Distances. *Philosophical Transactions of the Royal Society of London, Series A*. 194, 135–174.

Oldham, R. D. (1906) The Interior Constitution of the Earth as Revealed by Earthquakes. *The Quarterly Journal of the Geological Society of London*. 62, 456–475.

Schweitzer, J. (2003) Early German Contributions to Modern Seismology. In W. H. K. Lee, H. Kanamori, P. C. Jennings, and C. Kisslinger (eds.) *IASPEI International Handbook of Earthquake and Engineering Seismology, Part B*. London: Academic Press.

Shan, Y. (2019) A New Functional Approach to Scientific Progress. *Philosophy of Science*. 86, 739–758.

Smith, G. E. (2014) Closing the Loop. In Z. Biener and E. Schliesser (eds.) *Newton and Empiricism*. Oxford: Oxford University Press, pp. 263–345.

Turner, H. H. (1915) Seismological Investigations – Twentieth Report of the Committee. *Report of the British Association for the Advancement of Science*. 85, 52–79.

Turner, H. H. (1926) Seismological Investigations – Thirty-first Report of the Committee. *Report of the British Association for the Advancement of Science*. 96, 267–273.

von Rebeur-Paschwitz, E. (1889) The Earthquake of Tokio, April 18, 1889. *Nature*. 40, 294–295.

Walker, G. W. (1913) *Modern Seismology*. London: Longmans, Green and Co.

Wiechert, E. and Zoeppritz, K. (1907) Über Erdbebenwellen. *Nachrichten von der Gesellschaft der Wissenschaften zu Göttingen, Mathematisch-Physikalische Klasse 1907*, 415–549.

Wood, R. M. (1988) Robert Mallet and John Milne – Earthquakes Incorporated in Victorian Britain. *Earthquake Engineering and Structural Dynamics*. 17, 107–142.

10 Progress in Psychology

Uljana Feest

1. Introduction

In 1991, the psychologist David Lykken (1928–2006) published an article in an edited volume honouring the eminent psychologist and methodologist Paul Meehl. The article, entitled 'What's Wrong With Psychology Anyway?', made the case that psychology as an academic field was in bad shape, identified a number of reasons for this and made some suggestions for improvement (Lykken 1991). The article invites the reader to imagine Linus Pauling travelling back in time to his dissertation defence in 1925, telling his captive audience about the mindboggling progress that has occurred in biochemistry in the intervening 60-plus years. He then imagines a similar thought experiment, involving Paul Meehl traveling back to his dissertation defence in 1945. 'What could he amaze his committee with? What wonders of new technology, what glistening towers of theoretical development, could he parade before their wondering eyes?' Lykken asks sarcastically.

> They will be interested to learn that Hull is dead and that nobody cares anymore about the 'latent learning' argument. He could tell them now that most criminals are not helpless victims of neuroses created by rejecting parents: that schizophrenia probably involves a biochemical lesion and is not caused by battle-ax mothers and bad toilet training; that you cannot fully understand something as complex as language by the simple principles that seem to account for the bar-pressing behaviour of rats in a Skinner box.
>
> (Lykken 1991, 13)

Lykken's worries resonate with crisis declarations that have haunted psychology since the late nineteenth century (see Sturm & Mülberger 2012), including recent debates about the crises of replication (Open Science Collaboration 2015), generalisation (Yarkoni 2020), validity (Schimmack 2019), and theory (Eronen & Bringmann 2021; van Rooij & Baggio 2021). The long history of such debates indicates that it is contested whether there has been real progress in psychology or, at the very

DOI: 10.4324/9781003165859-13

least, that there are some obstacles to making true progress in psychology. What is lacking so far are more general reflections about the very notion of progress, as it applies to psychology. This concerns a number of different questions, such as what is the subject matter of psychology, what are aims of psychology, and what are criteria by which to judge whether there has been progress in psychology. In this contribution, I will try to make some headway with regards to these issues.

The underlying contention of this chapter is that whereas traditional accounts of progress have typically focused on scientific *theories* (evaluating them with regards to their truthlikeness or their explanatory and predictive success), we should pay closer attention to efforts of forming and developing scientific *concepts* (evaluating them with regards to how well such concepts serve the respective aims of research, which can include, but are not limited to, explanation and prediction). If we focus on concepts rather than theories, our attention is drawn to the questions of (a) what is the status of the subject matter *of psychological research* and (b) what are basic requirements for *conceptual progress* with regards to this subject matter.

Section 2 begins by narrowing down the time frame I will be focusing on (roughly, 1900 to today) and argues that one requirement for progress is a minimal continuity of subject matter. Section 3 will raise the question of whether well-known theoretical and methodological shifts that occurred in twentieth-century psychology (e.g. from behaviourism to cognitivism) pose a threat to this continuity. I will conclude that they do not. My argument for this will be that both, the units of analysis (individuals) and the objects of research (capacities, typically individuated by means of folk-psychological concepts) remained relatively stable throughout these changes. Section 4 will argue that eliminativist critiques of the use of folk-psychological concepts in scientific research misconstrue an important aim of much psychological research, namely, to obtain and adequate taxonomic and descriptive accounts of their objects. Given this aim, Section 5 will argue, the progress we should be interested in is conceptual progress (i.e. progress in the accuracy or usefulness of our concepts). Section 6 raises a complication to this view, namely, that it is not clear that there is a natural fact of the matter as to what the psychological objects 'really' are. Section 7 clarifies that it doesn't follow that psychological kinds are not real, though I will adopt the proposal that psychological kinds are historical kinds. Section 8, finally, articulates the main thesis of this article, namely, that we should broaden our notion of progress to include the accumulation of epistemic resources that researchers can draw on in their attempts to track their objects conceptually.

2. 'Modern Psychology' and the Question of Progress

First of all, we should specify what time frame we are talking about when considering progress in psychology. There is a long-standing narrative, according to which philosophical reflections about mind and behaviour go back to early Greek philosophy (e.g. Robinson 1995) but that the beginnings of psychology as a properly scientific endeavour can be dated to the mid-nineteenth century (e.g. Boring 1950; Mandler 2007), which saw experimentalists like Fechner (1860) conduct systematic experimental manipulations with the aim of formulating mathematical laws. In turn, these experimental efforts were heavily influenced by developments in experimental sensory physiology, for example by figures like Johannes Müller and Hermann von Helmholtz (Ash 1980; Fancher & Rutherford 2013). This push culminated in the founding of institutionalised psychological laboratories (e.g. Wundt's lab in Leipzig in 1879) and, a few decades later, the establishment of the discipline of psychology as separate from philosophy (e.g. in Berlin and Harvard).

A lot of ink has been spilled about whether this is indeed an accurate description of when psychology became scientific, with scholars pointing out that conceptual (i.e. non-experimental and non-quantitative) work can be scientific, too (Hatfield 1995), that people were already discussing quantification and experimentation in the eighteenth century (Sturm 2006), that there was a close connection between the development of scientific psychology and the applied needs of pedagogy already in the early nineteenth century (Gundlach 2004), etc. Regardless of details and exact dates, and regardless of how one evaluates the changes in question, it is fairly clear that some relevant shifts occurred around the turn of the nineteenth/twentieth century. 'Since that time', the historian Kurt Danziger remarks, 'psychology has had a social presence and methods of inquiry it did not have before' (Danziger 2002, 1). He goes on to say that '[t]hese novelties provide us with usable criteria for deciding what forms part of the history of *modern* psychology' (Danziger 2002, 1).

This suggests that when thinking about progress we should distinguish between (a) the question of whether the shift towards 'modern' psychology constituted progress vis-a-vis earlier reflections, and (b) whether there has been progress *within* the (quantitative and experimental) framework of modern psychology. The first of these suggestions boils down to the idea that psychology's progress towards becoming modern and scientific was coextensive with it becoming a *natural* science and that this should be considered the right move. While this notion is widely endorsed among contemporary psychologists, there have long been critical voices as well. Two lines of response have been (1) to make the *descriptive and historical* point that there are areas of psychology that do not fit into this natural scientific mould and (2) to question the *normative* assumption that psychology can and should be a purely naturalistic discipline (see Richard et al. 2014; Smith 2007).[1] Present-day critiques of psychology focus on various aspects of this latter suggestion, including the assertions that quantification in psychology is misguided because the psychological subject matter is not 'quantitative' (e.g. Michell 1999) and the experimental method cannot capture the fact that the psychological subject matter contains irreducibly experiential and intentional features (e.g. Wertz 2018). Speaking more broadly, some scholars have worried that within modern psychology, presumptions about scientific methods (concerning, e.g. the use of inferential statistics) can sometimes dictate specific ways of conceptualising the subject matter, rather than the other way around (e.g. Danziger 1985, 5).

While this latter concern is, of course, well taken, it raises the question of what the subject matter of psychology 'really' is. Depending on how we answer this question, we are likely to have divergent responses to the question of what progress might look like and by what means it can be (or has been) achieved. There is another question here, namely, whether the subject matter of psychology over the past 160 years (or the last 2,000 years, for that matter) has exhibited enough stability and continuity in order for the science of psychology to have been able to make progress about 'it'. One might wonder, for example, whether the Cartesian notion of a *res cogitans* was even remotely concerned with the same subject matter as (say) today's cognitive psychology. A similar worry is also pertinent to *modern* psychology because of the common assumption that conceptions of the relevant subject matter have shifted several times since the middle of the nineteenth century, ranging from conscious (sensory) experience in the late nineteenth and early twentieth centuries (spearheaded by figures like Wundt, James, and Titchener) to behaviour, roughly from the 1920s to the late 1950s (as notoriously associated with figures like Watson and Skinner) and cognition, roughly from the early 1960s to today (associated with the works of George Miller and Jerome Bruner, Chomsky, and the beginnings of AI). Again, the question is whether these shifts should be regarded as progress (on the assumption that the kind of mechanistic theorising associated with cognitivism allows for better explanations), or whether the ways in which they are entangled with methodological and conceptual commitments makes it very difficult to even judge them by the same standards.

3. Developments in Psychology: Methodological and Substantive

The issue just raised amounts to the question of whether twentieth-century developments should be regarded as paradigm shifts in a Kuhnian sense such that the progress question cannot even meaningfully be asked. One observation in support of this proposition is that the changes mentioned earlier can be construed as involving both substantive/ontological and methodological/epistemological elements, which are closely intertwined. For example, behaviourists presented themselves as making both an ontological claim (that the true subject matter of psychology is behaviour) and a methodological claim (that the only admissible data in a scientific psychology are behavioural data). Relatedly, behaviourists rejected conscious experience as an area of study in part because it required (introspective) methods, which they regarded as unscientific. Likewise, the cognitivist appeal to cognitive states and mechanisms violated the behaviourist rejection of internal explanatory entities. With this in mind, it might seem that the twentieth century has seen a significant degree of discontinuity regarding both subject matter and method. However, I will argue in this section that this characterization overstates the discontinuities in twentieth-century psychology. I will (1) begin by questioning the notion that the *methodological shifts* were as radical as they are sometimes portrayed. I will then (2) think more closely about what we mean, precisely, when talking about the *psychological subject matter*.

First, both the shift from 'introspectionism' to behaviourism and the shift from behaviourism to cognitivism constituted neither radical nor universal methodological ruptures. For example, even though it is true that the 1920s saw a decline in research that focused on conscious experience, it is by now fairly well established that this was not (just) because of growing distrust of introspective methods as such but rather had to do with a declining interest in the study of consciousness as well as the fact that one particular research programme (the structuralism of Edward Titchener) was no longer viewed as plausible (e.g. Beenfeldt 2013; Hatfield 2005). Moreover, when we look at Titchener's methodological writings at the time (e.g. Titchener 1902, 1905), we find that many of the issues he was concerned with pertained to the problem of how to control for potentially distorting factors. In other words, as an experimentalist, he was concerned with very similar issues as experimentalists today.

Likewise, even though it is true that the 1960s saw a shift of interest towards cognitive processes, the basic methodology of manipulating and observing behaviour remained similar to the behaviourist era (e.g. Leahey 1992). Moreover, the appeal to 'internal' explanatory entities had been anticipated by proto-cognitivist behaviourists, such as Tolman and Hull, as early as the 1930s (Smith 1986), thereby suggesting that the reign of a radical behaviourism was less pervasive and dominant than the standard story might have it. This is further backed up by the fact that some versions of introspection continued to be practiced throughout the so-called behaviourist era (e.g. in questionnaire research) (Gundlach & Sturm 2013). Lastly, even though the early twenty-first century has seen a push towards cognitive neuroscience (in contrast with the cognitive-behavioural approach of traditional psychology), I argue that here, too, we see no major ruptures in how the subject matter is construed.

To develop this point more clearly, I propose that we distinguish between the following two aspects of the psychological subject matter, namely (a), the question of what are the *objects* of psychological research, and (b), the question of what are the *units* at which psychologists direct their attention when studying those objects. By 'object of psychological research' I mean the kinds of attributes that psychologists subsume under a specific term.

For example, memory, attention, perception, and personality can all be objects of psychological research. Now, with regards to the units that are typically assumed to display these attributes, I claim that throughout the history of modern and pre-modern psychology, the focus has mostly been on the human (and animal) *individual*.[2] If we conceptualise the *unit of psychological analysis* in this way, it is obvious that the different foci chosen by different paradigms of twentieth-century psychology (consciousness, behaviour, cognition) are ultimately geared at different (i.e. experiential, behavioural, and cognitive) aspects of *the same unit of analysis*. This is supported by the fact that it is typically the individual that is put in a laboratory situation and exposed to experimental stimuli or questionnaires, with the goal of drawing inferences about their conscious experience, behavioural principles, or cognitive mechanisms. With this understanding of the unit of psychological analysis, we can account for a certain degree of continuity not only across different approaches within the twentieth century but also between modern and 'pre-modern' psychology.

With regards to *objects of psychological research*, that is the capacities under investigation that are exhibited by the units of psychological analysis, I would like to highlight two things, namely, that (a) they are often individuated by means of folk-psychological vocabulary, and (b) at the time of their scientific investigation they are typically not yet fully grasped (which is why researchers direct their attention at them), yet are intuitively understood as involving experiential, behavioural, and cognitive phenomena. The notion of an object of research as 'not fully grasped' picks up on Rheinberger's notion of an *epistemic thing* (Rheinberger 1997). The notion that such objects involve not only cognitive, but also behavioural and experiential phenomena, makes room for investigative efforts that integrate these phenomena in order to get at a better understanding of what the objects are (see also Feest 2011, 2017). If we look at memory, for example, relevant questions are not only what are the mechanisms that explain 'it' but also how items in memory are consciously represented, and what kind of behaviours should be regarded as 'memory' behaviours in the first place. Taken together, these points suggest (a) that the continuity of psychology – in addition to being allowed by the historical stability of the unit of analysis – is also tied to the continuity of our folk-psychological vocabulary, while (b) our uncertainty about the objects in the extension of our folk-psychological concepts, and indeed the fact that these concepts can shift over time, raises the question by what standards such conceptual shifts might constitute progress.

4. Folk Psychology and the Aims/Objects of Psychological Research

The fact that objects of psychological research are individuated by folk-psychological concepts has not gone unnoticed in the history and philosophy of psychology. In this vein, Kurt Danziger observes that in the history of psychology '[c]ertain key terms from everyday language became part of disciplinary language without any very profound change of meaning' (Danziger 2002, 2). One important issue that has been debated in the philosophy of mind in the last 40 years is whether this is an impediment to progress in psychology. For example, Paul Churchland famously argued that '[t]he story [of folk psychology] is one of retreat, infertility, and decadence' (Churchland 1981, 74). As he puts it, '[t]he FP of the Greeks is essentially the FP we use today, and we are negligibly better at explaining human behaviour in its terms than was Sophocles' (ibid.). To Churchland, this suggests that there is something wrong with the explanatory categories of folk psychology, and hence it is questionable that they should be used for our scientific endeavours. In turn, this verdict gave rise to various 'eliminativist' proposals in the philosophy of science, which have pointed out that our folk psychological categories are often too coarse to figure as genuine explanatory constructs (cf. Ramsey 2021).

While this assessment throws severe doubt on the possibility of making scientific progress by means of folk psychological concepts, it rests on a specific assumption about what psychological research is all about, namely, 'to explain human behaviour', and that mentalistic terms refer to psychology's theoretical entities. However, a competing view has long been that folk-psychological concepts do not describe explanatory entities (such as beliefs or desires), but rather refer to the very explananda of psychology, namely, *behavioural capacities* (for the classic references, see Cummins 1983, 2000). In the current contribution, I am adopting and extending this perspective on the psychological subject matter. Specifically, I am adopting the view that targets of psychological research are *capacities*, and I am extending it in stressing (in accordance with my notion of objects of psychological research) that this involves not just *behavioral* but also *experiential* and *cognitive* capacities. In accordance with my account of units of psychological analysis, we can also specify that objects of psychological research are sets of capacities *of individuals*, as they are individuated by folk psychology and other practical (e.g. applied) contexts.

If we endorse the notion that folk-psychological terms refer not to theoretical entities that explain behaviour but rather to the very capacities that we hope to delineate, describe, and explain by functional analysis, eliminativist worries about whether folk-psychological concepts have what it takes to be theoretical concepts seem misplaced. Folk concepts, on this conception, individuate *research objects*, not *explanantia*. With this, I do not mean to deny that one of the aims of psychological research will be that of providing explanations. However, the use of folk-psychological concepts to single out objects of research does not automatically commit us to the notion that scientific explanations of those objects need to invoke folk-psychological concepts. In other words, we can individuate an object of psychological research (e.g. memory) at the level of interest to folk psychological contexts. But the explanatory posits of memory research need not contain the word 'memory' but can refer to more basic functions assumed to be active when the capacity of interest is actualised.[3]

While the notion of explanation as functional analysis was still committed to the idea that the main aim of research is to provide explanations, my claim that objects of psychological research are (by definition) not yet fully grasped suggests that the very question of what we want to explain is itself under investigation and that (accordingly) at least part of the aim of psychological research is to *describe and delineate* the very capacities under investigation. For example, it is one thing to individuate a specific capacity (e.g. the ability to store and retrieve information) by using the folk-psychological term 'memory', but it's quite another to have an adequate empirical and conceptual grasp of this capacity. I argue that much psychological research is geared towards this latter type of question. In this vein, researchers might inquire about the capacity and duration of memory, they might ask what factors memory functions are moderated by, and what are phenomenological features of the ways in which items are represented in memory. Asking and investigating these questions amounts to asking what memory (as individuated by our folk vocabulary) really is, that is what we mean (or ought to mean) by words like 'memory'.

If we take seriously our folk categories as individuating the objects of scientific research, our attention is drawn to two more things. First, the choice of a folk-psychological vocabulary as individuating objects of psychological research highlights that this research is expected to be responsive to questions and issues that arise in folk-discourse. This suggests that the progressiveness of conceptual development should be evaluated (at least in part) relative to how well such developments answer to folk-psychological concerns. Second, if we acknowledge that folk-psychological categories typically describe capacities that are exhibited by sensing, behaving, and cognising human (or animal) individuals, there is no reason to privilege descriptions of explanatory mechanisms over other (e.g. experiential) features of the objects (Feest 2017).[4]

5. Progress as Conceptual Development?

In the previous sections, I have argued that the aims of psychological research include the delineation, description, and explanation of features of psychological objects of research, such as memory, intelligence, perception, personality, and attitude, and that such objects are typically individuated as sets of experiential, behavioural, or cognitive capacities. One way of paraphrasing the aim of psychology, then, is to say that psychology aims at gradually formulating more and more accurate and refined *concepts* of the objects of psychological research. Intuitively, on this account, progress occurs when the conceptual development with regards to a specific object meets certain criteria. This raises the question of what such criteria might be.

Even though the extant philosophical literature about scientific progress has not really focused on conceptual progress (preferring, instead, to talk about explanatory theories), we can nonetheless turn to this literature to identify the following possibilities: Conceptual progress can be said to have occurred when a concept develops in such a way that (1) it allows for an increase of *true statements* about a given research objects (this is also sometimes referred to as the semantic approach and is commonly attributed to Niiniluoto 1980), (2) it allows for an *increase in knowledge* about a given research object (also known as the epistemic approach, Bird 2007), and (3) it allows for an *increase in problem-solving abilities* (This is also referred to as the functionalist approach, which is associated with Larry Laudan 1977, 1981). The main difference between the semantic and the epistemic approaches is that the latter, but not the former, require not just an accumulation of true beliefs but of *justified* true beliefs. By contrast, the functionalist takes to heart the Kuhnian insight that truth is relative to a given paradigm and thus prefers not to tie progress to truth. In turn, both the semantic and the epistemic approaches assume that success indicates truthlikeness (e.g. Niiniluoto 1980), and that to leave truth out of an analysis of progress amounts to a change in subject (Bird 2007).

Leaving aside the issue of truth (and reference) for now, I want to begin by considering the question of what problem-solving abilities amount to. To a first approximation, we can distinguish between (1) problem-solving within the context of basic research activities and (2) problem-solving as pointing to realms of application. Douglas (2014) has rightly pointed out that many analyses of scientific progress have too narrowly focused on basic research. Here, my focus is on the fact that any kind of problem-solving, including the epistemic problem of ensuring conceptual progress, occurs within scientific traditions and contexts, using specific methods, instruments, background assumptions, and truth conditions (Massimi 2018). If follows that we cannot decontextualise conceptual progress but rather have to take into account the factors that make such progress possible. In other words, we cannot leave instrumental and methodological developments out of the picture when considering conceptual progress. This is crucial because, as Shan (2020) points out, such instrumental and methodological features are important not only with regards to problem-*solving*, but also problem-*defining*. As I have argued in Feest (2010), in psychology, such problem-defining features of research approaches include the very conceptualisation and operationalisation of the research objects in question. My analysis, according to which operational definitions of research objects can be regarded as tools in the very research that can potentially push the concept ahead, aligns with other accounts of conceptual development that have emphasised that scientific concepts often contain resources for their own development (e.g. Andersen et al. 2006; Bloch-Mullins 2020).

One obvious problem with the functionalist account is its 'internalism', that is that it seems to define progress relative to a given context or paradigm, making it difficult to account for progress across paradigm-shifts. One strategy with which one might respond to this problem is to formulate two necessary conditions for conceptual progress, that is that the conceptual developments in question (1) can be construed as rational and (2) require a minimal continuity of reference. The possibility of meeting the former requirement in general has long been discussed affirmatively in the philosophical literature on conceptual change (e.g. Nersessian 2008; Andersen et al. 2006; Arabatzis & Kindi 2013, and more recently Haueis 2021). Speaking specifically to psychological concepts, I have discussed the latter requirement in Section 3 above, where I have distinguished between (1) the basic unit of psychology (the sensing, behaving, and cognising individual) and (2) the objects of psychological research (the experiential, behavioural, and cognitive capacities of individuals, as delineated by our folk-psychological language). As I argued earlier, these two facts can provide a certain degree of stability and continuity to scientific research in psychology. For example, consider again the case of memory: if we follow our rough folk-psychological notion that memory has to do with the capacity of individuals to learn, store, and retrieve information, this is compatible with very different research programmes of what this amounts to, while also allowing for a common reference as to what these different programmes are about (Danziger 2002, 2003).

6. Conceptual Progress in Psychology, and the Question of 'Psychological Kinds'

Summing up what was said thus far: I argue that an important aim of psychological research is to delineate and describe objects of research, and that our account of progress in psychology should therefore capture conceptual progress with regards to such objects. As just reiterated, a necessary condition of conceptual progress is a minimal stability and continuity of the objects in the extension of the concepts. But this hardly seems sufficient since, intuitively, conceptual progress also requires an *improvement* of the relevant delineations and descriptions.

This latter idea is best captured by the epistemic account of progress, which would suggest that conceptual progress reflects an increased degree of knowledge about the precise contours and descriptive features of the object under investigation. One problem raised by this, however, is that the notion of true belief about a given object of research seems to lock us into a realist position about this object. For example, in order to make conceptual progress about memory, it would seem that we need it to be the case that there is (a) a relatively stable concept that individuates memory as an object of psychological research but also (b) a natural fact of the matter as to what memory 'really' is. In the following two sections

I will argue that this is too strong a requirement for psychology. This will prompt me to propose a weaker version of an epistemic account of progress, according to which progress in psychology should be understood as an accumulation of epistemic resources that can not only aid in an understanding of objects of psychological research but can also accommodate changes in those objects.

One way of conceptualising realism about the objects of psychology is to say that they are mind-independent kinds. To make conceptual progress about them, then, might mean to increasingly 'zoom in' on them with the aim of having our concepts align more closely with those kinds. In recent years, there have been critical voices with regards to such an understanding not only of scientific kinds in general (e.g. Reydon 2016; Brigandt 2020), but also of the kinds of the social sciences in particular (e.g. Mäki 2011). Here I will focus my attention on the kinds of psychology, a topic that is sometimes discussed under the heading of 'cognitive ontology', though I prefer to talk of the ontology of *psychology*, since (as I argued earlier) psychology studies not only *cognitive* capacities, but also behavioural and experiential ones. As Janssen et al. (2017) point out, two interrelated issues here are (a) what is the *domain* of psychology (i.e. the sets of entities in the scope of the relevant concepts) and (b) what are the *categories* that might settle the correct taxonomy for this domain. Applying this to my previous analysis, I argue that the objects in the domain of psychology are experiential, behavioural, and cognitive capacities of individuals. The crucial question, then, is whether our folk-psychological concepts can be said to provide adequate categories in this domain.

While it is tempting to assume that the identity conditions of psychological kinds, such as memory, are settled by neural structures (e.g. Craver 2004), other scholars have pointed to the role of social practices as determining (at least some) 'human kinds'. In this vein, Mallon (2018) distinguishes between 'social role kinds' and 'bio-behavioral kinds'. Well-known examples of the former are race and gender. Prima facie it seems reasonable to assume that kinds like memory fall more on the bio-behavioural end of the spectrum. This is reinforced by the fact that we have neuroscientific knowledge that identifies specific brain regions, such as the hippocampus, as instrumental in the realisation of memory functions. However, it is important to recognise the distinction between (1) the existence of neuroscientific mechanisms and structures as underwriting or realising specific memory functions and (2) the much stronger claim that neuroscientific localisations provide the identity conditions of memory as a mental faculty or natural kind, that is as the entity that our scientific accounts of memory are *about*. This stronger claim runs into several difficulties. First, already nineteenth-century writers like Herbart and Wundt were wary of the notion that there are biological categories directly corresponding to our historically contingent folk concepts. Their concern was that this assumption would reinforce outmoded faculty-theoretical assumptions (Danziger 2002) (see also Anderson 2014 for a more recent articulation of this kind of worry). A second problem is that if we hope to settle the identity conditions of psychological kinds by appeal to neuroscientific kinds, this requires the possibility of settling the identity conditions of neuroscientific kinds in non-conventional terms. However, as Craver (2009) points out, there is no one uniquely correct way of individuating (say) the hippocampus. I argue that these challenges can be met if we distinguish the role of folk psychological concepts as individuating objects of research from that of singling out natural objects. The former role does not necessarily imply the latter. And an object does not have to be natural to be real, or to be worthy of scientific investigation.

7. The Historicity of Psychological Objects

The considerations in the previous section imply that progress in psychology should not be understood as the accumulation of true beliefs about mind-independent objects, whose identity conditions are settled by non-psychological facts. I would like to emphasise that this does not mean that non-psychological (e.g. neuroscientific or genetic) facts play no role in the tasks of delineating, describing, and explaining phenomena related to psychological objects of research. It only means that those facts might not unambiguously settle the one correct way of individuating the objects in the domain of psychology.

There are (at least) two interpretations of this, namely (1) that there are no facts of the matter as to what folk-psychological concepts 'really' refer to or (2) there are facts of the matter, but these are socio-historically contingent and not uniquely determined by mind-independent neuroscientific facts. It is this latter possibility that I pursue here, if (unfortunately) only very programmatically. My basic idea is that folk-psychological concepts play an important role in our conceptual and causal practices, which in turn confer reality to the corresponding objects (see Feest forthcoming). However, given that our folk-psychological practices and interests may change over time (and/or exhibit cross-cultural differences), the position developed here seems to imply a certain degree of conventionalism with regards to psychological taxonomies.

This adds another element to be taken into account when considering progress with regards to objects of psychological research, namely, that we are not only looking at *conceptual development* (asking under what conditions such development can be regarded as progress) but also at the possibility that *the objects themselves* can change over time. The literature about scientific objects having 'biographies' (e.g. Daston 2000) does not always distinguish the notion that concepts have biographies and that the objects described by concepts have biographies. However, Danziger (2003) clearly articulates the idea that in psychology not just our scientific concepts, but also the very subject matter can undergo developments.

In his words, objects like memory (and other objects of psychological research) are 'discursive objects' because their identity conditions are at least partially constituted by the contexts in which we become interested in them. Danziger is careful to emphasise that it does not follow that they cannot be studied objectively: 'Certainly the targets of the scientist's activity are objects, but that does not mean that they are necessarily natural objects that have no history' (Danziger 2003, 12).

While such pronouncements may sound quite radical at first sight, it should be noted that similar ideas have gained traction through Ian Hacking's work on the looping effect (e.g. Hacking 1995), according to which the very kind under investigation can change as a result of the investigation when it feeds back into the ways people discursively conceptualise themselves or their (experiential, behavioural, and cognitive) properties. We might also add that recent interest in ecological and extended approaches to cognition, by widening the scope of psychological inquiry to include contextual and environmental features, opens the door to the idea that objects of psychological research change because contexts and environments change. Notice that this broader environmental focus is compatible with the notion that the objects of psychological research are the (experiential, behavioural, and cognitive) capacities of individuals.

Changes in contexts and environments affect not just the objects we turn our scientific attention to but inform also the very questions and aims of our scientific interest in them. This connects with Brigandt's thesis that scientific concepts incorporate *aims* (e.g. Brigandt 2009). This is particularly evident when we consider objects of applied scientific interest, or the applied contexts in which we become interested in specific objects. The entire field of clinical psychological research, for example, is motivated by the aim of explaining, intervening, and controlling certain disorders. Likewise, the entire field of testing has practical aims of providing predictions or diagnoses. While I have not specifically addressed applied psychological research in this chapter, I would like to stress that applied questions typically arise in discursive and practical contexts that invoke folk-psychological vocabularies.

8.   The Accumulation and Utilisation of Epistemic Resources as Progress

If we follow the argument of the previous section, it is clear that conceptual progress in psychology cannot simply be a gradual accumulation of true beliefs about (ahistorical) natural objects. This makes both the semantic and the epistemic conception of scientific progress problematic for an understanding of conceptual progress in psychology. However, there are important insights worth preserving about the epistemic view, in that conceptual progress should track the objects of psychology in their historical trajectories and in their cultural and social variability, thereby (potentially) providing us with knowledge. Notice that on this account conceptual progress is epistemic progress and that such progress can be had even if it turns out that there are no stable objects about which we can accumulate justified true beliefs.

Does my account of conceptual progress imply that 'accumulation' has no place in thinking about progress in psychology? I argue that it does not. And this is where *epistemic resources* come in. Epistemic resources are the totality of existing theories, instruments, experimental paradigms, methodological reflections, measurement instruments, etc., that have been developed throughout the history of modern psychology and allow us to describe, reflect upon, measure, intervene on, etc., objects of psychological research in the service of conceptual development. In this vein, my thesis is that *progress in psychology importantly involves, and builds on, the accumulation of such resources.* Importantly, even though such resources underwrite conceptual and propositional progress, they are not themselves conceptual and propositional as they include things like experimental techniques, statistical methods, and experimental effects. By emphasising that the accumulation of such resources should be regarded as progress, I can capture the intuition that scientific knowledge should not be reduced to propositional knowledge (see also Shan 2020).

Now, there is an obvious objection to this suggestion, namely, that it is both too permissive and begging the question: if we treat every aspect of past psychological research as progressive, merely by virtue of (potentially) contributing to conceptual development, this does not provide us with criteria as to (1) which resources are actually relevant to psychological objects and (2) how to use these resources correctly. Moreover, if we simply assume that every addition to the toolbox of existing epistemic resources is progressive, this seems to be question-begging, at least if we want to measure progress in terms of actual successful applications of those resources. This objection addresses two concerns. The first is that according to my account 'anything goes', that is there are no criteria for high-quality resources. The second is that my analysis is counterintuitive by virtue of including unused resources as 'progressive'.

My response to the first concern is twofold: I do indeed argue that it constitutes progress to have a range of resources in the toolbox of psychological research, even if they are not all currently in use. However, this does not amount to an 'anything goes', given my previous analysis of the psychological subject matter (cognitive, behavioural, and experiential capacities of individuals), which puts some constraints on the resources that can reasonably be regarded as relevant. Moreover, something can also be a resource for conceptual development if scientists engage with it critically. In other words, something can also be a resource by virtue of being regarded as interestingly wrong. Regarding the second objection, I recognise that my claim needs to be qualified. Obviously, there is a difference between the mere existence of epistemic resources and their successful application in the service of conceptual development. That being said, it is also the case that resources can only be applied if they exist in the first place. My analysis attempts to capture this intuition while also making the normative suggestion that in order to make conceptual progress, psychologists would be well-advised to engage with the wealth of resources provided by the history of psychology (see Araujo 2019, and Smith, 2016 for similar points).

As already emphasised earlier, epistemic resources can include a range of things, from theories and methodologies to specific (e.g. statistical, experimental, or phenomenological) methods and particular experimental effects (such as, e.g. the chunking effect of short-term memory). I argue that such diverse entities can be epistemic resources even when the research contexts in which they were first developed have changed or the objects they first thought to be relevant to have been discarded. To briefly mention two examples of what I have in mind: 19th-century psychology, with its focus on consciousness as a mark of the mental, saw sustained debates about the nature and epistemic function of introspection. These debates, which were largely forgotten in the second half of the twentieth century, can still provide us with sophisticated sources of insight and reflection in the course of a renewed interest in the conscious mind, even if the general framing of the psychological subject matter has shifted (see Feest 2014). To consider a completely different example: even if psychology were to abandon the concept of short-term memory, the chunking effect (as an instantiation of a behavioural capacity) would remain in the repository of resources.

9.   Summary and Conclusion

As the theoretical psychologist Lisa Osbeck rightly remarks, psychology as '[a] field of practice is a system that must progress or face stagnation' (Osbeck 2019, 190). Surely, stagnation is the opposite of progress. But (1) what does it take for psychology to progress? (2) When does such 'progression' constitute *progress*? (3) And do we have any reason to believe that there has, in fact, been progress in modern psychology since the late 1800s? While Osbeck (2019) addresses the first of these questions, I have mostly focused on the second and third question here. I began by positing that one intuitive notion of progress in psychology holds that there is progress when there is an increase in knowledge about the *objects* of psychology. In turn, these objects are often individuated in terms of folk psychological concepts and scientific investigations of those objects often answer to folk psychological and practical concerns. After arguing that there are no 'natural' facts about what the objects of psychology really are and that folk-psychological and practical concerns (and with them psychology's objects) can shift, I argued that conceptual progress should be measured by how well conceptual changes can track such shifts.

In a second step, I asked what needs to be the case in order for such shifts to occur. In response to this question, I argued for a broader, and more permissive, account of progress in psychology, which takes into account not only actual conceptual progress but also the accumulation of epistemic resources that are available to researchers in their quest for conceptual progress. Such resources are not only limited to scientific theories or concepts but also include methods, methodological reflections, tacit knowledge, and experimental effects. I argued that the increasing availability of these resources should itself be viewed as progress. My account, thus, tries to do justice to two facts: first, not all knowledge is propositional. Therefore, our account of epistemic progress needs to be able to accommodate progress in non-propositional knowledge. Second, the history of psychology provides us with a rich (though often under-appreciated) trove of resources. I argue that having those resources constitutes progress over not having them. Given the open-endedness of psychological research, we don't know which aspects of past psychological research might help psychology to progress. Whether such progression constitutes progress can often only be determined in hindsight.

Dellsén's recent (2018) analysis of progress as an increase in understanding serves as a useful counterpoint to bring my basic commitments into sharper relief: first, while I am sympathetic to the suggestion that progress has something to do with understanding, I argue that the very availability of epistemic resources that can increase understanding should be regarded as progress (as opposed to the idea that we need an actual increase of understanding). Second, while I agree with Dellsén that understanding is factive, the epistemic resources that enable an increase of understanding (i.e. the aforementioned tools and paradigms) need not be (and typically are not) factive. Third, while Dellsén assumes that the relevant factive knowledge pertains to explanation and prediction, I argue that, at least in psychology, descriptive and taxonomic knowledge is relevant, too.

In conclusion, we can now return to Lykken's verdict that twentieth-century psychology has not made much progress with regards to specific research objects (Lykken singles out intelligence). It will be instructive to look at his own evaluation of the situation. After laying out the numerous problems that stand in the way of productive research in psychology, he remarks that he has never regretted his choice of career, stating that 'I am a rough carpenter rather than a finisher or cabinetmaker and there is need yet for rough carpentry in Psychology's edifice. This is a field in which there remain many simple yet important ideas waiting to be discovered and that prospect is alluring' (Lykken 1991, 37). The imagery of psychology as 'rough carpentry' resonates with my analysis of psychology being in the business of figuring out the shapes and descriptive features of their research. It is also compatible with the idea that many of the required tools are already available, though not always made adequate use of.

Notes

1. These two critiques are logically independent of each other.
2. With this, I do not mean to deny that the existence of psychological approaches that have focused on group-level phenomena. However, historically, I take these to be the exception rather than the rule.
3. Notice that this account of is not committed to the notion that capacities are fixed, localised, or innate. (This is in response to a reviewer who worried that my account of objects of research as capacities might come with faculty-psychological baggage.)
4. In turn, this means that we should be open to integrative and pluralist approaches, an issue that has attracted some attention (e.g. Kirschner 2006).

References

Andersen, Hanne, Peter Barker, & Xiang Chen (2006), *The Cognitive Structure of Scientific Revolutions*. Cambridge: Cambridge University Press.

Anderson, Michael L. (2014), *After Phrenology: Neural Reuse and the Interactive Brain*. Cambridge, MA: MIT Press.

Arabatzis, Theodore & Kindi, Vasso (2013), "The Problem of Conceptual Change in the Philosophy and History of Science." In Stella Vosniadou (ed.), *International Handbook of Research on Conceptual Change*. New York: Routledge, 247–372.

Araujo, Saulo de Freitas (2019), "A Role for the History of Psychology in Theoretical and Philosophical Psychology." In T. Teo (red.), *Re-Envisioning Theoretical Psychology. Diverging Ideas and Practices*. Cham: Palgrave Macmillan, 111–130.

Ash, Mitchell (1980), "Experimental Psychology in Germany Before 1914: Aspects of an Academic Identity Problem." *Psychological Research* 42, 75–86.

Beenfeldt, Cristian (2013), *The Philosophical Background and Scientific Legacy of E. B. Titchener's Psychology: Understanding Introspectionism*. New York: Springer.

Bird, Alexander (2007), "What is Scientific Progress?" *Nous* 41(1), 64–89.

Bloch-Mullins, Corinne (2020), "Scientific Concepts as Forward-Looking: How Taxonomic Structure Facilitates Conceptual Development." *Journal of the Philosophy of History*, 1–27.

Boring, Edwin (1950), *A History of Experimental Psychology*. New York: Appleton-Century-Crofts.

Brigandt, Ingo (2009), "The Epistemic Goal of a Concept: Accounting for the Rationality of Semantic Change and Variation." *Synthese* 117(1), 19–40.

Brigandt, Ingo (2020), "How to Philosophically Tackle Kinds without Talking About 'Natural Kinds'." *Canadian Journal of Philosophy*, 1–24.

Churchland, Paul (1981), "Eliminative Materialism and Propositional Attitudes." *The Journal of Philosophy* 78(2), 67–90.

Craver, Carl (2004), "Dissociable Realization and Kind Splitting." *Philosophy of Science* 71(5), 960–971.

Craver, Carl (2009), "Mechanisms and Natural Kinds." *Philosophical Psychology* 22(5), 575–594.

Cummins, Robert (1983), *The Nature of Psychological Explanation*. Cambridge, MA: MIT Press.

Cummins, Robert (2000), "'How Does it Work?' Versus. 'What are the Laws?': Two Conceptions of Psychological Explanation." In F. Keil & R. Wilson (eds.), *Explanation and Cognition*. Cambridge, MA: MIT Press, 117–144.

Danziger, Kurt (1985), "The Methodological Imperative in Psychology." *Philosophy of Social Science* 15, 1–13.

Danziger, Kurt (2002), "How Old is Psychology, particularly Concepts of Memory?" *History and Philosophy of Psychology* 4(1), 1–12.

Danziger, Kurt (2003), "Where Theory, History and Philosophy Meet: The Biography of Psychological Objects." D. B. Hill & M. J. Kral (eds.), *About Psychology: Essays at the Crossroads of History, Theory and Philosophy*. Albany, NY: SUNY Press, 19–33.

Daston, Lorraine (ed.) (2000), *Biographies of Scientific Objects*. Chicago, IL: University of Chicago Press.

Dellsén, Finnur (2018), "Scientific Progress: Four Accounts." *Philosophy Compass* 13(11), e12525.

Douglas, Heather (2014), "Pure Science and the Problem of Progress." *Studies in History and Philosophy of Science* 46, 55–63.

Eronen, Markus & Bringmann, Laura (2021), "The Theory Crisis in Psychology: How to Move Forward." *Perspectives on Psychological Science* 16(4), 779–788.

Fancher, Raymond & Rutherford, Alexandra (2013), *Pioneers of Psychology*. 4th edition. New York: W.W. Norton & Co.

Fechner, Gustav Theodor (1860), *Elemente der Psychophysik*. Leipzig: Breitkopf & Härtel.

Feest, Uljana (2010), "Concepts as Tools in the Experimental Generation of Knowledge in Cognitive Neuropsychology." *Spontaneous Generations: A Journal for the History and Philosophy of Science* 4(1), 173–190.

Feest, Uljana (2011), "Remembering (Short-Term) Memory. Oscillations of an Epistemic Thing." *Erkenntnis* 75, 391–411.

Feest, Uljana (2014), "The Continuing Relevance of 19th-Century Philosophy of Psychology: Brentano and the Autonomy of Psychology." In M. C. Galavotti & F. Stadler (eds.), *New Directions in the Philosophy of Science*. Cham: Springer, 693–709.

Feest, Uljana (2017), "Phenomena and Objects of Research in the Cognitive and Behavioral Sciences." *Philosophy of Science* 84, 1165–1176.

Feest, Uljana (forthcoming), "(Folk-) Psychological Concepts and the Investigation of Psychological Kinds." In C. Bloch-Mullins & T. Arabatzis (eds.), *Concepts, Induction, and the Growth of Scientific Knowledge*.

Gundlach, Horst (2004), "Reine Psychologie, Angewandte Psychologie und die Institutionalisierung der Psychologie." *Zeitschrift für Psychologie* 212(4), 183–199.

Gundlach, Horst & Sturm, Thomas (2013), "Ursprünge und Anfänge der Kognitionswissenschaft." In A. Stephan & S. Walter (eds.), *Handbuch Kognitionswissenschaft*. Stuttgart: J.B. Metzler, 7–21.

Hacking, Ian (1995), "The Looping Effect of Human Kinds." In Sperber et al. (eds.), *Causal Cognition. A Multidisciplinary Debate*. Oxford: Clarendon Press, 351–383.

Hatfield, Gary (1995), "Remaking the Science of Mind. Psychology as Natural Science." In C. Fox, R. Porter, & R. Wokler (eds.), *Inventing Human Science. 18th Century Domains*. Berkeley, CA: University of California Press.

Hatfield, Gary (2005), "Introspective Evidence in Psychology." In P. Achinstein (ed.), *Scientific Evidence. Philosophical Theories & Applications*. Baltimore and London: Johns Hopkins University Press, 259–286.

Haueis, Philipp (2021), "A Generalized Patchwork Approach to Scientific Concepts." *British Journal for the Philosophy of Science*. https://doi.org/10.1086/716179.

Janssen, Anneli, Klein, Collin, & Slors, Marc (2017), "What is Cognitive Ontology Anyway?" *Philosophical Explorations* 20(2), 123–128.

Kirschner, S. R. (2006), "Pluralism and Psychology: Towards the Psychological Studies." *Journal of Theoretical & Philosophical Psychology* 26, 1–17.

Laudan, L. (1977), *Progress and its Problems: Toward a Theory of Scientific Growth*. Berkeley: University of California Press.

Laudan, L. (1981), "A Problem-solving Approach to Scientific progress." In I. Hacking (ed.), *Scientific Revolutions*. Oxford: Oxford University Press, 144–155.

Leahey, T. H. (1992), "The Mythical Revolutions of American Psychology." *American Psychologist* 47(2), 308–318.

Lykken, David (1991), "What's Wrong With Psychology Anyway?" In Dante Cicchetti & William Grove (eds.), *Thinking Clearly about Psychology. Volume 1: Matters of Public Interest. Essays in Honor of Paul E. Meehl*. Minneapolis: University of Minnesota Press, 3–39.

Mäki, Uskali (2011), "Scientific Realism as a Challenge to Economics (and Vice Versa)." *Journal of Economic Methodology* 18(1), 1–12.

Mallon, Ron (2018), "Constructing Race: Racialization, Causal Effects, or Both?" *Philosophical Studies* 175, 1039–1056.

Mandler, George (2007), *A History of Modern Experimental Psychology: From James and Wundt to Cognitive Science*. Cambridge, MA: MIT Press.

Massimi, Michela (2018), "Four Kinds of Perspectival Truth." *Philosophy and Phenomenological Research* XCVI(2), 342–359.

Michell, Joel (1999), *Measurement in Psychology. A Critical History of a Methodological Concept*. Cambridge: Cambridge University Press.

Nersessian, Nancy (2008), *Creating Scientific Concepts*. Cambridge, MA: MIT Press.

Niiniluoto, Ilka (1980), "Scientific Progress." *Synthese* 45, 427–462.

Open Science Collaboration (2015). "Estimating the Reproducibility of Psychological Science." *Science* 349(6251), DOI: 10.1126/science.aac4716.

Osbeck, Lisa (2019), "Vertical and Horizontal Development in Theoretical Psychology." In T. Teo (ed.), *Re-Envisioning Theoretical Psychology. Diverging Ideas and Practices*. Cham: Palgrave Macmillan, 189–207.

Ramsey, William (2021), "Eliminative Materialism." In Edward N. Zalta (ed.), *The Stanford Encyclopedia of Philosophy* (Fall 2021 ed.). <https://plato.stanford.edu/archives/fall2021/entries/materialism-eliminative/>.

Reydon, Thomas (2016), "From a Zooming-in Model to a Co-creation Model: Towards a More Dynamic Account of Classification and Kinds." In C. Kendig (ed.), *Natural Kinds and Classification in Scientific Practice*. London and New York: Routledge, 79–93.

Rheinberger, Hans-Jörg (1997), *Toward a History of Epistemic Things. Synthesizing Proteins in the Test Tube*. Stanford: Stanford University Press.

Richard, T., Walsh, G., & Theo, T. (eds.) (2014), *A Critical History and Philosophy of Psychology: Diversity of Context, Thought, and Practice*. Cambridge: Cambridge University Press.

Robinson, Daniel (1995), *An Intellectual History of Psychology*. Madison: The University of Wisconsin Press.

Schimmack, Ulrich (2019), "The Validation Crisis in Psychology." HYPERLINK <https://replicationindex.com/2019/02/16/the-validation-crisis-in-psychology/>.

Shan, Yafeng (2020), *Doing Integrated History and Philosophy of Science: A Case Study of the Origins of Genetics*. Cham: Springer.

Smith, Laurence (1986), *Behaviorism and Logical Positivism: A Reassessment of the Alliance*. Stanford, CA: Stanford University Press.

Smith, Roger (2007), *Being Human: Historical Knowledge and the Creation of Human Nature*. New York: Columbia University Press.

Smith, Roger (2016), "History of Psychology: What for?" In S. H. Klempe & R. Smith (eds.), *Centrality of History for Theory Construction in Psychology*. Cham, Switzerland: Springer, 3–28 (Annals of Theoretical Psychology 14).

Sturm, Thomas (2006), "Is There a Problem with Mathematical Psychology in the Eighteenth Century? A Fresh Look at Kant's Old Argument." *Journal of the History of the Behavioral Sciences* 42(4), 353–377.

Sturm, Thomas & Mülberger, Annette (2012), "Crisis Discussions in Psychology – New Historical and Philosophical Perspectives." *Studies in History and Philosophy of Biological and Biomedical Sciences* 43, 425–433.

Titchener, Edward B. (1902), *Experimental Psychology: A Manual of Laboratory Practice, Vol. I: Qualitative Experiments, Part 1: Student's Manual* (2nd ed.). New York and London: Macmillan and Co.

Titchener, Edward B. (1905), *Experimental Psychology: A Manual of Laboratory Practice, Vol. II: Quantitative Experiments, Part 1: Student's Manual*. New York and London: Macmillan and Co.

van Rooij, Iris & Baggio, Giosuè (2021), "Theory Before the Test: How to Build High-Verisimilitude Explanatory Theories in Psychological Science." *Perspectives on Psychological Science* 16(4), 682–697.

Wertz, F. J. (2018), "Beyond Scientism: Toward a Genuine Science of Psychology." In R. N. Williams & E. E. Gantt (eds.), *On Hijacking Science: Exploring the Nature and Consequences of Overreach in Psychology*. New York: Routledge/Taylor & Francis Publishing, 107–119.

Yarkoni, Tal (2020), "The Generalizability Crisis." *Behavioral and Brain Sciences*, 1–37. https://doi.org/10.1017/S0140525X20001685.

11 Progress in Sociology?

Stephen Turner

The question of whether sociology cumulates, or even progresses, in any meaningful way, and in what sense, has been an on-going concern within sociology itself (Bell 1982; Collins 1981; Deutsch et al. 1971; Rule 1997). It is a question complicated by both intrinsic features of sociology as a subject matter itself and technical issues. Strictly applying to sociology the traditional notions of progress from the philosophy of science is largely fruitless, however, because sociology has no theories in the sense that accounts of progress have traditionally focused on; its 'mechanisms' do not have the same character,[1] and prediction, for reasons to be discussed, is not easily distinguished from description. Some of the earlier discussion does focus on the question of the establishment of true beliefs or at least the disproof of false ones, but as will become apparent, claims about this are problematic as well: simplicity, a traditional cognitive value in science, applies to predictive theories, but in dealing with complex social situations is often simply reductive. The criteria of puzzle solving appear more promising, but this turns out to be problematic as well: the puzzles are externally generated by policy issues or culturally generated, rather than simple products of the failures of theories or of the internal puzzle-generating processes of science.

There are, nevertheless, good reasons to consider sociology in relation to the question of progress, because, as will become apparent, much of the rest of science, especially in biomedicine, is coming to resemble sociology in crucial respects as a result of the expanded role of statistical models. Model building in sociology is the strategy that most resembles 'science', and in which 'progress' might be expected, so its main form, causal modelling, will be the major focus of the chapter.

Statisticians have promoted the idea that scientific theories are themselves merely statistical models that are true or false, and that statistical tests can give effective certainty. This is a half-truth. It is true in the sense that general scientific theories can be tested by turning them into statistical models which then make predictions under given assumptions. But it is false if it is taken to mean that theories read off of successful models will necessarily generalise, either in the form of more general theories

DOI: 10.4324/9781003165859-14

or in the form of mechanisms that have general applicability. One way in which sociology and the social sciences differ is with respect to the generalisability of its models. The problem of generalisation, or what I will call semantic instability, is the core of the issues I will discuss here, but to see its significance requires a substantial detour into the history of methodological thinking in sociology.

I will divide this chapter into three parts. The first will be a general introduction to the two divergent but complementary methodological traditions in sociology and their limitations. After some discussion of the problem of 'theory' and its relation to the problem of generalisation, I will turn to the main focus of conventional quantitative sociology, the problem traditionally known by the title of a classic methodology text, *Causal Inference in Non-Experimental Research* (Blalock 1961), that is, the construction of statistical models with some sort of causal significance. There are of course many applications of quantitative methods in the social sciences, but the difficulties with them have close family resemblances. In 1961, Blalock listed the 'shortcomings of social science research' as 'substantial measurement errors, the inability to randomise and replicate, a high percentage of unexplained variation, and numerous unmeasured variables and unclear concepts' (1961, viii). A few examples will show why this is still the case. The 'qualitative' or interpretive tradition in sociology associated with case studies, the second tradition, addresses the problem of what we may roughly call the cultural specificity of social processes. This issue has complex implications for the concept of progress, which will be discussed in terms of two examples of research areas in sociology which may be said to have progressed. The final section will discuss the applicability of some philosophical accounts of scientific progress to this material and to sociology generally.

1. Two Approaches and Their Relations

Sociology as a discipline has sources in two divergent traditions. The 'quantitative' tradition derived from Comte, Quetelet, and Mill, but evolved into a statistical tradition defined by the methods of Pearson and Edgeworth centred on correlation, which were applied creatively to sociology by Giddings and his students. The 'qualitative' tradition is associated with the concept of *Verstehen*, historicism, Dilthey, phenomenology, and somewhat misleadingly with Weber, who attempted to reconcile the two traditions. The differences between the two follow the explanation-understanding distinction: for the qualitative tradition explanation is ultimately understanding; for the quantitative tradition, understanding is a source of testable hypotheses and necessary background knowledge for model construction and measurement. A central characteristic of both approaches is that they depend on categories derived from the understandings of the agents whose activities are being explained, rather than purely theoretical concepts. The subject, the social world, is by definition in a sense already 'known' to its participants. Sociology is necessarily, in some sense, a second-order form of inquiry. This has profound implications for accounts of progress: the agents' categories are relative to historical and social context, unlike the categories of natural science.

Sociologists have laboured under the model of physics since Comte and Quetelet, but its 'theories' have never approximated the physics model, despite both concentrated effort and various reconstructions of the physics model that allowed for comparisons. The way in which current sociology retains the original aspiration to discover laws and derive them within a deductive structure is the aspiration to identify causal correlations that are reasonably robust over different contexts when subject to standard statistical controls. The similarity between this project and general theories comes down to this: just as a law in physics retains its form no matter how its instances are divided up, the kinds of correlations regarded as genuinely causal do not change their form when they are partialled, that is when the data points are divided into groups based on causal variables that have the potential to affect the statistical distribution.[2]

This standard is negative, in that it is an assertion that a supposedly causal correlation is 'not affected by' any other potential causes, and thus not provable. There is of course no way to exclude the possible existence of a causal variable which has such an effect. Consequently, causality is always a matter of assumption (see Pearl 2000, 136–137, 139, 174–182). But it is nevertheless usable for the purpose of causal inference, and may indeed resemble actual non-scientific ordinary causal inference, if one can plausibly assume that the model contains the relevant variables for the processes it represents. This appears less problematic when the causal situation being represented is, so to speak, local, and one can rely on local knowledge about potential causal relations. The issue of the status of local knowledge will be discussed shortly: it is the primary concern of the qualitative tradition.

We can use the example of a model of the causal determinants of teen pregnancy in the United States (Rodgers and Rowe 1993) to illustrate the basic situation of model construction, what it aspires to do, and what its limitations are. The same considerations apply throughout the quantitative tradition. The casual relations are modelled by assembling the variables that are expected to influence both the outcome and one another, and by eliminating those relations which are not possible – effects that go backwards in time, for example – and subjecting the reduced set of relations to statistical analysis, determining how much variation in the outcome can be predicted on the basis of the arrangement of variables that produces the best predictions. One can then apply the same model to new settings, such as other populations, to see whether it is 'robust' or merely local. The creators of this model describe their results in this way:

The parameters that were estimated by our model-estimation exercises were consistently interpretable and plausible. Furthermore, the tests of adaptations demonstrated a number of features of how the world would work under the simplified assumptions of this model and about how it does work to the extent that these assumptions are reasonable.

(Rodgers and Rowe 1993, 503)

The 'goal' is 'reproducing the basic patterns in the prevalence data', that is causally relevant patterns. But it is admitted that 'other approaches could probably do as well at reproducing the data'. There are acknowledged trade-offs, in this case between a model 'which has less theoretical richness' and 'reproduced the data slightly better' and alternative statistics which 'also be fit to these data' (Rodgers and Rowe 1993, 503).

The advantage claimed for their model depends on 'the interpretability of the parameters and the substantive knowledge that comes out of testing the model adaptations' (Rodgers and Rowe 1993, 503). 'Substantive knowledge' here means that given the plausible assumptions, we know something more than we did prior to fitting the model about the empirical case at hand. The authors concede that

[W]e would be irresponsible to suggest that we have found the correct model of how sexual behavior develops in adolescents, on several grounds. First, the particular simplification of reality that we describe here, based on the social interaction and influence processes among adolescents, is only one of dozens or hundreds that are reasonable and appropriate.

(Rodgers and Rowe 1993, 503)

They note that there are potential confounders and common causes not included in the model:

parental influence, . . . the way individuals respond to sex education in school, explicit hormonal variables, or . . . many other processes that we know to be important in the development of adolescent sexuality. It simplifies the reality we are trying to explain by projecting that reality into one particular domain.

(Rodgers and Rowe 1993, 503)

They of course also assume without needing to state it that these results are context bound: the practices, attitudes, and so forth that are the background to the causal mechanisms. The 'basic patterns' they are attempting to isolate vary culturally and historically in ways that are outside the model.

This is the basic logic of causal modelling. There is a standard of improvement, or progress: the reduction of error in prediction. There is, however, a gap between prediction and explanation. A good predictor in the case of teenage pregnancy is smoking, which is not a cause. So interpretation, or making causal assumptions, is necessary to go from the raw statistical facts to causal conclusions. What grounds these assumptions? The usual strategy is *not* to ground them, but to make innocuous assumptions which are unlikely to be challenged, such as that smoking does not cause pregnancy, and to let the data eliminate others. But this normally leaves, uneliminated, many 'possible' causes, which get included in the model by virtue of not being eliminated. Many of these 'causes' are problematic: they may suffer from causal arrow ambiguity, for example, and typically have only an anecdotal or folk basis for regarding them as causes. They may also be suggested by larger scale social theories, such as theories about class and stratification, which provide categories and variables. But they cannot be eliminated as impossible.

The same issues arise for measurement. Regarding a particular scale based on answers to a questionnaire, even if it has been used widely as a 'measure', depends on some intuitive sense that the scale represents the concept that appears in the theory, or in the relevant conceptualisation. This in turn depends on contextual, and largely 'local', knowledge about what the items on the questionnaire might mean to the respondent. A famous and much ridiculed and copied example is F. Stuart Chapin's living room scale (Chapin 1928), designed to allow an interviewer in a subject's residence to place the person in a social category by assigning index numbers to items of furniture and other objects present in the room. The history of measurement in sociology, briefly, is this. Prior to Giddings' *Inductive Sociology* (1901), social researchers either relied on official statistics or collected information based on conventional categories. Giddings pioneered the idea of inventing quantitative surrogates for concepts like 'labor unrest'. His students, who produced the basic framework of quantitative sociology, invented more – this opened up social life to quantification based on concepts in social theory as well as policy concerns. But in each case, the measure depended, as the living room scale did, on local, contextual knowledge of the 'meaning' of the things in the scale. Chapin carefully validated his scale by showing that the items correlated with relevant measures of status. But the items were themselves unstable indicators, and their meaning and value as predictors would change as fashions changed.

The implication of instability – the lack of anything like the principle of the uniformity of nature – is that while these correlations may provide grounds for causal claims, they are subject to the limitation that one does not know the range or conditions of their application, and that the relations both of causality and the correlation itself may cease to apply. Put differently, these are not general claims: they are models that apply when they apply. The same goes for measures. Rather than having a set of fixed, highly generalisable measures that derive from one another, measures in sociology are *ad hoc* and related to theoretical concepts by local knowledge.

2. Local Knowledge and Understanding: The Qualitative Tradition

Quantitative sociology depends on local knowledge. But what is local knowledge, and how does it relate to sociology as a knowledge producing system, and in what sense does *it* progress? Much could be said about this, but the key feature is that it is 'local', in the sense that it differs from other settings, which makes the difference a problem or puzzle to account for, and that it is knowledge, in the extended sense of the term that includes the practices that underlie the beliefs, attitudes, experiences, and intentions that arise within a given setting.

Mary Morgan and Albert Hirschman have recently described this situation of comparison between local settings in a discussion of case studies, the 'method' for this kind of sociology, usually labelled 'ethnographic' or qualitative, and associated with the 'method' of participant observation:

> Easy comparison is not there for the taking in case study work, because what is found in the field of study are not simply facts, observations or phenomena, let alone ready-made descriptions of phenomena awaiting the social scientist's re-description, but something much less obvious. Clifford Geertz, as anthropologist, described the problem thus: the field presents the social researcher with a knot of puzzles:
>
>> a multiplicity of complex conceptual structures, many of them superimposed upon or knotted into one another, which are at once strange, irregular, and inexplicit, and which he must contrive somehow first to grasp and then to render (Geertz 1973, 10).
>>
>> (Morgan and Hirschman 2017, 5)

As they add, 'case study work in social sciences is a mode of enquiry that most fruitfully focuses on the exploration of puzzles within a single case, not on re-describing materials to fit a general hypothesis or theory' (Morgan and Hirschman 2017, 5). For our purposes, several things must be taken from this: the puzzle is implicitly or explicitly comparative; the solution is to be found in putting together elements of the context, such as related beliefs or practices, which cannot be identified as relevant in advance (indeed, if they could there would be no puzzle), and the solution is to make the initially puzzling practices, and therefore the beliefs, experiences, and the rest of the 'local knowledge' intelligible.

Thus for case studies, or ethnographies, 'improvements' take the form of providing context that allows for greater intelligibility. Typically this involves making tacit features of the local knowledge being studied, such as cultural norms, explicit. This might come from a novel comparison, for example, between one culture of poverty and another, which points to additional similarities and differences that are relevant to, and serve to modify the original interpretation. In these cases the new comparisons create a new puzzle, with a new solution. In the strict sense this amounts to changing the subject, but changing it to one that swallows up the original puzzle and interpretation, and in this sense it is progressive. There is, in short, a process of criticism whose form can be made explicit and exemplified. There are various accounts of this process (Geertz 1973; Roth 1989; Turner and Carr 1978; Turner 1980). The process is not always conclusive, however, and is prone to ideologisation.

The motives for this kind of inquiry are often political or ideological: the thing that is puzzling and needs to be understood is something that offends a local sense of social justice or rightness. In the case of cultures of poverty, for example 'cultural' accounts have been accused of 'blaming the victim', and there is an insistence that the 'context' of 'capitalism' be blamed instead (Wacquant 2002). For this kind of critic, the explanations are incomplete. But it is because what is not a puzzling practice for one audience – normal capitalist behaviour – is puzzling or somehow problematic for audiences with a particular political viewpoint. 'Understanding', in short, requires an understander with a particular cognitive agenda and particular epistemic needs to make something intelligible.

The qualitative and quantitative approaches are, in some sense, complementary: the 'understanding' approach supplies the knowledge necessary to make judgements about plausibility; the quantitative approach determines the strength of causal relations and how they are connected to one another to produce causal structure. There is a famous model of explanation called Coleman's boat, which works like this. If we are to account for a causal relation between two 'macro' or societal attributes, such as Protestant religious doctrine and early capitalism, we need more than a correlation. We need a mechanism to make the causal relation plausible. We cannot rely on local knowledge here: we need to get this by understanding the early capitalists and early Protestants, who are intellectually alien to us. We do this by reconstructing the mental world that connects religious motivations to capitalist actions, at the individual level. This is the 'mechanism' that makes the causal claim plausible. In other areas, such as economics, the mechanisms might simply be rational choice reconstructions.

On the surface, such 'mechanisms' seem more or less similar in function to the chemical process mechanisms of biology. Biological mechanisms are supposed to be robust across multiple instances, such as species or classes of organisms. They are also supposed to more or less fully represent some causal transformation, normally one that can be reproduced under controlled experimental conditions. They might, in epidemiological contexts, for example, be regarded as contributory to a statistical outcome, just as a mechanism in the social sciences would. There are grey areas in which a statistical connection is poorly understood but assumed plausibly to be causal in both social science and medicine – what physicians call 'empirical' relations, meaning that the mechanism is not understood properly, but that there probably is one, and one can, as a practical matter, intervene in it. The difference between the fields is that our understanding of such relationships can be improved in medicine (i.e., such things as the causal chemical relationships fully identified), while in the social sciences they remain at the level of plausible local knowledge.

The possibility of verifying mechanisms by other means, rather than merely attributing them to correlations based on local knowledge, is a significant difference. Although one could say that experimental economics and social psychology experiments perform the same function in warranting mechanisms, judgements of plausibility are bound to considerations of applicability in local contexts to a greater extent. Nevertheless, there are strong similarities to the biological sciences with respect to variability.

It might appear that making conduct intelligible, at least to the point of providing plausibility to attributions of causality to correlations, is sufficient as a substitute for experimental evidence of the regularities needed for 'mechanisms' so that the only deficiency is epistemic: the inability to do experiments is just an accidental feature of the situation. Intelligibility is not the same as causality, but in many cases is only a poor but irrelevant guide to it. Reasons are causes only in, as Donald Davidson puts it, a low probability sense, and there are many causes that go into social activity and human action, which are not 'reasons' (Davidson 1963). The case study or ethnographic approach attempts to be close to the material of social life, as it is experienced and articulated by the agents. This excludes those potentially large parts of the causal basis of action that are outside the limits of these modes of expression. Self-understanding may be largely self-deception, and the deeper sources of our desires and mental processes are occluded to us, and also to observers. So while one might aspire to a complete representation of a complex biological process and its conditions, 'understanding' can never be more than a partial representation of the causally relevant processes. What one finds in social science is abstraction. The most intelligible mechanisms, such as those of decision theory, are abstractions that are mathematically convenient rather than representative of actual reasoning processes, but are nevertheless predictive for the highly aggregated data being studied.

This is a case of purpose relativity, which raises important problems. In the case of social science generally, it is a commonplace that social science should be 'emancipatory'. This implies the possibility of some sort of intervention that would resolve a normatively defined policy problem. The causal model sought in this case is purpose- or comparison-relative, in the sense that there is a problem it is trying to address which can be solved by varying some condition or set of conditions. To talk about progress in model construction is to talk about it in relation to these purposes. Greater predictivity is progress. But what we want to predict is defined by the aims of inquiry themselves.

In the case of policy-oriented causal models, the aim is to identify main effects and enough causal structure for the purpose of intervention, or, sometimes, to rule out proposed interventions. But all these 'causal structure' terms are purpose or comparison relative, which I will call, for convenience, problem relative. A 'main effect' is relative to a comparative, counterfactual question and a comparison with whatever situation has no such main effect. The terms used in discussing these models, such as endogeny/exogeny (or caused within the system or an external causal condition for the system) and identification (meaning specifying causal relations), and terms like over and under identification (meaning insufficient causal links to identify the relevant structure or the inclusion of too many intervening or confounding variables, which obscure the relevant structure), all relate to this quest for the effects relevant to the problem. These are not intrinsic categories – one correlation is as good as another in the eyes of God.

These considerations are confusing, but they can be roughly summarised in this way. There are four separate problems here: causality, predictivity, intelligibility, and intervention. In the traditional philosophical literature, there have been arguments defining each of these in terms of the others, and in the model case of physical laws they do seem to be the same thing. In the case of statistical models of human activity, however, they pull in different and often conflicting directions. Smoking is a good predictor of pregnancy, but not a cause. However, it might also stand in for a large set of sub-rational dispositions that are outside of intelligibility, but do have a causal role in relation to action. Intelligible actions are low probability causes, and the parts of the causal process revealed by *Verstehen* are a small subset of the total causal situation. Moreover, as in economics, abstractions from the complexity of action and decision may predict well despite being poor representations of the causal processes. Intervention requires isolating a system to intervene in – an endogamy-exogamy distinction – from the vast ramifying world of correlations and causes, and the intellectual act of isolating a system depends on controversial judgements of plausibility, as well as on the intervention-related purposes of constructing a model of the system. These features of model building all bear on the problem of progress, which we can better understand in terms of examples.

3.   Progress in Two Topics in Sociology

It is useful to bring this discussion down a few levels of abstraction to some actual empirical research to illustrate some of the issues with the concept of progress that these features of sociological research produce. I will consider two examples of research areas in which there was a trajectory, with an eye to the kind of progress each represents. One is a 'theory', the other a group of causal models. Both relate to the core issue of sociology, social hierarchy: the first to power differentials, the second to differences in status. The first example will involve two related problems in elite theory: the problem of a ruling class and the problem of community power. The second, the status attainment literature, generally regarded as the great success story of quantitative methods in sociology.

The basic idea of elite theory is that the rule of the few, governed by an 'Iron Law of Oligarchy', as Robert Michels put it, holds regardless of the ideology, including the egalitarian ideology, of the society. His own study of political parties showed that the German Social Democratic party was itself oligarchic ([1911]1915). The classic statements of this theory by Gaetano Mosca ([1896]1939) and Vilfredo Pareto (1916, 1966) focused on the ruling class and applied to political societies. Mosca's claims were universal: all organised societies were divided into a ruling class and a ruled class, and this is a result of 'constant facts and tendencies that are to be found in all political organisms' (Burnham 1943, 98). These were, principally, the psychological drive to pre-eminence, and with respect to the composition of the ruling class, the need to adapt to prevailing (and changing) social forces. Pareto added considerations about the circulation of elites, that is the problem of absorbing the potential leaders and contributors from the non-elite and pushing the weak out of the elite, a consideration he took to be a key condition for elite stability. These too were universal considerations and processes. But to explain them it was necessary to see how the same processes operated in different social and political orders, and that different elite qualities were necessary in different orders, that is to show the similarities despite the apparent differences.

Did this paradigm progress? It developed in two directions. On one side was a continued interest in what C. Wright Mills, in the 1950s, called 'the power elite' and described as an interrelated group of leaders in the military, business, politics, and so on (Mills 1956). Its most recent manifestations are two books, one which presents itself as a general theory of elites (Milner 2015) and the other is an account of elite stability and instability (Shipman et al. 2018). Partly independent of this, a vast literature developed on community power structures aimed at identifying elites and elite structure. This literature dead-ended in the 1970s with an unresolved conflict between multiple models of elite structure and community politics, after a long debate over methods – the two methods being 'reputational', meaning that informants were asked which people in the community were powerful, and 'decision-making', meaning what might now be called process-tracing, which worked by following specific decision-making processes through to their outcome. There were many interesting findings in this literature, such as the long-term trend for elected community leaders to not come from the category of 'economic dominant' (Clelland and Form 1964, 511). So, on the empirical level, something was achieved. Nothing like theoretical consensus emerged, however. What emerged instead was the recognition that there were many different types of community power structures, which is to say that the models of community power did not generalise well. That was progress, as was the proliferation of models. It was not progress to a new general theoretical account. But each new model allowed for the inclusion of new kinds of facts.

At the national level in the United States, a great deal of data was collected showing the personal relations and connections among members of the elite (Domhoff 1967). This work is highly specific to the current elite and the American case. At the other extreme are two other recent works (Milner 2015; Shipman et al. 2018), one of which argues theoretically, based on historical examples, that elites rule by cohesion and by allying with either the middle classes or the poor, and that alliances with the middle class are more stable, but indicates that the main cause of elite instability is internal division within elites over which group to ally with.

The work of Shipman et al. may be seen as a case of progress within the general 'problem' of elite stability and instability, or change, which has a long prior history. The novelty is in adding the consideration of class alliances and relating changes in class alliances to divisions within the elite, and noting that alliances with the middle class based on such things as secure property rights have proven more stable. This can be taken as a novel theory with a novel prediction, within the general 'problem' of elite stability, which is intrinsically about comparisons: between stable and unstable elite regimes.

Milner attempts a highly general 'model' of elites – a synthesis of the elite literature which identified different components of the elite and differentiated types of elite power, which is then applied to three historically disparate cases (2015). The key to this model is a list of attributes of elites. He distinguishes three types of elite, distinguished by the primary source of their elite character: political power, economic means, or status (2015, 27). This is also predictive, in the sense that the attributes are intercorrelated. This 'general' model, however, also provides conceptual clarification, and in a sense prediction about the necessary elements of elite power or the normal components of elites. But in this case the model veers into the category of 'definitional': having the properties is also what the analyst defines as 'elite'. And when we consider the other account, there is an element of this as well.

That 'theories' tend to become 'definitional' is a normal result of what in sociology are called issues of 'scope' or applicability. For some societies these concepts will seem less clearly applicable, and to affirm the generality of the theory requires a suspension of these concerns and a loss of the particularity of the historical or comparative situation. Put differently, these represent standardisations of comparisons or comparative problems that elide much of the substance that an interpretive approach and different comparisons would reveal. Weber ([1904] 2012) famously discussed such models under the heading of 'ideal types', and took the view that their goal was to facilitate comparison, that they provided conceptual clarity through their one-sidedness, and that they applied where they applied and helped reveal explanatory lacunae when they didn't, but should not be regarded as general theories. Rather, they should be taken as tools for interpretation to be applied when they were useful for this purpose. Weber rejected the identification of ideal types with 'reality' and considered theories such as Marx's to have made that error, and were better understood as ideal types. In the terminology I have used here, they are purpose- or comparison-relative and need to be evaluated in terms of their success in solving the puzzles that arise in the comparisons.

With this we can identify an important issue relating to 'progress': when is a redefinition of a problem progressive, and when is it a matter of changing the subject? In this case and the following example, we face a similar issue. The aims of the 'theories' or models differ, as do their definitions or terms, what they take to be causally internal or endogamous and what they leave out as exogamous, and what they take to be their scope or the range of comparisons they regard as empirically relevant or open to intervention. Nevertheless, they fall within more or less the same problem domain. In the case of elite theory, we can identify 'progress' in the sense of replacing one problem definition with another with respect to a specific common issue – elite instability. But this does not make that problem definition superior or progressive in a universal sense; it does not help us with the question raised by Milner of the necessity of some kind of transcendental grounding. But Milner also is considering a different set of 'elites' as his empirical base than community power studies.

We see an analogous issue with the tradition of 'status attainment' research, which is a case of causal modelling. The topic grew out of the combination of status research and a prior issue of 'rural depletion', which was concerned with the fact that as populations urbanised through migration to cities, the people left behind were less educated, intelligent, and able than those who left for the city. This was part of a large international phenomenon, and manifested itself in different policy responses in different countries, reflecting somewhat different concerns. In the United States, the policy aim was largely egalitarian – to make country life as attractive as city life, leading to policy-oriented research producing a country life commission and an academic field of rural sociology oriented to rural amelioration and needs. The empirical research in the United States relating to 'depletion' focused on such things as intelligence, and this developed into a general programme designed to explain 'status attainment'. The introduction of Path analysis, a structural equation modelling method, in the 1960s, enabled a massive and influential series of projects that assigned causal weights to a wide range of variables affecting ultimate socio-economic status. This body of work 'was viewed as exemplary' (Grusky and Weeden 2006; cited in Sakamoto and Wang 2020, 1), and exemplary of progress: 'a leading example of successful cumulative knowledge development' (Hout and DiPrete 2006; cited in Sakamoto and Wang 2020, 1)

The basic strategy in this literature was to look at inter-generational occupational mobility. This fit a basic set of intuitions: status was mostly a matter of the job you had, something long entrenched in the literature on status (Goode and Hatt 1949); moving to a higher status job in the next generation was the paradigm of upward mobility. 'Progress' here took the form of adding variables and improving predictive models by the kind of statistical adjustment discussed earlier. Causality is assumed in the models: the predictive importance of the variables is what is assessed, and improvement consists in changing the weights of different causal variables on the basis of the predictive power of the model in different configurations of causal order assumptions. But the modelling adds nothing to our knowledge about the causal mechanisms themselves: they are part of the 'assumptions' of the model.

We can call this internal progress. The models get 'better' in various ways: they include more potentially undermining variables and increase the variance explained – partly by making them more sophisticated, meaning tweaking assumptions. Strictly speaking, models are not comparable: they each make their own assumptions, which conflict. But if we think of this literature more broadly, and acknowledge that the assumptions are purpose relative or comparison relative, we can make the same kind of intuitive or common-sense judgements, perhaps even based on some qualitative knowledge of how people think about occupations and their desirability, about which models are an improvement.

The problem of progress gets posed in a different way when the models are sufficiently different that, even in relation to the same purposes, 'progress' begins to look more like 'changing the subject'. In recent years, this tradition has been supplanted by an economic approach, which avoids occupational categories – the surrogate for status – in favour of lifetime incomes as the outcome variable. The reasons for abandoning occupation are these:

> [O]ccupational categories typically harbor a great deal of underlying job variability, firm heterogeneity, and earnings inequality. Using some sort of occupational average to indicate an individual's earnings is highly imprecise. Occupational mobility tables are unable to discern substantial economic immobility that persists within any given occupational category.
>
> (Sakamoto and Wang 2020, 4)

The use of parental lifetime income has its own intuitive case for changing the subject, because this is what we often think of as wealth, as well as an empirical reason for thinking it is better. 'The parent-offspring income correlation can be high even when the two generations have different occupations because most earnings inequality derives from other variables', such as the ability to invest in high status education. 'The economic approach avoids this shortcoming of contextual imprecision by measuring incomes directly at the individual level for both parents and offspring' (Sakamoto and Wang 2020, 4).

I leave aside the many technical complexities here to make a simple point: the occupational status approach assumed the stability of a set of causal relations, allowing some to be ignored as irrelevant or exogamous, and for occupational status to stand in for a more generic notion of life-time status. Changes in the labour market that destabilise these relationships, such as frequent career changes and greater differentiation in status within occupations, have the effect of making the models both less predictive and a worse fit with intuitions about such things as the variations within occupational categories. The earlier models took status within occupations to be more or less equal. Changes in the work world made status attributable to the organisation doing the employing (Jencks et al. 1988) – a programmer for Microsoft is not the same as the one for the local municipality. The new approach is 'better' not because it is a more comprehensive and complex model, which it is not, but because it is a better fit to the changed conditions, given the thing shared between the approaches – what I am calling 'the problem' – a general concern with the promotion of social mobility.

The changes here are continuous. Models apply when they apply. But the conditions in which they apply are in constant flux, and it is only by assumption that we can treat them as predictive and as something other than a snapshot of a particular dataset. From a purely technical point of view, this is changing the subject – the models are not strictly speaking comparable, as they involve different and conflicting assumptions about cause – both about endogeneity and identification of the causal links. And determining causal strength, the point of the model, is something internal to the model. But these changes can be better understood in a positive way: these models are purpose relative. And there is a larger purpose here, vaguely associated with the possibility of intervention for the purpose of promoting social mobility or removing barriers to it. From this larger point of view, the point of view of the users of the model, whose purposes are larger than the model itself, and consistent with many models, it is an improvement.

4. Problem Exogeneity and Progress

How does this relate to concepts of progress in the philosophy of science literature? Begin with this: the 'problems' or puzzles of social science are largely defined by common-sense concerns. The 'problems' are often initially constituted by normative or policy concerns. The core *explana-tory* problems of sociology arise when either outcomes or practices vary or change for unknown reasons. Answering such questions sometimes involves moving to a higher level of abstraction, for example, by thinking about rural-urban relations historically, in order to provide comparisons, or general claims. Weber did this by turning to the agricultural history of ancient Rome and the evolution of social and property relations in rural estates during its decline. Claims over putatively comparable cases require abstract concepts. Comparisons of cases that differ on multiple dimensions – Prussian and Roman estates, for example – require ignoring most of these dimensions and producing a simplified model. Weber called these ideal types, and understood them to be tools that provided concep-tual clarity, but at the cost of specificity, and were inherently distortive, but also necessary. But he presented them only as 'useful' and therefore tied to the purposes of the investigator. These purposes were, ultimately, a product of our own cultural concerns, and varied historically: the salience of problems and therefore also puzzles changes as our concerns change.

In his *Wissenschaftslehre*, which influenced Karl Popper and other philosophers of science, Weber provided a comprehensive answer to the question of the historicity of social science, and the differences between social science and natural science, as well as the relation to normative or dogmatic sciences such as law and theology. He made the semantic point that while the sciences could replace and discard concepts, the concepts of the social sciences were tied to ordinary language as meta-concepts, or concepts about concepts. Because ordinary language concepts themselves change historically and vary culturally, the explanatory objects of social science themselves change, and the meta-concepts, such as 'estate', are abstractions that may become, or be, less applicable as the actual practices they describe change.

In short, there is a kind of semantic instability in the terms of social sci-ence based on what was called 'the language of life', or what I have called here 'local knowledge', and indeed the 'puzzles' or problems of social science derive to a significant extent from this fact: terms cease to apply in the same way, acquire different associations, and come to be used differ-ently. The apparent solution to this is to appeal to highly general concepts.

We can think of accumulating truth and increasing verisimilitude in terms of adding variables and increasing explained variance (or alter-natively in terms of specifying mechanisms). If we do, the dilemmas become apparent. Increasing explained variance generally has a price in robustness or generalisability: the models work best in confined settings with limited instability. Adding models with new variables amounts to changing the question. And because each variable requires some 'plau-sibility', typically adding a variable amounts to adding a less plausible 'cause', with less plausible grounds for thinking it is unaffected by other causes, or is not a confounder. But it produces new results, and thus new knowledge, subject to the limitations of the model-building process itself, which requires assumptions about identification or causal mechanisms and exogeneity, or what to include. If the new models are relatively stable over time and are more robust or generalisable, which is to say they keep applying in new settings, they are cases of 'progress' in the form of replacing puzzles. If the proliferation of models with some empirical basis is analogous to the proliferation of established experimental hypotheses, this is accumulation of truth.

The fact that the models are non-comparable and purpose relative because they employ incompatible assumptions makes this process different from the kind of accumulation normally understood in traditional models of science. The status attainment literature provides many examples of this. Within the general 'problem' one can add variables, such as the effect of less educated mothers' further education during early childhood on children's educational performance in adolescence. This creates not only a new model, but also a new comparison. It is motivated by policy concerns, which have generally come to focus on early childhood interventions as a means of improving educational outcomes (Esping-Andersen et al. 2012; Heckman 2012; Winther-Lindqvist and Svinth 2020). There is a fine line here between progress and changing the subject, or extending the problem in a novel direction. There is learning involved in these changes: learning that the exogenous conditions for the older model have changed, or that the particular interventions not contemplated in the earlier model (and motivated by different perspectives, in this case human capital theory from economics) might be effective. But this kind of progress is closely bound up with the issue of problem-relativity. The changes in the problem are themselves exogenous: rural depletion is out, and inner-city education is in. Moreover, as noted, the change of subject also requires a change in 'assumptions', and not only statistical assumptions, but in what is taken to be plausible or reasonable.

What is taken to be plausible or reasonable reflects purposes or comparisons. In the case of comparisons of the kind found in interpretive social science, this is apparent in the fact that the meta-language in which an interpretation is given is in the 'language of life' of a different community. There is no generic language of 'science' to translate into. We describe a 'culture of poverty' in terms of our concepts, in order to make sense of it for our audience. Improvement, in the internal sense, is making more sense. But, as with quantitative models, there can be the 'improvement' of changing the subject to such things as the context of oppression. These changes are purpose relative, so one would need to claim that purposes themselves progress rather than merely change. One could make such a case in biomedicine on pragmatic grounds, and perhaps in some contexts in social science. But this still implies that the models are purpose relative.

Does this result translate into the problem of progress in the philosophy of science? Following Bird's account of the alternatives into semantic, epistemic, or functional (Bird 2007; cf. Shan 2019), we get results like this. Distinguishing the semantic and epistemic cases would require us to find an example of an unreliable method that had produced true beliefs in the past being replaced by a sound method. The situation in sociology is that there have always been competing methods, which generate different kinds of results, which are at least partially complementary: the quantitative results require plausible assumptions, while the qualitative 'methods' make the assumptions plausible. So the distinction between semantic and epistemic is not relevant: results do accumulate; past consensual methods do not get replaced by new consensual methods. 'Truth', in a realist sense, or any sense beyond the observational, is trickier: there are continuing and extensive 'paradigmatic' differences between basic theoretical approaches, because different approaches constitute their explanatory problems in different ways. To decide on an elite theory, for example, is to decide first what the subject of the theory is supposed to be.

In the case of causal models, something analogous occurs: we define the problem that the model is intended to solve, and apply these definitions, together with plausible assumptions about system endogeny and about causality. Changing a model to make it more predictive, or improving it in terms of plausibility or value for intervention, changes the question, so the models are not strictly speaking comparable. But there is a sense in which there is internal improvement through the replacement of one model by another in relation to a 'problem' or question that is defined less strictly. So one can say that our models of status attainment improve. When (occupational) 'status' was replaced by lifetime income as a dependent variable one could claim improvement – but this improvement was a result of instability in the assumptions, which no longer applied, and were replaced by ones that were more plausible for a changed occupational structure, rather than a failure of the earlier model to apply to its own historically bound problem.

This leaves us with the alternative of understanding progress as internal to methods or perspectives, in relation to a problem domain. The last clause is the problematic one: in the usual view of progress as puzzle solving, the puzzles are generated by the internal failures of a dominant theory. In sociology, the theories are vocabularies that describe problems, such as the problem of elite rule, differently, and answer different questions. In the case of causal modelling, the 'problems' are defined externally: by such things as a desire for greater social mobility or decreasing adolescent pregnancy. When Laudan comments that 'A problem need not accurately describe of affairs to be a problem: all that is required is that it be an actual states of affairs' (Laudan 1977, 16), he is describing an odd case in the history of science. In sociology this is the normal case: to be a 'problem' is to be a problem within the limits of a form of local knowledge. One can construe these cases as arising naturally out of the failures of prediction and understanding that arise in encountering novel or different social settings (cf. Turner 1980). But here too 'the problem' depends on our starting point: it is a problem from the point of view of some local knowledge that has failed to make sense of the new setting. Progress dead ends with the fact of our on-going dependence on local knowledge. We do make progress relative to the problem defined by our local knowledge, but we never escape our dependence on it, and it varies not through theoretical advance, but because social life is changeable.

A better way of putting this would be to vary Laudan's title, *Progress and Its Problems*, to read *Do Problems Progress?* In the internal science model they can be said to progress in the following way: puzzles arise out of the theories that are the solutions to previous puzzles. In some limited way this kind of internal progress does occur in sociology. We get understandings of action in novel settings that improve on our original understanding, and we replace models with better models, with different assumptions which are better at serving particular purposes. But this does not produce a chain of puzzle-solving. Rather, new puzzles arise for external reasons. And in this respect sociology is like medicine dealing in a practical way with new diseases as they arise. But medicine can often depend on an expanding base of causal knowledge, which is a kind of progress. This is not analogous to dependence on local knowledge and judgements of plausibility based on it.

The implications of this difference can be clarified by comparing these theories of progress to an image made famous by Michael Polanyi, in which progress in science is analogised to the fitting together of a large puzzle in which scientists contribute pieces that fit with other pieces – the established hypotheses. For Polanyi, this was fundamental; it was the form of discipline that adjacent areas of science imposed on one another. Medicine does this in the face of novel diseases. But the adjacent areas in question are generalising forms of knowledge, such as chemical models of biological mechanisms.

Science as statistical modelling tacitly abandons the idea of the parts of science fitting together into a coherent whole for the simple technical reason that the simplifying assumptions of models, which are also the means but which they generate the predictions that are their 'results', are not consistent with one another. This extends the applicability and reach of 'science', and its marketability to governments. But it accepts as inevitable a plurality of licit models between which one must decide on non-empirical grounds, such as the plausibility or practical relevance of assumptions, on which opinions will vary, or in terms of extrinsic purposes. In this respect, the situation of sociology represents what is becoming a normal state of science.

Notes

1. In the classic formulation, 'Mechanisms are entities and activities organised such that they are productive of regular changes from start or set-up to finish or termination conditions' (Machamer et al. 2000, 3). Very few if any of these mechanisms are 'regular', much less a series of regular change-producers organised in a series.
2. These are the cases that produce Simpson's paradox.

References

Bell, Daniel. 1982. *The Social Sciences since the Second World War*. New Brunswick, NJ: Transaction Books.

Bird, Alexander. 2007. "What Is Scientific Progress?" *Noûs* 41(1): 64–89.

Blalock, Hubert. 1961. *Causal Inference in Non-Experimental Research*. Chapel Hill: The University of North Carolina Press.

Burnham, James. 1943. *TheMachiavellians: Defenders of Freedom*. Chicago: Henry Regnery Company.

Chapin, F. S. 1928. "A Quantitative Scale for Rating the Home and Social Environment of Middle Class Families in an Urban Community: A First Approximation to the Measurement of Socio-Economic Status." *Journal of Educational Psychology* 19(2): 99–111.

Clelland, Donald, and William H. Form. 1964. "Economic Dominants and Community Power a Comparative Analysis." *American Journal of Sociology* 69(5): 511–521.

Collins, Randall. 1981. *Sociology since Midcentury: Essays in Theory Cumulation*. New York: Academic Press.

Davidson, Donald. 1963. "Actions, Reasons and Causes." *Journal of Philosophy* 60: 685–700.

Deutsch, Karl W., John Platt, and Dieter Senghaas. 1971. "Conditions Favoring Major Advances in Social Science." *Science* New Series 171 (3970): 450–459.

Domhoff, G. William. 1967. *Who Rules America?* Englewood Cliffs, NJ: Prentice Hall.

Esping-Andersen, G., I. Garfinkel, W. J. Han, K. Magnuson, S. Wagner, and J. Waldfogel. 2012. "Child Care and School Performance in Denmark and the United States." *Children and Youth Services Review* 34(3): 576–589.

Geertz, Clifford. 1973. "Thick Description: Toward an Interpretative Theory of Culture." Pp. 3–36 in *The Interpretation of Cultures*. New York: Basic Books.

Giddings, F. H. 1901. *Inductive Sociology: A Syllabus of Methods, Analyses and Classifications, and Provisionally Formulated Laws*. New York/London: The Macmillan Company.

Goode, William, and Paul Hatt. 1949. *Methods in Social Research*. Oxford/New York: Prentice Hall.

Grusky, D. B., and K. A. Weeden. 2006. "Does the Sociological Approach to Studying Social Mobility Have a Future?" Pp. 89–108 in *Mobility and Inequality: Frontiers of Research from Sociology and Economics*. Edited by S. L. Morgan, D. Grusky, and G. Fields. Stanford: Stanford University Press.

Heckman, James J. 2012. "Promoting Social Mobility." *Boston Review* 1(September). HYPERLINK "http://bostonreview.net/forum/promoting-social-mobility-james-heckman" (accessed 21 July 2021)

Hout, M., and T. A. DiPrete. 2006. "What We Have Learned: RC28's Contributions to Knowledge about Social Stratification." *Research in Social Stratification and Mobility* 24: 1–20.

Jencks, Christopher, Lauri Perman, and Lee Rainwater. 1988. "What is a Good Job? A New Measure of Labor-Market Success." *American Journal of Sociology* 93(6): 1322–1357.

Laudan, Larry. 1977. *Progress and Its Problems: Toward a Theory of Scientific Growth*. Berkeley: University of California Press.

Machamer, Peter, Lindley Darden, and Carl F. Craver. 2000. "Thinking about Mechanisms." *Philosophy of Science* 67(1): 1–25.

Michels, Robert. [1911]1915. *Political Parties: A Sociological Study of the Oligarchical Tendencies of Modern Democracy*, trans. Eden Paul and Cedar Paul. New York: The Free Press.

Mills, C. Wright. 1956. *The Power Elite*. Oxford: Oxford University Press.

Milner, Murray, Jr. 2015. *Elites: A General Model*. Cambridge: Polity Press.

Morgan, Mary S., and Albert O. Hirschman. 2017. "Narrative Ordering and Explanation." *Studies in History and Philosophy of Science* Part A 62: 86–97.

Mosca, Gaetano. [1896]1939. *The Ruling Class*, ed. Arthur Livingston, trans. Hannah D. Kahn. New York: McGraw-Hill.

Pareto, Vilfredo. 1916. *The Mind and Society*. New York: Harcourt, Brace and Company.

Pareto, Vilfredo. 1966. *Sociological Writings*, selected by S. E. Finer, trans. Derick Mirfin. Totowa, NJ: Rowman and Littlefield.

Pearl, Judea. 2000. *Causality: Models, Reasoning, and Inference*. Cambridge: Cambridge University Press.

Rodgers, J. L., and D. C. Rowe. 1993. "Social Contagion and Adolescent Sexual Behavior: A Developmental EMOSA Model." *Psychological Review* 100(3): 479–510.

Roth, Paul. 1989. "Ethnography without Tears." *Current Anthropology* 30(5): 555–569.

Rule, James B. 1997. *Theory and Progress in Social Science*. Cambridge: Cambridge University Press.

Sakamoto, Arthur, and Sharron Xuanren Wang. 2020. "The Declining Significance of Occupation in Research on Intergenerational Mobility." *Research in Social Stratification and Mobility* 70. https://doi.org/10.1016/j.rssm.2020.100521

Shan, Yafeng. 2019. "A New Functional Approach to Scientific Progress." *Philosophy of Science* 86(4): 739–758.

Shipman, Alan, June Edmunds, and Bryan S. Turner. 2018. *The New Power Elite: Inequality, Politics, and Greed*. London/New York: Anthem Press.

Turner, Stephen. 1980. *Sociological Explanation as Translation*. Rose Monograph Series of the American Sociological Association. New York/Cambridge: Cambridge University Press.

Turner, Stephen, and David Carr. 1978. "The Process of Criticism in Interpretive Sociology and History." *Human Studies* 1: 138–152.

Wacquant, Loïc. 2002. "Scrutinizing the Street: Poverty, Morality, and the Pitfalls of Urban Ethnography." *American Journal of Sociology* 107(6): 1468–1532.

Weber, Max. [1904] 2012. "'Objectivity' in Social Science and Social Policy." Pp. 100–138 in *Max Weber: Collected Methodological Writings*, trans. Hans Henrik Bruun. Edited by H. H. Bruun and S. Whimster. London/New York: Routledge.

Winther-Lindqvist, Ditte Alexandra, and Lone Svinth. 2020. "Early Childhood Education and Care (ECEC) in Denmark." *Oxford Bibliographies*. http://doi.org/10.1093/obo/9780199756810-0093

12 Progress in Economics

Marcel Boumans, Catherine Herfeld

1. Introduction

In this chapter, we analyse how economics progresses by discussing not only when progress occurs in economics but also when it does not. As in most disciplines, economics as a field is a patchwork of subfields, many of them with their own methodology, and is not as unified as has sometimes been claimed. Therefore, to discuss progress in economics, we first have to choose the subfield in which progress is to be described. In the following, we focus on those subfields in economics that predominantly use models. Because modelling is the most common methodology in economics, we are able to cover most parts of the field of economics.

Since the middle of twentieth century, economics has become a modelling science (Morgan 2012). Economic models have become 'endemic at every level' (Morgan 2012, p. 2). The reliance on models also introduced a new way of reasoning to economics: 'As models replaced more general principles and laws, so economists came to interpret the behaviour and phenomena they saw in the economic world directly in terms of those models' (ibid., p. 3).

Economists use a variety of models. Economic models can be mathematical, statistical, or diagrammatic, and even include physical objects that can be manipulated in various ways. To discuss whether progress occurs in economics, we also have to choose a subset of these and consequently limit ourselves to a discussion of mathematical models. The development and use of mathematical models is indeed representative of what large parts of economics does as a modelling science.[1]

By analysing progress in economics enabled by mathematical models, we also attempt to understand when progress is altered and how new methodological developments are considered to be potentially progressive. To do so, we draw on the recent literature on model making and application in philosophy of science to look at a case study from economics. We argue that progress in model-based economics occurs when a model is considered sufficiently useful to be reapplied to define and solve pending or new problems. Progress is disrupted when models are

DOI: 10.4324/9781003165859-15

considered useless for those purposes and advances again when a model is developed that proves useful in this respect.

2. Concepts of Scientific Progress and Scientific Modelling

To fruitfully discuss how large parts of economics progress, we choose the functional approach to progress because it best captures the kind of progress accounted for in this chapter (see Part I of this volume for a detailed description of the four distinctive approaches to characterise scientific progress). Economics is chiefly a problem-solving science, in particular where model making and application predominate. We can even observe instrumentalist attitudes in economics which can best be captured by the functionalist approach. This is not to say that all progress in economics is functional, even in those subfields that draw heavily on models. Progress can also be theoretical. Indeed, other accounts, such as epistemic, semantic, and understanding-based approaches to scientific progress, are focused on how theoretical achievements enable progress. However, although economics has long made extensive use of theory, theoretical progress is no longer what economists primarily aim to achieve.

We assume that there is a consensus among economists that economic theory, that is to say general equilibrium theory combined with utility theory, has reached a stage at which no further refinement is needed. However, an uncritical view of economics as exclusively focused on progress produced by models is overly simplistic. General equilibrium theory and utility theory have changed considerably during the twentieth century. For instance, utility theory was augmented by prospect theory, introduced to economics by the psychologists Daniel Kahneman and Amos Tversky in 1979; this shows a continuous commitment to theory in economics and a desire to develop the field further by developing new theory. However, as Robert Lucas (2004, p. 22), one of the most influential economists of the twentieth century, claims, the progress one can find in economics is mostly technical:

> We got that view from Smith and Ricardo, and there have never been any new paradigms or paradigm changes or shifts. Maybe there will be, but in two hundred years it hasn't happened yet. So you've got this kind of basic line of economic theory. And then I see the progressive – I don't want to say that everything is in Smith and Ricardo – the progressive element in economics as entirely technical: better mathematics, better mathematical formulation, better data, better data-processing methods, better statistical methods, better computational methods.

We consider Lucas' view a representative and credible description of the overall disciplinary development of model-based economics from the mid-twentieth century onwards.

Lucas' understanding of progress shifts the focus from the role of theory to that of models so that he sees a theory as 'an explicit set of instructions' for making a model (Lucas 1980, p. 697). Yet, theorising and modelling should be distinguished from each other. In economics, theories are general accounts assumed to provide truths about economic phenomena, whereas models function primarily as instruments of investigation (Morgan and Morrison 1999) and hence are built for specific purposes. A model is therefore validated by assessing how much or in what respect it can reach its purposes. These purposes can be meaningfully reconstructed as problems that models are meant to solve.

We consider Shan's (2019) 'new functional account of progress', to provide a particularly useful framework for thinking systematically about model-based progress in economics.[2] One major advantage of Shan's account over traditional functional approaches is that instead of emphasising only the problem-solving dimension of progress (Kuhn 1970; Lakatos 1970; Laudan 1981), he defines progress in terms of the usefulness of a particular scientific practice for problem defining and problem solving. To achieve progress, a scientific practice is useful when it repeatedly defines and solves problems. Consequently, this practice should provide a 'reliable framework' for solving and defining more problems in other areas or disciplines (Shan 2019, p. 746). The repeatability and reliability of the process of defining and solving research problems is thus a prerequisite for recognising its usefulness.

Shan characterises progress as the successful application of frameworks that guide the process towards defining and solving new problems within and across fields and disciplines. He explicates the process behind problem defining as a sequence of problem proposal, problem refinement, and problem specification. He emphasises what he calls a 'common recipe' for a useful solution to indicate that the solutions provided by this framework do not have universal and stable characteristics, but that a solution consists of the following nonexhaustive list of components: a set of concepts employed in problem and solution; a set of practical guides specifying the procedures and methodologies as means to solve a problem; a set of hypotheses proposed to solve a problem; and a set of patterns of reasoning indicating how to use other components to solve a problem (Shan 2019, p. 745). Usefulness is thus defined as the repeated successful application of a common recipe to define and solve new problems. Hypotheses should not be narrowly construed as statements or propositions; they can also take the form of models. This is how we can think of them when we apply this account to economics.

In this chapter, we start from the idea of thinking about problem definitions and solutions as common recipes to address how we can consider the parts of economics that progress through mathematical models. To specify the components of a problem's definition and solution in economics using Shan's idea of a common recipe, we combine ideas from the modelling literature to capture not only what economic models are but also how they are built, revised, and reapplied. These aspects help us specify the procedures and methodologies for solving problems in economics, the framework proposed by economic models to solve a problem, and patterns of reasoning indicating how to use other components to solve that problem.

We start from Paul Humphreys' account of understanding models as based on what he calls 'templates', which are mathematical or computational forms that are flexible enough to be applicable to various kinds of target systems (Humphreys 2002, 2004, 2019). Generally, a template constitutes 'a pattern that can serve as a common starting point for the development of a product but that can be adapted for the purpose at hand' (Humphreys 2019, p. 116, fn. 20). Humphreys uses his concept of a template to clarify how using such templates within one domain, such as a discipline or subdiscipline, aids 'the acquisition of explicit knowledge and understanding, and to explore similar questions when they are applied across domains' (ibid., p. 114). Thinking of modelling as grounded on templates allows us to see how their application to other target systems within economics might ensure a repeatable process that leads to the definition and solution of new problems in the respective domain.

On Humphreys' account, templates are in the first instance subject independent, albeit to differing degrees. Whereas the models that are ultimately made from such templates represent specific target systems, the underlying template is general in that it lacks a specific interpretation other than that which is given by a general theory accepted in that domain. Therefore, templates are transferrable and applicable to radically different target systems, sometimes studied in distinct domains. While Humphreys has introduced various notions of a template, we draw on his recent definition of a 'formal template' (Humphreys 2019). A formal template is described as a mathematical form without interpretations beyond a mathematical one and of which the construction assumptions only have mathematical content.[3] Formal templates can be used to model phenomena only after undergoing a process of construction and specification. As examples of one-equation templates, Humphreys mentions the Laplace's equation, Poisson's equation, and the diffusion equation. Other examples of formal templates are statistical distributions such as the Gaussian, Poisson, and binomial.

A well-known example of a formal template from economics is the two-variable homogeneous function of the first degree $f(x,y) = Ax^{\alpha}y^{1-\alpha}$, also known as the Euler function. In economics, this function is better known as the Cobb-Douglas function, because it was originally used by Charles Cobb and Paul Douglas to define a production function representing the relationship between inputs for production and outputs generated on their basis.[4] This function became a template because it possesses the attractive advantage of displaying a direct relationship between its 'elasticities' and 'returns to scale'. Elasticities in economics refer to the proportional change of some economic variable in response to a change in another, which can be expressed as $\left(\frac{\partial f}{\partial x}\right)\Big/\left(\frac{f}{x}\right)$. For example, the elasticities of the Cobb-Douglas function variables x and y are represented by α and $1 - \alpha$, respectively. They show how much an increase in the input factors influences the output or return.

The template account of models can capture how formal structures such as the Cobb-Douglas function are transferred within and across domains and used to define and solve several problems in economics beyond the problem they were originally meant to address. As we show in the following, many such mathematical structures are used in economics, which is why the template account applies well to the kinds of economic models we discuss here. However, to consider not only when functional progress in economics occurs but also when it does not, we must first understand why a specific template is chosen at all. Two additional aspects of templates are relevant. The first aspect is the frequently observed intertwining of the conceptual and mathematical dimensions of a template, which rests on the premise that templates are not neutral, contrary to Humphreys' account. The second aspect is the set of prerequisites for initial adoption of a template. This second aspect is particularly relevant to understanding progress when old templates are replaced by new ones. Both aspects are largely neglected in Humphreys' account of templates but are relevant to the integrative aspect of modelling in economics.[5]

Building on Humphreys' account, Tarja Knuuttila and Andrea Loettgers (2016, 2020) capture the first element, the intertwining of conceptual and mathematical dimensions in a template, by proposing the concept of a 'model template' as an alternative to the formal template account. They argue that a model template is a mathematical structure that is '*coupled with a general conceptual idea that is capable of taking various kinds of interpretations in view of empirically observed patterns in materially different systems*' (2016, p. 396; italics in original). To illustrate their concept of a model template, they studied the cases of the Lotka-Volterra system of nonlinear coupled differential equations and the Ising model. Their approach suggests that Humphreys' account should acknowledge the aspect of analogical reasoning to explain how templates are reapplied and even transferred across domains (Knuuttila and Loettgers 2020). They include this aspect because they consider the conceptual dimension coupled with the mathematical structure to specify some characteristics of potential target phenomena. Without taking analogical reasoning into account, it will be difficult to explain what drives the application of a formal template both within and across domains; in their view, scientists can do so because they identify these characteristics as shared across target domains.

More specifically, according to Knuuttila and Loettgers (2020, p. 128), useful templates embody a vision of the phenomenon exhibiting a general pattern for the study of which the template, including the underlying conceptual idea, offers appropriate tools. For instance, in the case of the Lotka-Volterra model, the model template provides a framework that 'renders certain kinds of patterns as instances of cooperative phenomena [i.e., the conceptual idea] coupled with associated mathematical forms and tools that enable the study of such phenomena' (p. 135). Seeing various kinds of systems as instances of a familiar general pattern captured by this conceptual idea enables scientists to apply the template to different systems. It is particularly the conceptual side of the model template that mediates between the mathematical form and its various empirical interpretations. According to Knuuttila and Loettgers (2020, p. 135), this combination that model templates embody, of a conceptual vision of the phenomenon and a mathematical form, is what makes them attractive for reuse.

When we view functional progress in economics from the template view, it appears largely to occur when such templates are reused as common recipes to define and solve problems. However, before we discuss whether we can find functional progress in economics through the reuse of model template, we address the second aspect, the use of a new template, to capture the conditions under which progress is disrupted in economics. Generally, as we show in the following, progress is disrupted when the conceptual vision of the phenomenon has changed in such a way that the mathematical form is no longer considered to be sufficiently representative of the phenomenon in question. We address this second aspect by presenting a few accounts of model-making that complement the template account by focusing on the process of template making and can partly explain why a new template is adopted.

Morgan (2012) distinguishes four accounts of how models are constructed. The first was proposed by Boumans (1999). In this account, the process of model-making can be understood as analogous to recipe-making. The notion of a recipe encapsulates two ideas. The first idea is that economists choose a model's ingredients; these include theoretical notions, metaphors, analogies, policy views, mathematical concepts, mathematical techniques, stylised facts, and empirical data. The second idea is that these ingredients are integrated to create something new: mixing, shaping, combining, and moulding them to create a new model. The final product may well not be what was envisioned at the beginning; recipe-making is a creative process. However, the resulting recipe can then be used for making other models to answer similar questions. For example, one wishes to find the business cycle mechanisms of two different economies. If we skip an ingredient or replace it with a different one, we obtain a new model that answers questions similarly to the old one. Some of the stylised facts may differ between the two business cycles. The cycles may have different periodicities. So, what makes the recipe useful is that it offers a template useful for developing other models.

The second account is Morgan's own (Morgan 2004). It rests on the view that model-making requires, first, imagination to hypothesise how the economy might work, and second, the power and skill to envisage that idea. A third account understands model-making as a process of idealisation: a process of picking out the relations of interest and isolating them from any disturbances which interfere with their workings. The final account is on Mary Hesse's (1966) work, who argues that model-making depends upon our cognitive ability to recognise analogies and our creativity in exploring them. However, Morgan suggests that the point of modelling is not to recognise analogies but to create them. According to Morgan (2012, p. 25), what these four accounts have in common is 'that forming models is not driven by a logical process but rather involves the scientist's intuitive, imaginative, and creative qualities'. Although the second and third accounts emphasise the construction of a common vision as a necessary part of model-making, it is the first and fourth accounts that prove useful in thinking about the usefulness of economic models understood as model templates. This is because they capture the process by which new templates are developed to define and solve new problems.

We propose to complement the template account with these model-making accounts that explain why templates are adopted. They also provide a systematic way to capture how template-building can be a progressive step. However, they capture only one aspect of functional progress. Of course, being able to build a satisfactory template, and thus model, can be considered some form of progress. Whether a model can be considered satisfactory depends on whether it meets certain aims or criteria, such as satisfying theoretical requirements, representing certain facts, and being useful for policy (Morgan 1988). However, what is relevant for our argument is whether these models are a constitutive element of a satisfactory framework and thus provide a representative practice for defining and solving new or unsolved problems. The first and the fourth of the accounts discussed earlier implicitly indicate how this may happen. Boumans' recipe account suggests that the recipes can be followed to make other models for similar aims. The analogy account suggests that these models can be used as formal analogies to create other models for similar purposes. But both accounts only clarify the application of model templates once some similarity has been ascertained and acknowledged between certain aspects of the problems. Therefore, these accounts cannot explain why some models have been fruitfully applied to defining and solving problems of very dissimilar kinds. Neither can these accounts explain how models are applied to cases for which no similarity has yet been ascertained.

To understand the progressive aspect of the reuse of models, we need to complement the model template account with an account of model making that captures the idea that templates are not only complemented by a conceptual vision but can also co-define or shape this conceptual vision. Take the example of the Cobb-Douglas function presented earlier. It has found applications in various domains not because of similarities between such aspects as the phenomena in different domains, but because of its directly displayed relation between elasticities and scale: because of its neat mathematical features. It is these mathematical features that are conceptually considered useful for defining and solving new problems. In this way, these mathematical features create the similarities between two patterns.

3. A Case Study From the History of Modelling the Business Cycle

To explore how the model template account combines with the model building accounts to help understand functional progress in economics, we discuss a specific case of modelling in economics, that for methodology phenomena such as the business cycle and growth initiated by Jan Tinbergen's methodology (Boumans 2004). Tinbergen's methodology included a vision of what business cycles are, a general template for modelling them, and a number of requirements prescribing how cyclical movements of the economy should be explained.[6] Tinbergen's methodology of modelling and the model template it provided can be understood as providing a common recipe for defining and solving business cycle problems in economics. This case shows how model templates are not only reused but also how scientists create models initially and how, once they are considered less useful, such old templates are replaced with new ones. Finally, this case illustrates that it is not only the mathematical structure that is reused but that such reuse also requires a shared conceptual vision of the phenomenon that underlies the phenomenon. If that vision is no longer shared, a model template can become useless and has to be replaced, sometimes against resistance, with a different one. This case underscores how the concept of a model template plays a key role in enabling functional progress in economics.

A good starting point for our analysis is Tinbergen's business cycle research in the 1930s (Boumans 2004). He was the first to introduce the methodology of modelling to economics by providing a common recipe for model building in business cycle modelling and later in economics in general. Tinbergen's methodology was based on James Clerk Maxwell's 'method of formal analogy' (Boumans 2004). According to Maxwell, 'we can learn that a certain system of quantities in a new science stand to one another in the same mathematical relations as a certain other system in an old science, which has already been reduced to a mathematical form, and its problems solved by mathematicians' (Maxwell [1871]1965, pp. 257–258). Tinbergen introduced this methodology to economics in a 1935 survey article on recent developments in business cycle theory because he was interested in explaining the observed dynamics of an economic system. According to Tinbergen (1935, p. 241), the aim of business cycle theory is to 'explain certain movements of economic variables. Therefore the basic question to be answered is in what ways movements of variables may be generated'. To answer this question, one had to find the underlying mechanism, which he specified as 'the system of relations [i.e., a model] existing between the variables' (p. 241). This system of relations 'defines the structure of the economic community to be considered in our theory' (p. 242).[7]

To find the mechanism underlying the dynamics of business cycles, Tinbergen emphasised that one needs to distinguish between the mathematical formulation of the mechanism captured by its defining equations, and the economic interpretation, or 'sense' as Tinbergen called it, of these equations. According to him, '[t]he mathematical form determines the nature of the possible movements, the economic sense [i.e., the economic interpretation] being of no importance' in the first instance. The mathematical form could be reapplied to various systems because, for instance, 'two different economic systems obeying . . . the same types of equations may show exactly the same movements' in variables (p. 242). However, whereas finding the appropriate 'mathematical form' was essential, it was not sufficient to develop a business cycle theory. Their economic interpretation would ultimately be crucial because only the combination of both would reveal the significance of this theory for economics, or as Tinbergen wrote: 'no theory can be accepted whose economic significance is not clear' (p. 242). As we show, this economic interpretation can be understood as the conceptual vision of the phenomenon accompanying a model template discussed earlier, and it did not come only after the mathematical form was determined, as Tinbergen suggested. We show that more is at stake than merely an interpretation of the model; it also includes an interpretation of the specific phenomenon, the business cycle, and its mechanism.

After surveying the mathematical forms available at that time to model what Tinbergen viewed as the mechanism underlying the business cycle, he arrived at the following template that captured his conceptual vision of the business cycle mechanism (Tinbergen 1935, p. 279):

$$\sum_{i=1}^{n} a_i p\left(t-t_i\right) + \sum_{i=1}^{n} b_i \dot{p}_t\left(t-t_i\right) + \sum_{i=1}^{n} c_i \int_0^{t-t_i} p(\tau)d\tau = 0,$$

where a, b, and c are coefficients; p represents the general price level; and t represents time. To ascertain that this general difference-differential-integral equation represents a business cycle mechanism, its parameters need to meet two 'wave conditions': the first condition is that the parameters should be such that the solution of this equation consists of a sine function; in this the time shape of $p(t)$ is cyclic. The second condition is that the cycle period should be much longer than the time unit and that its amplitude is constant. These conditions taken together imply that the summation of the coefficients c should be zero, so that $\sum c_i = 0$, and thereby reduce the kinds of mechanisms that are acceptable.

At first sight, Tinbergen's equation seems to be a good example of a template as proposed by Humphreys. It constitutes a mathematical structure without any a priori economic interpretation attached to it. However, Tinbergen developed this template with a clear conceptual vision of the main characteristics of the business cycle, as reflected by the mechanism captured by the model. Thus, his contribution is actually a good example of how the conceptual vision of the phenomenon, the mechanism coupled with the view that cycles should be endogenous and permanent, shapes the mathematical structure constituting the template. In other words, Tinbergen's template can best be described as a model template as proposed by Knuuttila and Loettgers.

Paul Samuelson turned Tinbergen's methodology into a more general modelling methodology for the field of macroeconomics, which became a new area in economics. Samuelson (1939, p. 78) considered Tinbergen's methodology as 'liberating' when he wrote that '[c]ontrary to the impression commonly held, mathematical methods properly employed, far from making economic theory more abstract, actually serve as a powerful liberating device enabling the entertainment and analysis of ever more realistic and complicated hypotheses'. In a four-page article, Samuelson discussed the dynamics of a simple four-equation model of an abstract economy to illustrate this methodology. As in Tinbergen's template, the dynamic behaviour of the economy could be expressed by the reduced form equation of this four-equation model, which was the following second-order difference equation:

$$Y_t = 1 + \alpha[1 + \beta]Y_{t-1} - \alpha\beta Y_{t-2},$$

where Y represents national income, and α and β are some coefficients. The 'qualitative properties' of the model, that is the dynamic characteristics derived from Tinbergen's conceptual vision, then depended only on the values of α and β. Samuelson showed that all possible combinations of the values of α and β lead to four types of behaviour of Y: Y will 'approach asymptotically' a specific value, 'damped oscillatory movements', 'explosive, ever increasing oscillations', and 'an ever increasing' Y (Samuelson 1939, p. 77).

With his equation, Tinbergen had not only proposed a model template to be applied to the business cycle. He had also introduced a whole methodology of mathematical modelling, which was further elaborated by Samuelson for macroeconomics and which later became the foundations of mathematical economics. This methodology provided a common recipe for defining and solving problems. Textbooks appeared in which this approach was demonstrated. One of the best-known textbooks was Alpha Chiang's *Fundamental Methods of Mathematical Economics*, first published in 1967. According to Chiang, mathematical economics is 'an *approach* to economic analysis, in which the economist makes use of mathematical symbols in the statement of his problem and also draws on known mathematical theorems to aid to his reasoning' (1974, p. 3, italics in original). Another important textbook was R.G.D. Allen's *Mathematical Economics*, first published in 1956. As Allen explained, '[t]he elements of macro-dynamic economic theory (Chapters 1, 2 and 3) show up the need for using difference and differential equations and for the description of oscillatory variation by means of complex variables and vectors (Chapters 4, 5 and 6)' (Allen 1959, p. xvi), clearly pushing Tinbergen's methodology. Besides the existing mathematical methodologies of static equilibrium analysis, comparative-static analysis, and optimisation problems, both textbooks also contained mathematical economics in the Tinbergen–Samuelson tradition: a methodology for dynamic analysis that includes difference equations and differential equations and their combination.

Against this background, Tinbergen's methodology, as generalised by Samuelson, can be understood as a common recipe for studying dynamic economic phenomena, including a template coupled with a conceptual vision of the main characteristics of those phenomena. Textbooks such as Allen's and Chiang's can then be considered as recipe books that provide the formal templates. When dealing, for example, with an optimisation problem, economists could use them to look for a recipe to tell them which kind of mathematical tools or concepts would be appropriate for which mathematical problem. For example, it became standard in economics to apply the Lagrangian equation to any optimisation problem. So, Tinbergen's and subsequently Samuelson's methodology exemplifies the idea of a template (from the recipe book) plus a conceptual vision to study the phenomenon in question as crucial ingredients of a common recipe for defining and solving a diverse array of economic problems.

Catherine Herfeld and Malte Doehne (2019) give a characterisation of the process of developing a scientific idea, which could be a new template, from its innovation to its application to concrete problems in a specific domain. They propose a typology of roles that scientific contributions can play in this process. Firstly, contributions can play the role of innovators, as Tinbergen's contribution does, which formulate a novel template. Secondly, they may be elaborators, as was Samuelson's contribution to generalising Tinbergen's template, that clarify, adapt, and sometimes modify the conceptual, theoretical, or empirical scope of the template. The third role is that of translators, such as Chiang's contribution, which adopt the elaborated template and modify it in such a way that it aligns with established frameworks and concepts in a specific domain so that it can be used for specific problems. Indeed, translator contributions sometimes provide concrete recipes for how an abstract template can be repeatedly applied to concrete and specific problems in various domains in what Herfeld and Doehne (2019, p. 43 f.) call a specialist contribution. We can understand the process sketched in our case study as following a similar sequence: the recipes resulting from elaboration and translation of a novel template are crucial in enabling the recurring application of the template to focused and specific problems and thus can lay the ground for functional progress in economics.

The case of Tinbergen's methodology shows how a template is applied to define and solve multiple problems following a common recipe. Furthermore, we can use this case to show that a template in this process is not reapplied on its own but that its reuse depends crucially on conceptual considerations and whether the underlying conceptual idea remains stable; in other words, it does not function merely as a formal template but as a model template. Although the methodology of using difference and/or differential equations to capture the dynamic properties of the business cycles can be understood as providing a useful common recipe for defining and solving problems in business cycle research, it also has some disadvantages that limits its use for defining and solving new problems in economics. A first disadvantage is that the stability assumed for the cyclical movement was too dependent on specific parameter values. Take, for example, a second-order differential equation. Whether the cycle is stable, gradually diminishes, or grows in amplitude is determined by the parameter of the first-order term. Tinbergen's wave conditions required that its value was zero. In discussing a business cycle model, where this issue played a role, Ragnar Frisch pointed out that 'since the Greeks it has been accepted that one can never say an empirical quantity is exactly equal to a precise number' (Goodwin 1989, pp. 249–250).

To further see how the template was connected to Tinbergen's conceptual vision of the business cycle, a second disadvantage has to be considered, namely, the assumption that the business cycle is endogenous, meaning that the existence of the business cycle is fully explained by the dynamic characteristics of the economic system alone. In economics, the dominant view has been that the economic system is a stable equilibrium system and that, as such, every position out-of-equilibrium would move back to equilibrium. This implied that the business cycle should be explained by exogenous factors. The disadvantage here is that the mathematics of difference and differential equations implied an endogenous explanation that was in conflict with the standard equilibrium view in economics and thus with an important methodological commitment of economics, which in turn called into question Tinbergen's conceptual vision of the business cycle. A third disadvantage of the Tinbergen approach is that observed business cycles do not resemble a smooth harmonic oscillator but less regular, that is to say more like a constantly changing harmonic, a cycle whose period lengths and amplitudes vary and whose shape is erratic.

These changes in the conceptualisation of the business cycle as erratic fluctuations instead of a smooth harmonic and in the explanatory requirements as neither deterministic nor endogenous exemplify how disadvantages can hamper or even halt the reuse of a template because the template does not capture mathematically the general properties of the target system. To overcome them, Frisch's (1933) influential paper 'Propagation Problems and Impulse Problems in Dynamic Economics' demonstrated that the business cycle can be generated by an iteratively disturbed stable equilibrium system. Frisch's starting point was Tinbergen's template, a mixed difference and differential equation. However, this equation did not show how the cycles are maintained: 'when an economic system gives rise to oscillations, these will most frequently be damped. But in reality, the cycles we have occasion to observe are generally not damped' (Frisch 1933, p. 197). Frisch introduced the concept of what he called 'impulses' to solve this problem. He explained the maintenance of the dampening cycles and the irregularities of the observed cycle by the idea of erratic shocks repeatedly disturb the economic system. These impulses are then propagated through the system, represented by the mixed difference and differential equation, creating a behaviour that resembled the observed business cycle.

By adopting a new conceptual vision of the characteristics of business cycles, Frisch replaced the model template of mixed difference and differential equations with a template that combined this deterministic system with a stochastic process. He assumed that $Q(t)$ denotes a damped oscillation that is the solution of the mixed difference and differential equation, t is time, and e_k represents the random shocks; then, the business cycle generated by Frisch's propagation and impulse model is:

$$y(t) = \sum_{k=1}^{n} Q(t - t_k) e_k,$$

where y is production. As Frisch noted,

> '$y(t)$ is the result of applying a linear operator to the shocks, and *the system of weights in the operator will simply be given by the shape of the time curve that would have been the solution of the determinate dynamic system in case the movement had been allowed to go undisturbed.*
>
> (Frisch 1933, p. 201; italics in original)

In other words, the business cycle was a summation of weighted shocks whose weights are determined by the economic system. Frisch had a conceptual vision of the economic system as a dynamic system that determined the dynamic properties of the cycle but did not explain its continuation. According to him, the continuation of the cycle was explained by a sustained series of external random shocks.

Besides the three disadvantages of Tinbergen's model template outlined earlier, another reason why Frisch had developed a new template, which became known as the 'rocking horse' model, is that Eugen Slutzky had suggested a 'deeply worrying possibility' that 'cycles could be caused entirely by the cumulation of random events' (Morgan 1990, pp. 79–80). In an article with the title 'The Summation of Random Causes as a Source of Cyclic Processes' (1933, originally published in 1927 in a Russian journal), Slutzky discusses whether it is 'possible that a definite structure of a connection between random fluctuations could form them into a system of more or less regular waves' (p. 106). Slutzky showed that it was indeed possible. He used a rather simple model, a 10-item moving summation of a random number series, the last digits of the numbers drawn for a Russian government lottery loan, e_k, in this equation: $y(t) = \sum_{k=1}^{10} e_k + 5$. As a response to this unsettling result, Frisch (1933) suggested that the weighting system of the shocks was determined by the economic system, which therefore could be thought of as functioning as a propagation system.

The three disadvantages of Tinbergen's model meant that a new template had to be created. Tinbergen's old model was not able to redefine and solve these problems. Because these problems were themselves created by a changing vision of the business cycle and its generating mechanism, the new template would have to be coupled with a new conceptual vision that could accommodate this change. Interestingly, Slutzky was not the only economist in the 1920s to doubt whether the business cycle was a regular harmonic movement and wonder whether it might better be approached as a stochastic process. Irving Fisher (1925, p. 191; italics in original) even questioned the very existence of a business cycle:[8]

> Of course, if by the business cycle is meant merely the statistical fact that business docs *fluctuate* above and below its average trend, there is no denying the existence of a cycle – and not only in business but in any statistical series whatsoever! If we draw any smooth curve to represent the general trend of population, the actual population figures must necessarily rise sometimes above and sometimes below this mean trend line. . . . In the same way weather conditions necessarily fluctuate about their own means; so does the luck at Monte Carlo. Must we then speak of 'the population cycle,' 'the weather cycle' and 'the Monte Carlo cycle'?

Fisher's considerations showed clearly that conceptual considerations about the general characteristics of cycles played a substantial role in deciding which template was best for explaining them. He argued that business fluctuations were not characterised by cyclical or regular patterns. He denied the possibility of tendencies towards regularity because 'these tendencies may always be defeated in practice, or blurred beyond recognition' (p. 192). Drawing on analogical reasoning to specify his conceptual vision, a physical analogue to the business cycle for Fisher would have been

> the swaying of the trees or their branches. If, in the woods, we pull a twig and let it snap back, we set up a swaying movement back and

forth. That is a real cycle, but if there is no further disturbance, the swaying soon ceases and the twig becomes motionless again. In actual experience, however, twigs or tree-tops seldom oscillate so regularly, even temporarily. They register, instead, chiefly the variations in wind velocity. A steady wind may keep the tree for weeks at a time, leaning almost continuously in one direction and its natural tendency to swing back is thereby defeated or blurred. Its degree of bending simply varies with the wind. That is, the inherent pendulum tendency is ever being smothered.

(Fisher 1925, p. 192)

Fisher's conceptual vision of the business cycle came closer to Slutzky's than Frisch's because the dynamics of the 'rocking horse' or 'pendulum' were 'smothered' by the erratic behaviour of the wind. However, Fisher was not the only economist who was struggling with what could be considered to be the best way of modelling the business cycle. As these considerations were closely connected to the choice of the most useful template, their views differed about what the best template was for modelling the business cycle.

In the decades that followed Frisch's contribution, economists increasingly argued that the economic system was not itself the mechanism that produced dynamic behaviour such as a business cycle. According to them, the only dynamic it produced was stable growth. Because the economic system was believed to be a stable growth equilibrium system, the business cycle could only be the product of exogenous shocks. The random generator of shocks could be represented by a probability distribution, most often a Gaussian distribution, and hence the dynamic properties of the business cycle could best be represented by the moments of this distribution. It was a conceptual shift away from locating the dynamic properties within the economic system towards locating them outside the system, as external disturbances. This view resulted in dynamic stochastic general equilibrium (DSGE) models, first developed by Fynn Kydland and Edward Prescott (1982) in the early 1980s. When these DSGE models grew in dominance in the late 1980s and 1990s, the use of mixed difference-differential equations increasingly moved to the periphery of economics. However, the general procedure underlying the core of the original Tinbergenian methodology, solving a problem in a mathematical way by finding an appropriate mathematical model template that could capture all aspects of the conceptual vision, did not disappear.

This condensed history of modelling the business cycle with a sequence of templates shows how the choice of model template is closely tied to specific conceptual visions of general characteristics of target systems and how strongly visions determine which templates are accepted. Consequently, a shared conceptual vision partly explains the choice of one template over another in the history of business cycle analysis. However, it also shows that a new template does not easily replace a so far satisfactory template just because the conceptual vision has changed. In our case, the replacement of one model template by another did not immediately follow the changes in the conceptual vision of the business cycle. This is nicely expressed by Slutzky (1937, p. 105) in the opening paragraph of his 'Summation of Random Causes' article:

> Almost all of the phenomena of economic life, like many other processes, social, meteorological, and others, occur in sequences of rising and falling movements, like waves. Just as waves following each other on the sea do not repeat each other perfectly, so economic cycles never repeat earlier ones exactly either in duration or in amplitude. Nevertheless, in both cases, it is almost always possible to detect, even in the multitude of individual peculiarities of the phenomena, marks of certain approximate uniformities and regularities. The eye of the observer instinctively discovers on waves of a certain order other smaller waves, so that the idea of harmonic analysis, viz., that of the possibility of expressing the irregularities of the form and the spacing of the waves by means of the summation of regular sinusoidal fluctuations, presents itself to the mind almost spontaneously. If the results of the analysis happen sometimes not to be completely satisfactory, the discrepancies usually will be interpreted as casual deviations superposed on the regular waves.

The general patterns observed are thus based upon what the economic system is believed to be.[9] To model the business cycle, a mathematical template was needed that mathematically reproduced the conceptual vision of the general characteristics of the economy. When it was believed that the business cycle is endogenous, difference and differential equations were useful in building the business cycle models. When it was believed that the business cycle is exogenous, it was mathematically represented as a stochastic process.

4. Model Templates and Progress in Economics

Progress in economics is functional: a model can be used repeatedly as a template that is part of a common recipe to define and solve new problems. The notion of a model template is thus fruitful for thinking about progress in economics because it shows how common recipes that rely on models are reused in economics to define and solve diverse problems and therefore how their reapplication leads to functional progress. We have also seen that the contribution of a model template to progress in economics depends crucially on a shared conceptual vision of the general characteristics of the phenomenon, such as the economic system, of which the business cycle is one particular element. In economics, progress through template use only occurs as long as this conceptual vision is shared and sustained.

That a template could only be reapplied as long as economists shared a conceptual vision about the business cycle becomes particularly clear when considering how the conceptual vision underlying Tinbergen's template changed within business cycle research. Mixed difference-differential equations ceased to be applied when economists changed their conceptual vision of the economic system from a cycle mechanism to an equilibrium system. In line with Knuuttila and Loettgers, our case study suggests that Humphreys' template view needs an important modification to explain the template's adoption and reapplication. A formal template can indeed be reapplied because it is in principle neutral to particular economic interpretations, but the application of a model template is not neutral; it is guided by a conceptual vision of the phenomenon to which the template is to be applied. Thus, the concept of a model template can capture the observation that models are reapplied only to those phenomena that share general characteristics. This is what makes a model template well-suited to study such phenomena. However, we have also noted that the mathematical structure in return can help shape the conceptual vision underlying the model template.

This is a different view of progress than Lucas's. Because Lucas argues that general theoretical ideas have not changed for a few centuries and progress is mainly 'technical', he observes a long history of continuous progress. Our contrasting view is that progress can be hampered by a change in the conceptual vision of the general characteristics of the phenomenon, in our case the economic system. The conceptual vision underlying a template presupposes a set of general characteristics of the target system. If the vision of the target system changes, it requires a different explanation and thus a different template to provide this explanation. A new model template will be chosen whose mathematical form and underlying conceptual vision can provide such an explanation.

Is a shared conceptual vision of the target system sufficient to explain the application of model templates? If so, progress in economics could be characterised as the reapplication of model templates. However, we have seen that the application of a template can also be explained by its strong initial influence on the conceptual vision, by inducing a specific mathematical form. Tinbergen influenced the conceptualisation of the business cycle in the 1930s by introducing a specific approach in which the business cycle was represented as generated by a mixed difference and differential equation. Another example is Slutzky's representation of the business cycle as a random walk. These were new conceptual considerations about the general characteristics of the phenomenon; new specific patterns or shapes of the phenomenon were formulated that changed conceptual visions. When adopted more widely, it enabled the application of that form as a template for new models. Thus, a template not only leads to progress because it helps define and solve new problems. It can also lead to progress because it mathematically co-creates the conceptual vision that makes it suitable for studying the phenomenon with these characteristics.

In the introduction to the special issue 'Knowledge Transfer and Its Contexts', Herfeld and Chiara Lisciandra (2019, p. 6) add to the characterisation of model transfer that 'templates are particularly apt to be transferred to those domains that either a) share a similar methodology with the source domain or b) deal with problems that can in principle be tackled with that methodology'. As our case study suggests, this is in part because a common recipe including a new model template and procedures for its creation can lead to functional progress. It was the introduction and acceptance of Tinbergen's methodology that allowed the business cycle to be modelled. This model template was then reapplied to various other problems in economics. Once the conceptual vision underlying the model template changed, the template was no longer considered useful.

5. Conclusions

In this chapter, we discussed a specific kind of progress in economics, progress involving the repeated use of mathematical models, which are part of the common recipes used in economics to define and solve new problems. We explicated such common recipes as including model templates, which are defined as mathematical structures that capture an underlying conceptual vision of the phenomenon's general characteristics. As such, these model templates are integrated combinations of conceptual visions of the phenomenon to be modelled and highly abstract mathematical forms that are needed for actually modelling them. We discussed the model template that Jan Tinbergen formulated to model the business cycle as a specific case study of formulating and applying a model template, and we argued that functional progress in economics occurs when a shared conceptual vision of the phenomenon allows a model template to be reused to define and solve problems. Progress can be hampered when model templates are considered useless for specific problems, for which a change in the conceptual vision of the phenomenon can be responsible. However, functional progress can continue, once a new model template is developed and the conceptual vision underlying it is accepted.

Acknowledgements

We are grateful for the useful and constructive comments of Anna Alexandrova, Paul Humphreys, Tarja Knuuttila, Andrea Loettgers, Mary S. Morgan, and Robert Northcott on an earlier version of this chapter.

Notes

1. For a recent philosophical account of progress in economics, see Northcott and Alexandrova (2009), which also concerns economics as a modeling discipline. Also note that in the following, we provide an internalist rather than an externalist account of progress in economics. Our goal is not primarily to offer an appraisal to answer the question of whether economics indeed progressed by some external standard but rather provide an account that is also descriptive of economic practices.
2. See also Chapter 3 'The Functional Approach: Scientific Progress as Increased Usefulness' in this volume and Chapter 6 'A Functional Account of the Progress in Early Genetics' in Shan 2020.
3. Formal templates also include forms in formal logic or programming languages. In model-based economics, so far, they do not play any significant role.
4. Although today it has many other applications, such as to define consumption, the Euler function is still called the Cobb-Douglas function after its first famous application. We follow this tradition in economics.
5. As Humphreys rightly pointed out to us, it should be noted that the extent to which the conceptual and mathematical dimensions of a template are intertwined will depend, however, on how abstract the interpretation is and whether the account given by us applies across changes in economic conceptualizations.
6. We understand a 'methodology' here along the lines of Laudan's definition as a broad set of procedures, strategies, and tactics used for the purposes of validation of scientific hypotheses as well as their heuristic advancement (Laudan 1984, pp. 3–5).
7. Instead of 'economic community', today we would use the term 'economy', but in the 1930s, this latter term did not yet have its current meaning. The same applies to the term 'model', which was not commonly used in the 1930s. Instead of this term, Tinbergen speaks of a 'system of relations'.
8. Fisher's view was neglected at the time he wrote this, but it became representative of the view that has predominated since the 1980s.
9. This is closely related to the idea that observation is theory-laden, defended for example by Norwood Russell Hanson's *Patterns of Discovery* (1958), in which Hanson argues that one observes the world through the lens of theory.

References

Alexandrova, Anna and Northcott, Robert (2009). Progress in economics. *In The Oxford Handbook of Philosophy of Economics* (pp. 306–337), edited by D. Ross and H. Kincaid. Oxford: Oxford University Press.

Allen, R.G.D. (1959). *Mathematical Economics*, 2nd edition. London: MacMillan.

Boumans, Marcel (1999). Built-in justification. In Morgan and Morrison (1999), pp. 66–96.

Boumans, Marcel (2004). Models in economics. In *The Elgar Companion to Economics and Philosophy* (pp. 260–282), edited by J.B. Davis, A. Marciano, and J. Runde. Cheltenham and Northampton: Edward Elgar.

Chiang, Alpha C. (1974). *Fundamental Methods of Mathematical Economics*, 2nd edition. New York: McGraw-Hill.

Fisher, Irving (1925). Our unstable dollar and the so-called business cycle. *Journal of the American Statistical Association* 20 (150), 179–202.

Frisch, Ragnar (1933). Propagation problems and impulse problems in dynamic economics. In *Economic Essays in Honour of Gustav Cassel*. London: Allen & Unwin.

Goodwin, Richard M. (1989). Kalecki's economic dynamics: A personal view. In *Kalecki's Relevance Today*, edited by M. Sebastiani. London: Palgrave Macmillan.

Hanson, Norwood Russell (1958). *Patterns of Discovery*. Cambridge: Cambridge University Press.

Herfeld, Catherine, and Malte Doehne (2019). The diffusion of scientific innovations: A role typology. *Studies in History and Philosophy of Science* 77, 64–80.

Herfeld, Catherine, and Chiara Lisciandra (2019). Knowledge transfer and its contexts. *Studies in History and Philosophy of Science* 77, 1–10.

Hesse, Mary (1966). *Models and Analogies in Science*. Notre Dame, IN: University of Notre Dame Press.

Humphreys, Paul (2002). Computational models. *Philosophy of Science* 69, 1–11.

Humphreys, Paul (2004). *Extending Ourselves. Computational Science, Empiricism and Scientific Method*. Oxford: Oxford University Press.

Humphreys, Paul (2019). Knowledge transfer across scientific disciplines. *Studies in History and Philosophy of Science* 77, 112–119.

Kahneman, Daniel, and Amos Tversky (1979). Prospect theory: An analysis of decision under risk. *Econometrica* 47 (2), 263–292.

Knuuttila, Tarja, and Andrea Loettgers (2016). Model templates within and between disciplines: From magnets to gases – and socio-economic systems. *European Journal of Philosophy of Science* 6, 377–400.

Knuuttila, Tarja, and Andrea Loettgers (2020). Magnetized memories: Analogies and templates in model transfer. In *Living Machines? Philosophical Perspectives on the Engineering Approach in Biology* (pp. 123–140), edited by S. Holm and M. Serban. London and New York: Routledge.

Kuhn, Thomas (1970). *The Structure of Scientific Revolutions*, 2nd edition. Chicago: University of Chicago Press.

Kydland, Finn E., and Edward C. Prescott (1982). Time to build and aggregate fluctuations. *Econometrica* 50 (6), 1345–1370.

Lakatos, Imre (1970). Falsification and the methodology of scientific research programmes. In *Criticism and the Growth of Knowledge* (pp. 91–196), edited by I. Lakatos and A. Musgrave. London: Cambridge University Press.

Laudan, Larry (1981). A problem-solving approach to scientific progress. In *Scientific Revolutions* (pp. 144–155), edited by I. Hacking. Oxford: Oxford University Press.

Laudan, Larry (1984). *Science and Values, The Aims of Science and Their Role in Scientific Debate*. Berkeley and Los Angeles: University of California Press.

Lucas, Robert E. (1980). Methods and problems in business cycle theory. *Journal of Money, Credit, and Banking* 12, 3–42.

Lucas, Robert E. (2004). Keynote address to the 2003 HOPE conference: My Keynesian education. In *The IS-LM Model*, edited by M. De Vroey and K.D. Hoover. *History of Political Economy* 36 (supplement), 12–24.

Maxwell, James Clerk [1871] (1965). On the mathematical classification of physical quantities. In *The Scientific Papers of James Clerk Maxwell*, vol. II (pp. 257–266). New York: Dover.

Morgan, Mary S. (1988). Finding a satisfactory empirical model. In *The Popperian Legacy in Economics* (pp. 199–211), edited by N. De Marchi. Cambridge: Cambridge University Press.

Morgan, Mary S. (1990). *The History of Econometric Ideas*. Cambridge: Cambridge University Press.

Morgan, Mary S. (2004). Imagination and imaging in economic model-building. *Philosophy of Science* 71 (5), 753–766.

Morgan, Mary S. (2012). *The World in the Model: How Economists Work and Think*. Cambridge: Cambridge University Press.

Morgan, Mary S., and Margaret Morrison (1999). *Models as Mediators*. Cambridge: Cambridge University Press.

Samuelson, Paul A. (1939). Interactions between the multiplier analysis and the principle of acceleration. *The Review of Economics and Statistics* 21 (2), 75–78.

Shan, Yafeng (2019). A new functional approach to scientific progress. *Philosophy of Science* 86, 739–758.

Shan, Yafeng (2020). *Doing Integrated History and Philosophy of Science: A Case Study of the Origin of Genetics*. Cham: Springer.

Slutzky, Eugen (1937). The summation of random causes as the source of cyclic processes. *Econometrica* 5 (2), 105–146.

Tinbergen, Jan (1935). Annual survey: Suggestions on quantitative business cycle theory. *Econometrica* 3 (3), 241–308.

13 Progress in Medicine and Medicines

Moving From Qualitative Experience to Commensurable Materialism

Harold Cook

When taking up the subject of progress in medicine, we might well begin by defining our terms. As a historian of a nominalist kind, however, I prefer not to work with categories but to try to imagine the real lives to which the words in the documents point and about which the secondary abstractions are partial generalisations. In considering 'progress', for instance, we might observe that in the time of Queen Elizabeth I she sometimes went on one, slowly travelling through the country with the court and hangers-on in tow; a few decades later a preacher, John Bunyan, adopted the term allegorically for his *The Pilgrim's Progress from this world, to that which is to come*. Today, the word is more commonly employed not to suggest a personal journey but a collective movement towards material or moral betterment: 'we are making progress'. Moreover, as indicated further in the following, the category of 'medicine', too, has shifted meaning over time. Even in the present, it remains difficult to confine within any well-defined boundaries. When we use the word, we may be thinking of expert bodily interventions; a collection of information and concepts; a professional, institutional, or financial system; aid in a healing process; wellness; childbirth; alternative or complementary ideas and practices; and so on. Although the bookshelves hold many histories of progress and many histories of medicine, our subject is so generous that establishing a precise definition of the topic would metaphorically threaten to cut off a living thing from the ever-shifting environment that sustains it.

Nevertheless, common usage indicates that when ordinary people speak of progress in medicine they usually mean 'improvements in human health'. The phrase implies that human health can be improved for persons and populations, the former often indicated by reports of grateful patients whose lives were saved by medical interventions and the latter often measured statistically as increased average length of life, which is often attributed to life-saving interventions but perhaps is due more to various determinants of population health, not least those of public health.[1] Students of modern culture have much more to say about

DOI: 10.4324/9781003165859-16

wide-spread discourses and counter-discourses related to popular assumptions about improvement. Here, I wish to consider a single thread in the progressivist understanding about what makes medical improvement possible: the world-wide research efforts that pursue questions of health and disease by examining their biochemical mechanisms, often referred to as biomedicine. I will take biomedicine to be an approach to human health and disease that is guided by the laboratory or experimental sciences, which in turn depend on impersonal methods of assessing physical evidence that create the possibilities for common agreement and action. Biomedicine is not a word used by ordinary people, but it is understood among researchers, professionals, administrators, charities, businesses, and political leaders.[2] Biomedical activities are usually connected to other forms of collective information-gathering and analysis, such as epidemiology, demography, and the like, but here we will keep our eye on the investigation, control, and use of tangible materials related to medical treatment. The accuracy of such methods is constantly improving while their effects are more powerful.

The methods of biomedical research and production also enable certain kinds of naming, instrumentation, and substance to move relatively easily across boundaries of cultural, linguistic, and national difference. Today, governments and funding bodies around the world commonly – although not universally – take biomedicine to provide a gold standard to which other practices can be compared. To put it simply, the benchmarking provided by materialistic assessment and the subsequent production of physical substances that act as intended in all appropriate instances, anywhere, is understood to flow from grasping an underlying impersonal 'nature', attention to which encourages certain kinds of disciplined human cooperation that do not depend on social, religious, ideological, or other personal or collective identities. In other words, pointing to the privileged respect given to improvements in the precision of materialistic assessments can enable us to recognise the means by which biomedical understandings and activities are widely thought to promote medical progress throughout the world despite diversity of cultures and polities. I will consequently refer to biomedical methods of assessment as a form of commensurable materialism.

Biomedical improvements can be put to various ends, however, so we should keep in mind the fact that the general association between such improvements and collective benefit does not itself rest on a self-governing natural order; governmentality channels the consequences to afflictive, utilitarian, profitable, humane, or other ends. The study of improvement of methods therefore leaves aside questions of moral and political philosophy. In disentangling material benefit from progress towards the good I wish to mark how the commensurable materialism of biomedicine often supports forms of political economy that require certain kinds of cooperative activity, but that other actionable values are necessary if it is to be directed towards sustaining and prolonging the lives of its beneficiaries, the ends commonly associated with progress. Moreover, other kinds of goods associated with health and well-being are made manifest by other kinds of human thought and action, and will also be set aside for the moment.

As a historian, I shall proceed by giving an account of the emergence of commensurable materialism in time and space, with attention to tangible medicines. The examples aim to clarify how the carefully disciplined procedures of evaluating questions of health and illness have increasingly moved from judgements rooted in qualitative bodily experience to physical investigations of substances using impersonal methods. Put another way, one can point to the formation of an intentionally universalisable set of disciplined activities that were intended to overcome distance and difference, activities which were not medical per se but which enabled biomedical investigation to take shape. If we look at the field of medicine as an example within the grand historical transformation writ large (currently often referred to as the history of capitalism), one notices a shift from qualitative evaluation to materialist assessment. In the process, various meanings and associations of powerful substances were stripped away so as to attend to the substances themselves. Examples of the process can be found in many times and places, allowing us to note that biomedicine is embedded in a lineage that commonly equates medicine with medicines.

A final brief remark may be helpful for situating this chapter within the genres of history and philosophy of science, medicine and technology, and history. In pointing to commensurable materialism this study shares some aims reminiscent of the position of the editor of this collection, Yafeng Shan, to present a 'new functional' framework for considering progress in science partly on the basis of taking the usefulness of any intellectual programme into account. Shan's aim is in part to rehabilitate important aspects of the positions held by Thomas Kuhn and Larry Laudan.[3] In my own case, I was introduced to Kuhn's *The Structure of Scientific Revolutions* (1962) as an undergraduate student shortly after the publication of its second edition of 1970, and as a graduate student in history, I continued to follow the debates within the field of history and philosophy of science provoked by it and other arguments (especially those of Michel Foucault about knowledge and power); I became somewhat acquainted with Laudan's views on research traditions, too. One of Kuhn's most famous arguments was of course this: that in the operation of 'normal science' the 'paradigms' of a field establish an intellectual consensus, the foundational assumptions of which are debated only during moments of disruption caused by the multiplication of anomalies. Kuhn drew further attention to the puzzle of consensus in his prefatory remarks to the second edition, stating that when he spent time with other research fellows at the Center for Advanced Studies in the Behavioral Sciences 'I was struck by the number and extent of the overt disagreements between social scientists about the nature of legitimate scientific problems and methods'. He doubted that the community of natural scientists in which he himself was educated understood the foundations of their fields any better, yet they did not engage in similar controversies, which intrigued him and led to his invocation of paradigms to describe the phenomenon of consensus.[4]

The question of how a consensus around knowledge claims might emerge, and whether commensurability has much to do with it, became a kind of base-line theme for others, too, such as those involved in movements like the Sociology of Scientific Knowledge or Actor Network Theory.[5] I owe much to the literature in those fields.[6] But my own reading of history and sources also drew on intellectual, cultural, social, political, colonial and post-colonial, economic, and global histories that raised both other questions about consensus and conflict and other problems about the reasons for the flourishing of 'modern science'. By limiting this examination to a few examples of the history of medicines, with commensurable materialism as a framework for containing certain kinds of historically grounded phenomena, I hope to encourage attention to the work done by commensurabilities within worlds full of diversity, and to do so by pointing to some of the underlying changes in political economies that suggest why the global reach of both commercial materialism and biomedicine flow from a common source. In this way, phrases such as 'progress in medicine' can be understood both by example and lineage.

1.   Afro-Eurasian Qualitative Medicine and Bodily Experience

Let us begin with an example of commensurable materialism from four centuries ago, at the end of the long period when qualitative assessments of the powers of nature were dominant. It sets up a helpful contrast between two widely accepted approaches to understanding the actions of medicinal substances, that of qualitative experience and that of physical description.

During the 1620s, Jacobus Bontius, a physician in the employ of a company-state, the United Dutch East India Company (VOC), made the half-year voyage from the Dutch Republic to the VOC's recently established and still contested headquarters at Batavia (now Jakarta). After his arrival, in the midst of two sieges and many duties, he set about recording information on the health, medical practices, and natural history of the region. He made a special effort to identify medicines yet unknown to his European peers that were used by Javans and by the groups of merchants from various parts of Asia who visited or settled in Batavia. He sometimes commented favourably on the people he met, who knew much about the God-given beneficial effects of many plants.

At the same time, however, Bontius was almost entirely silent about the meanings his interlocutors attached to the plants or their methods of using them, favouring bald description of the things themselves as the source of efficacy. He gave a hint of his self-conscious method in a favourable comment on Indian verbena, going on to remark that 'This herb is considered sacred among the old [East] Indian women (which they have in common with our own old women)'. In then apologising for setting down even this much context, he drew a line between his ways and theirs: they 'demonstrate[d] the foolish habit' of mind that considered the efficacy of herbs to be due to sacred powers while he himself was 'not one of those who has a propensity to superstitious belief about the natural powers of medicines'.[7] Since natural powers were at work, he could dispense with rituals and preparatory methods in favour of simply recording the medicinal effects of a plant and – so that it could be correctly identified – its physical characteristics. In his work on natural history, Bontius also visually depicted many plants and a few animals to which he added brief descriptions to allow ease of identification. While indicating his gratitude to many of his informants, he also self-consciously stripped away any unnecessary baggage – that 'context' that historians would now wish to have – so as to reduce his work to a materially descriptive report about things, the objects of his 'objectivity'.[8]

Let us also note a very important implication of such reductionist reporting: that Indian verbena and all the rest would be of use to any person suffering from the conditions it relieved, no matter their personal situation. In other words, Bontius' kind of reportage implied that even the bodies of his audience in The Netherlands would respond to illnesses and their treatments in similar ways, without regard to their unique personal characteristics. A similar set of expectations lay behind the trials of a century earlier when in 1524 an antidote against poison was offered to Pope Clement VII and he had it tested on two criminals condemned to death by empowering his physicians to administer a deadly aconite to them, with the one who did not receive the antidote dying in great agony while the other lived.[9] Remedies that worked in all cases, whether the persons were popes or destitute urban wayfarers or, as in the case of Bontius' report, people from other places in the world – all of whom were of various diets, habits of life, customs, ages, sexes, shapes, skin colours, and all the rest – could help anyone else in similar difficulties. What was dropping out of descriptions like Bontius' were not only the 'superstitious' meanings people ascribed to the living organisms they depended on but other qualitative characteristics of the persons who might be beneficiaries.

We should not take non-qualitative, descriptively materialistic approaches to medical knowledge as a given. Most of us have been conditioned to accept as 'natural' the relationships between tangible things and bodily responses, usually speaking about the process of cause and effect as located in a physical world. The biomedical model presumes that certain kinds of carefully specified substances, such as particular cells, genes, enzymes, molecules, and elements, interact with other such substances, which cause the intended bodily effects in a statistically significant numbers of cases (which we interpret as 'all but the exceptions'). But for most of human history the operating expectations, rooted in robust lived experience, were quite otherwise. Attention focused on felt experience rather than some sort of underlying set of propositions about a material substrate. In the common world of felt qualities, individual difference counted heavily, while general explanations of cause and effect made sense only in association with intellectual lineages that were also enormously diverse throughout Afro-Eurasia.[10] The emergence of commensurable materialism therefore requires explanation.

While today we might think of medicine as related to swallowing mass-produced pills that have no particular taste and act on our bodies

Figure 13.1 An ill man seated by a fireplace vomiting into a bowl. Pencil drawing, William Hogarth; courtesy of Wellcome Collection.

insensibly, in earlier traditions bodies and senses were meant to be aware of the workings of medicinal preparations made for the occasion. In fact, most medical practices aimed to produce quite noticeable powerful bodily effects such as evacuating the stomach or bowels upward or downward, or releasing blockages by needling or bleeding, or inducing sweating or blistering, or evoking a feeling of inner cooling or heating. Potencies were found in almost all things, but some were stronger than others. The Greek term *pharmakon* pointed to the variety of powers to be found, since under that heading could be listed not only remedies but poisons, intoxicants, sacraments, talismans, pigments, cosmetics, and perfumes. *Medicamentum* might also refer not only to a remedy but also to a dye or a poison.[11] In China, too, powerful medicines possessed *du* 毒, toxicity or potency, bringing poisons into the healing armamentarium if they were taken in the proper amount at the proper time.[12] In fact, throughout Afro-Eurasia the changing temporal character of any illness – which was commonly understood not to be an entity so much as a process – meant that attention to what might be necessary at the moment was critical. In other words, like tinctures, small amounts of powerful substances could produce strong bodily effects, which were in turn carefully noted and treated. Pharmacy therefore emerged as a kind of special kind of culinary art that dealt with a much larger number of ingredients, implements, and methods of preparation than ordinary cooking but could be relied on to be felt as required by the shifting situation at hand.

Medical effects were associated not only with plants, animal parts, minerals, and their mixtures per se, but also – as with foods – with the qualities discerned in them that would interact with the qualities of the protean sick body. When you taste, for example, you know immediately whether something is sweet or bitter, salty, and so on. The great culture hero of China, Shennong, who is associated with the origins of both agriculture and pharmacy, was therefore reputed to have tasted plants in order to determine their powers.[13] In Europe, too, a Latin word for knowledge, *sapientia*, derives from *sapio*, to savour or taste.[14] Older meanings of the Germanic 'proof' are also associated with taste, as in the English expression 'the proof of the pudding is in the eating' or modern Dutch usages such as *wijnproef*. (The more common English term 'proof' as a standardised measure of alcoholic content became common only in the eighteenth century.)[15] We might today think of a wine 'connoisseur' – a knower – as having a fine discernment of taste and a rich language in which to speak about it to others who are equally experienced, although such taste might be impossible to convey to a novice. For most of us, 'having good sense' is all we can aspire to.

A genre of early documents identifying the substances and their qualitative effects, called *materia medica* in the eastern Mediterranean, came to be associated with the name of Dioscorides and saw moments of subsequent flourishing, but there is a more continuous documentary trail in East

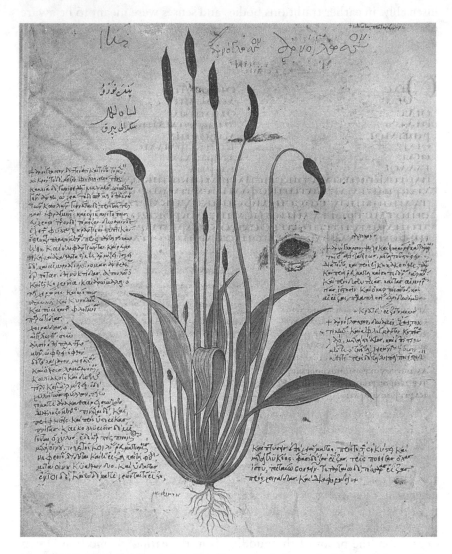

Figure 13.2 Plantago lanceolata (lamb's tongue), from the Vienna Dioscorides, c. 515, fol. 29v., public domain.

Asia: *bencao* texts. They flourished especially from the period of the Tang (618–907 CE), which is generally recognised as a period of technological innovation. The contents of eastern Eurasian *bencao* are similar but not quite the same as western Eurasian *materia medica* since the former include more information on many kinds of things found in nature, and while those are mainly medicinal minerals or herbs, non-medical creatures and their powers are included, too, even dragons.[16] Formularies, with

Figure 13.3 Drug preparation, *Buyi Lei Gong paozhi bianlan* (Supplement to Lei Gong's Guide to the Preparation of Drugs), 1591; courtesy of Wellcome Collection.

recipes and descriptions of processes and instruments, were also being produced from an early period, as is made plain by the example of Sun Simiao (581–c. 682) and his *Beiji qianjin yaofang* ('Essential Formulas for Emergencies [Worth] a Thousand Pieces of Gold') and its supplement, *Qianjin yifang*. By end of the Tang, an Imperial Medical Office (*taiyi shu*) was in operation for the imperial court, indicating the esteem in which medical remedies were held; during the succeeding Song dynasties (960–1279) imperially authorised pharmacists began publishing official guidebooks suggesting the authority of this branch of learned expertise. The *Shengji zonglu* (1111–1117), for example, contained 20,000 prescriptions (including demonological and astrological ones).

The example of the *Shengji zonglu* also indicates that by the tenth century pharmacists also were trying to correlate the effects of the medicines with the natural principles found in the medical literature on how to live well, such as *yin* and *yang* and the five phases (*wuxing*). The natural principles could be found in some of the oldest forms of canonical medical texts, from about two-and-a-half millennia ago – such as the Greek Hippocratic and Galenic corpus, the Sanskrit *Charaka*, or the Chinese *Huangdi neijing* – which might be termed a genre of wisdom literature exploring the relationship of health and illness to the regularities found in the universe. Happenstance or the ill will of other persons and beings were yet other ways of understanding bodily misfortunes, but the medical wisdom literature offered to correlate bodily experience with the general governing processes of the universe. While they also included recommendations about remedial help when necessary, they generally focused on advice about how to maintain proper relationships among the qualities typifying any person (one's 'temperament') and the situation in which those persons found themselves in order to preserve health and prolong life.

Put another way, vigorous efforts were being made to associate the assessments of potent substances with the underlying qualitative causes of change that were identified in the medical wisdom literature. But each cultural and linguistic textual lineage offered its own preferred grounding in an understanding of the foundations of the world and the qualitative expressions of it that so affected human lives. There were differences of opinion within any group drawing on their own preferred canonical sources, and there is certainly much evidence of interaction and borrowing among and between the major traditions. Paul Unschuld noticed traces of explanations for the natural power of things in texts like Sun Simiao's *bencao* that reflect contemporary Ayurvedic and Buddhist views from South Asia, and even Mediterranean understandings of the humours, while C. Pierce Salguero examined the movement of peoples and religions that translocated Sanskrit medical texts and domesticated them in China through processes of translation.[17] Yet even when in competition with one another, or defending their own assumptions or translating or appropriating selected ideas from other medical canons, the grand medical traditions pointed to qualitative experience as the foundation for understanding how the potencies of the world affected bodies.

In East Asia, therefore, drug actions would often be attributed to commonplaces about the activating movements of the world such as *yin* and *yang, qi,* and the five phases of transformation (*wuxing*); in South Asia such commonplaces included the three *dosha*s of *vata, pitta,* and *kapha,* which were in turn composed of the five elemental qualities, *panchamahabhutas*; in Unani and other dominant forms of Islamic medicine as well as in Europe preferences would be for the explanatory four qualities (hot, cold, wet, and dry), which combined in doublets (hot-wet, hot-dry, cold-wet, and cold-dry) to make up the four elements and four humours. On many occasions, however, in all these medical cultures at least some qualities had to be attributed to 'occult' causes, since their sources were unknown; other actions might be attributed to season of the year, astrological moments, and so on. The five *wuxing* were not the same as the three *dosa*s or the four qualities. The diverse customary ways might share the goal of helping humans live well and long, but they had their own ways of experiencing the world and hence different understandings of how to move towards that end.

2. Commensurable Description

By the time of Bontius, however, he stood among some of the physicians who were beginning to drop the categorisation of drug action according to qualities, striving for descriptive sufficiency rather than explanatory consistency. The reasons grew both from the huge influx of new information generated from robust global maritime commerce and from the related power of materialist calculation, which was more and more strongly affecting ordinary people and governments. Commercial materialism had two aspects that particularly bear on the changes in medical frameworks: an interest in the discovery, use, and marketing of effective new remedies; and the growing commensurability of common weights, measures, currencies, and other descriptive instruments. Precisely described and disciplined material description provided the ideal for conveying information about commodities being imported from or exported to far distant places, which in the case of drugs were most effectively recommended when they acted as expected on anyone in any place to which they were conveyed. That encouraged the kind of reductionist reporting we have noted in the case of Bontius, which in turn depended on careful physical description and depiction. With impersonal physical measures and descriptions circulating commensurably and taking precedence over qualitative experience, elements of an information economy were beginning to emerge.

Medicinal drugs, like spices (among which they were often classed), were among the early commodity forms traded over long distances throughout Afro-Eurasia. Sugar, for instance, was valued more for its medicinal value than as a food until its mass production in the later seventeenth-century Caribbean: it appeared very early in texts like the Ayurvedic *Charaka-Samhita* and *Susruta-Samhita* as well as Chinese and Greco-Roman sources, known for its physiological effects and preservative properties.

To Pliny the Elder, sugar was imported to Rome from the far reaches of the Indian Ocean and used as a medicine although it fell into the general category of spices and had as high a value as pepper – also imported from the same region.[18] In the Red Sea and Mediterranean, gifts of special resins used as medicine, perfume, and incense – like the famous frankincense that originated in southern Arabia, or like the myrrh of Ethiopia – could signify great wealth and honour. In East Asia, too, the early *Xinxiu bencao* (*Newly Revised Bencao*; also known as the *Tang bencao*), compiled by Su Jing or Su Gong (fl. 656–660 CE), contains at least 30 drugs from distant sources.[19] An example is a medicine/spice/plant from Central Asia that came to be widely known and used from antiquity to the present: in English it is called asafetida (or stinking *azā*, a Persian word for mastic), while in South Asia it goes by *hing*, and in Chinese it is *awei* (probably adopted from a word in now-defunct Central Asian Tokharian B).[20]

Merchants who specialised in the handling of powerful medicines and spices from distant places could therefore be found in ports and bazars throughout the archipelagos and coasts of East and SE Asia, the Bay of Bengal and Arabian Sea, along the Swahili coast, and across North Africa and the Mediterranean into northern and western Europe, while caravan-saries and market towns supported the overland trade as well. In European languages the merchants of valuable spices and 'drugs' (the word indicated dried substances) tended to be called apothecaries, from the Greek term for dock or wharf. Of course, finding a way to sail directly to SE Asia, where many of the fine spices and medicines originated, became one of the chief goals for European rulers and merchants who hoped for riches from long-distance trade. Once they reached those places in the late fifteenth century, and the Americas, they also took great care to look for additional medicines and spices that could be turned into helpful and valuable commodities.

Just as Europeans adapted quickly to the complex long-distance economic networks of the Indian Ocean, they also drew on traditions of commensurable mensuration that had long enabled the conditions of market exchange in Afro-Eurasia. Much recent attention has been given to the organisational forms used by itinerant traders when they wished to do business at distances that required them to engage other agents to conduct business on their behalf. Those forms might be personal contracts or, from the ancient period, 'sea loans', with later developments of the funduq, caravanserai, commenda, qirad, ortoy, and other arrangements that supported honest dealing among merchants and their agents. In the fifteenth century and early sixteenth centuries, the rulers of China (in the form of tribute 'trade' that for a time reached into the Indian Ocean) and Spain and Portugal (in the form of royal *Casa de India* and *Casa de Contratación*) engaged in long-distance trade themselves; a century later, the Dutch, English, French, and other governments developed the form of the chartered business corporation.[21]

But as important as these organisational forms were, other procedures made exchange in the markets possible: agreed standards of weights, measures, and currencies. From ancient times onwards, in places where commercial exchange among strangers was common, increasingly robust customs developed about physical standards of evaluation and agreed rules of evidence for markets, shaping larger patterns of behaviour in merchant cities. Throughout the cities of the eastern hemisphere, from Osaka and Guangzhou to Cochin, Aden, Mombasa, Cairo, Venice, Valencia, and Antwerp, commerce was subject to the physical standards of length, weight, and currency overseen by the local governors, while personal and commercial disputes were settled by resort to law courts, where the 'matters of fact' were determined before a judgement was rendered.[22] While you might like a piece of cloth for its colour but argue with the merchant about its value before committing to a deal, you would agree to use the public standards of the market for verifying the length, width, and number of threads per inch in the bolt, or for verifying any weight according to local scales. As Emanuele Lugli has noticed, the moral authority of the 'promise of sameness' is demonstrated by examples of measures of length carved into the stone fronts of churches overlooking Italian markets.[23]

Similarly, the quality of sameness in the making of coins – each batch meant to produce identical instances – was associated with the Roman temple of Juno Moneta, as in our word 'money'. In fact, the vocabulary of assaying metals remains deep within our languages of scientific truth: 'assaying' is from the French 'to try', *essayer*; 'testing' is from the French for the metaphorical inverted skull, *tête*, in which the materials for assaying were ground or poured;[24] and in minting, sample struck coins were 'proofs'. Joel Kaye showed how currencies of markets provided 'a common measuring continuum for all commodities in exchange', forcing late medieval scholastic philosophers to consider how the 'production of order and equality out of willed inequality [due to each merchant trying to make exchange work to his own advantage] violated the essence of the traditional metaphysical and physical understanding of the *ordo rerum*'. He has also drawn attention to how around 1260–1360 in Europe the market balance's indisputably determined equality of weights – still taken as a symbol of blind justice – became a powerful metaphor for new kinds of ideas about the common good. In short, the socio-political disciplines of any local market allowed several physical traits of tangible goods to be moved to the realm of impersonal physical verification, which did not require two or more parties to have a common language or common qualitative responsiveness.[25] As lawyers might say, 'for the avoidance of doubt' one specifies: indeed, our words 'specie' and 'species', like 'spice', come from Late Latin to indicate a type of marketplace ware.[26] Careful impersonal description derived from common measures manifested the power of sameness.

In Europe, at least, the associations between methods of commerce and the valuation of medicinal qualities gained traction, although the attempt ultimately came to naught. The attempt can be seen in the attempts of learned physicians to disaggregate each of the four classical qualities into quantitative degrees indicating how strong the quality was expressed, usually using a numerical scale from one to four.[27] In principle, if the degrees of the qualities of each simple could be specified, then the qualitative effects of compound medicines could also be unravelled. The method was similar to 'alternate allegation', a mathematical method otherwise used for assessing and mixing metallic alloys.[28] But disputes arose about how to determine the quantities of the qualities. For example, opium had been collected in Central Asia and distributed far and wide since at least the time of the Pharaohs. A Portuguese-Jewish physician in sixteenth-century Goa, Garcia de Orta, not only reported that he had encountered 'Arabs [who] cure all dysenteric diseases with opium rectified with nutmeg', but also pointed out that taken in small quantities it was very helpful in all kinds of complaints and was therefore in great demand as a medicine, with properties such as making you drowsy or causing you to feel no troubles.[29] Whether the effects of opium were to be ascribed to heat or cold, and in what degree, were nevertheless entirely unclear. How could the elemental qualities explain such effects in a range of very distinct diseases and conditions of life? A similar debate about explaining the action of medicinal agents was occurring in China in the same period.[30] The famous botanist Carolus Clusius therefore simply dropped the qualitative aspects of describing medicinal simples when translating works on the subject for wider circulation.[31]

Materialistically commensurable description therefore proliferated with regards to objects of exchange, yet the bodily effects of illness and its remedies were felt qualitatively. Different methods of description and explanation could be used in different situations and for different explanatory purposes. But from the merchant's point of view, when more and more medical substances came to be widely traded throughout Afro-Eurasia – many of them entirely new, originating in the Americas – they circulated past many cultural borders, raising questions about how they were to be specified. Walking into a local pharmacy might therefore be an overwhelming experience for the sense of smell, as well as rich in variety of objects for examination, but the intricacies of accumulating and preparing the substances occurred under conditions of highly disciplined and publicly scrutinised conditions of training, working, and vending.[32] Put another way, the collective processes of assessing tangible goods in markets, including currencies – processes that could be found throughout Afro-Eurasia – produced accepted rules and formal laws that focused on materialistic description rather than qualitative experience, even when it came to understanding the effects of medicines. Commercial trust was rooted in methods of material verification that created the conditions for commensurability.

Figure 13.4 Apothecary's shop, Switzerland: reconstruction of a late seventeenth-century shop, showing pharmacy equipment and decorations. Photograph, ca. 1920. Courtesy of Wellcome Collection.

3. Biomedicine and Modern Pharmacy

The more recognisable foundations for biomedicine are usually identified in the new fields of nineteenth-century chemistry, experimental biology, and microbiology, but we find material commensurability there, too. From the point of view of any modern clinical speciality it would be difficult to imagine how one could treat age-old diseases like plague, tuberculosis, and septicemia without the 'sulpha' drugs and antibiotics developed from laboratory efforts of the 1930s and brought into wide-spread use due to war-time innovations in precision pharmaceutical manufacturing. Huge numbers of diseases are now successfully treated or prevented by the application of specialised chemical, biological, and molecular substances that have flowed from research laboratories and controlled pharmaceutical facilities around the world. In Europe and North America, Japan, Brazil, South Africa, India, China, and now even more widely, this 'translational' industry – crossing not only between basic research and clinical applications but also between state, university, non-profit, and high-intensity capital-investment organisations – has applied sameness for the benefit of the lives of countless people. Transplantation of organs, treating for cancer or diabetes or asthma, sustaining the lives of victims of burns or gun violence, relieving pain or depression, and on and on:

without the 'drugs', treatments for whole classes of acute and persistent illnesses would be impossible.

Rather than once again charting that well-known history, however, let us consider how commensurable materialism enabled those developments. We can do so by examining one long-term example, which draws on a number of excellent recent studies related to cinchona bark, its transformation into quinine, and other anti-malaria drugs and campaigns. It shows us that understandings of diseases and medicines developed alongside one another from continued technical advances in processes of materially assessing and stabilising aspects of the more than human world.[33]

Seventeenth-century Jesuit missionaries took an active interest in medical substances they encountered among the people with whom they lived, and established pharmacies around the world to support their activities.[34] Some of them in the Viceroyalty of Peru noticed that in villages north of Lima, a bark from a certain tree was being used in cases of severe intermittent fevers with good results. They and others in the viceroyalty thought it so successful that they began to send samples of it to Rome – another region where severe intermittent fevers were common – from where the reputation of the 'Jesuits bark' quickly spread.[35] By the 1650s, it had become well known among elite Europeans as an excellent infusion for relieving the symptoms of 'agues', now usually identified as malarial fevers, then a widespread endemic problem.

Physicians like the famous Thomas Sydenham in England, and his associates Robert Boyle and John Locke, came to accept the 'Jesuits bark' as a 'specific': that is, a remedy against a particular disease entity that would almost always relieve its symptoms if properly administered. Robert Tabor (or Talbor), from the Essex marshes, 'perfected' the use of the medicine to such an extent that in 1678 Charles II granted him a knighthood and a reward of 5,000 pounds, then sending him on to the royal court of France to treat members of it, too. Tabor unsuccessfully attempted to corner the French market in the bark, but as a reward for his services to members of the royal family and for making his secret method of preparation public, Louis XIV bestowed on him the title of chevalier and 48,000 livres plus a pension of 2,000. Other physicians of the French court were concerned enough to publish a work arguing that the bark required administration under the care of a physician's direction; given considerations of such qualitative indications as the age of the patient, whether they have 'a hot constitution', whether they are 'much accustomed to Wine', and so on, the dose would need to be altered and the bark compounded with other ingredients to be safe and effective. Jesuit missionaries even offered it to the emperor of China as a remedy for his ills after trying it on some of his courtiers to be certain that it would have no harmful effects, and he was much pleased with the result.[36] Moreover, the bark continued to develop the reputation of a specific: by 1692, another English physician, Robert Morton, was so confident about the consistent actions of it that he diagnosed cases of malaria according to whether they responded to treatment with the bark; if they did not, they must be other kinds of intermittent fevers.[37]

The specific had itself become a kind of touchstone or litmus test.[38] Many of the European medical chemists and empirics of the previous century had begun arguing that diseases were indeed things a bit like spirits rather than processes, sometimes using the word *archeii* to indicate the forcible presence that their own medical preparations were meant to counteract.[39] In this case, however, provided the bark was authentic and potent, it would on its own cure cases of a particular class of fevers in anyone, acting as a diagnostic method as well as a remedy.

Quality control of the material bark was therefore critical. And indeed, in trying to enforce a monopoly on the supply of the bark, officials in the Spanish empire put considerable effort into assessing the bark and its varieties for purposes of quality control. Bitter arguments arose among officials in Spain and New Spain about whose expertise was a better guide to assessing its quality, which were not settled before the outbreak of the Napoleonic wars that created possibilities for the independence movements that expelled Spain from most of the Western Hemisphere.[40]

Disputes about the quality of the bark settled not politically but otherwise, by developments that furthered the exacting chemical methods of Antoine and Marie-Anne Paulze Lavoisier.[41] In 1809 François Magendie carried out a series of ingenious experiments on animals – even animal bodies could now reveal how human bodies would respond – to study the toxic action of several botanical drugs, allowing him to compare similar effects produced by drugs of different origin.[42] His work allowed him to argue that a long-standing ambition of medical chemists was now possible: to isolate in pure states the chemicals that create the effects. Building on Magendie's approach, Pierre-Joseph Pelletier used mild solvents to push alkaloidal chemistry forward, in 1817 isolating the emetic substance in ipecacuanha root, which he called 'emetine', then going on to identify and name brucine, caffeine, colchicine, narceine, strychnine, veratrine and, from cinchona bark, quinine and cinchonine, all of which were introduced into wide circulation by Magendie in his *Formulaire* of 1821. Experimentation on animal bodies and a new form of precision in chemical procedures now held out the possibility of isolation of the active substances so as to allow the elimination of impurities and invariable dosages. In practice, however, the new assessments were used mainly as a materialistic check on the quality of methods of producing 'extracts' from the bark, allowing Europeans to engage in malaria prophylaxis at home and in the colonies by taking preparations of quinine, often in the form of bitter 'tonic' waters (as in gin and tonic).

Control of quinine rather than bark per se now became the chief focus. The bark had been gathered in the woods by *cascarilleros*, but worries developed about their destructive methods and the future sufficiency of wild production, while other empires wished to break into the Latin American monopoly. In the early 1850s, Dutch colonial officials ordered cinchona trees to be smuggled across the Pacific to the Dutch East Indies; carried in a warship, 75 plants survived the crossing in 1854, and by 1862, more than a million young trees were in the ground on Java. The first results did not yield a very potent bark, however, so in 1858 a chemist was sent out to work with planters on the project. In what Arjo Roersch van der Hoogte and Toine Pieters have called 'colonial agro-industrialism', different growing environments were tried as well as selective breeding of trees to try to increase the quinine content in their bark.[43] By mid-1860s, the British were also successfully growing cinchona trees in Bengal, guided by a chemist working at the centre of their own imperial economic botany, at Kew Gardens.[44] Aiming to 'own' the market, working under the direction of a collaborative 'state scientist', the Dutch redoubled their efforts in 1864 in order to achieve standardisation and 'purity' of the bark. Following the identification in 1872 of a variety of the tree with a very high quinine content, and laboratory-based methods of seed selection, robust enough 'mother trees' were used to establish a line of high-yielding plantations. Initial processing occurred locally with export to Germany in order to supply its rapidly developing pharmaceutical industry, which in turn employed the latest methods to check inputs and outputs. Given increased profits, the planters eagerly exploited local labour regimes to produce large quantities of good bark, making the Dutch East Indies the supply centre for the world's industrial quinine production.[45] Supplies of quinine were consequently available during Europe's 'scramble for Africa', which divided most of the continent among their empires during the last two decades of the nineteenth century.

With the outbreak of war in the Pacific, however, Japanese occupation of the Dutch East Indies created a crisis of supply for the powers allied with Britain. The government of the United States, for example, obtained seeds from the Philippines and began operating cinchona plantations in Costa Rica, but could not produce enough for wartime demand, making chemical substitutes necessary. An antiprotozoal developed by Bayer in the early 1930s, quinacrine, better known by the brand names Mepacrine (in England) and Atabrine (in the United States), became widely used in the North African and Pacific theatres. Screening in the United States and Australia of over 13,000 compounds, 100 of which were tested at clinical level, led to the further development of chloroquine, which was not ready for mass production before the end of the war but later widely used by US forces in the war in Vietnam.

Given the identification at the turn of the century of mosquitoes as the insect vector transmitting the malarial parasite in colonial North Africa, India, and Cuba, other strategies were used to reduce the number of insects.[46] In 1942 an American military attaché in Berne forwarded a German formula for a new compound called DDT, which was produced in large quantity and from 1944 was used to spray invasion beaches during preparatory shelling in order to protect landing troops from disease. Anti-malarial campaigns were launched in many places after the war, too, to encourage material national development, often spraying massive amounts of DDT in attempts to suppress mosquitoes. The so-called vertical campaigns to eliminate diseases – organised with a narrow focus on eliminating the micro-organisms that are identified as responsible for the harmful bodily effects – often proved ineffective because of failing to take into consideration multiple environmental and socio-cultural factors.[47] In due course, the devastating consequences of widespread use of DDT caused it to be abandoned, although other kinds of chemical and biological attacks on mosquitos continue to be active in many places.

Given that malaria continues to be a dreaded endemic presence in many places, there remains a keen interest in 'front-line' drugs that can help those threatened by it. A recent well-regarded example of such a drug is Artemisinin, extracted from *Artemisia annua*, made famous by the 2015 award of a Nobel Prize in 'Physiology or Medicine' to Youyou Tu 'for her discoveries concerning a novel therapy against Malaria'. She was part of a Chinese army programme that screened the older *bencao* texts for potential agents useful against fevers in order to help their North Vietnamese allies in the war against the Americans. After the political reforms following the death of Mao Zedong, the extract came to wider attention, with further innovations in the production of the active molecules introduced by the pharmaceutical firm Novartis, which made it a 'drug of choice' since 2006.

Bioprospecting through historical pharmaceutical texts is used in other places too. But bioprospecting is mainly carried out by asking questions of people in regions that have rich botanical environments, or simply attempting to identify medically useful chemicals in anything that seems likely to yield results, continue to be the most common practices now, as in earlier periods.[48] Appropriating the active chemicals in 'traditional' medicines raises pressing questions of 'biopiracy', as in the case of neem, a tree that had been used for medical purposes for millennia by Unani, Siddha, and Ayurvedic practitioners but whose useful molecules were patented in 1995 by a US company, W. R. Grace. In successfully attacking the patent with the help of the Indian government Vandana Shiva helped to make the term 'biopiracy' common. But many governments also hope to turn local medical resources into bio-prosperity, too, investing in pharmaceutical research.[49] The disciplined activities that lie behind such hopes, fears, and border-crossing marketing are clearly centred on processes of describing material substances exactly, which makes them commensurable.

To sum up very briefly: if we look forward from the first age of global commerce, we see the emphasis on commensurable materialism only growing. Commerce was provoking similar changes in East Asia (and elsewhere) as well as in Europe.[50] By the seventeenth century, all kinds of new chemical preparations were appearing from the furnaces and glass-ware and porcelain vessels of workshops, with 'patent medicines' identi-fied by a maker's mark circulating through towns, and along post roads, and aboard ships.[51] Admirals and generals concerned about manpower were very supportive of singular and effective medicines.[52] Needless to say, the search for potent material substances led to a proliferation of discoveries of new 'elements' and theories about them and their combi-nations, leading to the 'chemical revolution' of Antoine Lavoisier's day. By the eighteenth century, pharmacists in some cities could buy many authenticated chemicals and compounds in common use from their own gilds, who together with some private licensed suppliers also provided medicines to the hospitals, commercial fleets, trading companies, and armed forces in bulk. Other bulk purchasers included planters with tens or even hundreds of enslaved or indentured workers under other forms of work discipline, who like servicemen would be subject to taking whatever the hired surgeon or physician prescribed.[53]

Improvements to transportation, particularly steam railroads, and the new forms of mass warfare stimulated the development of larger manu-facturing firms producing mass markets. During the US civil war, for instance, the government purchased medicines in bulk from Frederick Stearns and Co. (in Detroit, MI) and E. R. Squibb and Sons (Brooklyn, NY), both of whom after the war advertised the purity, reliability, and uniformity of their output; their wares were marketed to the public in popular sized packages, with printed directions for use and a list of names and quantities of the ingredients to reassure their customers about any questions of quality control. In the 1880s, Louis Pasteur's innovations in producing 'artificial' vaccines by inducing reactions in animals led to a new class of drugs, the biologicals, which increasingly drew the notice of national governments and led to the establishment of administrative bodies to provide further oversight of the foods and drugs manufacturing industries.

In other words, as commercial medicines grew dominant throughout the world, many firms developed methods to maintain the goal of being trusted for the material consistency and reliability of their medicinal prod-ucts, often with the oversight of government regulators who helped to reduce error and fraud.[54] Commensurable materialism supported the shift from hand-made medicines to mass-produced pharmaceutical goods.

4.   Biomedicinal Progress?: Medicine as Medicines

To grapple with biomedicine is to grapple with only a subset of the medi-cal world. The techno-scientific commonalities of it will be enacted in somewhat different ways in different places, and there are great worlds of human experience and healing intention that lie outside the carefully disciplined boundaries of commensurable materialism. The persons who might be called on for help will often be many and various, and go by a variety of names and titles. On the other hand, in conformity with modern expectations, the field of activities associated with biomedicine draws a line between its objective (object-oriented) knowledge and the rest, placing all bets on the former. It draws on other lineages of expertise that have to do with the powers of other natural bodies to effect human bodies, but judges them according to the standards of material substance rather than qualities or powers. For example, in arguments about the efficacy of acupuncture, the term *qi* has no commensurably materialistic meaning and so cannot be used to persuade biomedical investigators that it produces whatever phenomena are at issue; but a plant now named *Artemisia annua* that was once described as useful in certain conditions in the *bencao* literature is acceptable because a molecule found in it can be shown in commensurably materialistic ways to be effective against malarial parasites. To patients seeking answers to questions about their own circumstances of life a biomedically educated doctor speaking only of physical and biological processes, risks and probabilities, or comparisons with demographic or other enumerated reference groups, can seem an inadequate adviser, incapable of answering the 'why me' question.[55] Biomedicine, however, was built not for the grand qualitative questions of human experience but for impersonal descriptions of substances, aiming to set the common measure.

A world of very thoughtful critique has nevertheless drawn attention to the ambition implicit in biomedicine's universal commensurability. Many point to processes of commodification that teach us to understand our worlds as composed not of relationships but of things, a process sometimes termed reification or fetishisation.[56] A subset of such critiques invokes 'medicalisation' to address the processes by which frictions among humans and between humans and their larger worlds are turned into bodily problems, which can (according to its advocates) be resolved by interventions mounted by powerful agents of medicine or public health.[57] Still others – not only secular intellectuals but also religious scholars – worry about how scientific reductionism narrows any vision of the world. More straightforward egalitarian criticisms point to the hierarchies of social control that are evident in biomedical institutions, and the ways in which forms of biomedicine are implicated in systems of domination, from slavery to colonialism to imperialism.[58] Medical expertise has shaped warfare in many ways, too, most obviously in planning for and against chemical, biological, and nuclear weapons, less obviously by ameliorating its effects and making it thinkable. Many instances of biomedical research have also approached humans as material objects of study with no regard to the consequences, as in experiments conducted by early twentieth-century biomedical pioneers on Africans,[59] public health doctors and nurses on men enrolled in the Tuskegee syphilis study, Nazi party loyalists on concentration camp inmates, Japanese military doctors on prisoners of war, or US public health officials on Guatemalan inmates. In all of those examples, the cruelty of the enactors was normalised by racial or ethnic hatreds. The pseudo-sciences related to racialised categories of human types may not be biomedical per se, but they have clearly contributed to systems of coercion and suffering, despite the intention that the 'experimental results' produced by such studies would benefit the 'race' of the people running the studies because human bodies respond similarly.

Scandals and inequities continue to emerge. In 2012, for example, GlaxoSmithKline was fined $3 billion under the US False Claims Act for its campaign to promote the use of 'Paxil' while failing to disclose risks. More recently, Purdue Pharma was fined $8.3 billion for overprescribing its painkiller, Oxycontin, which has fuelled a fresh opioid addiction crisis in the United States and elsewhere, although as of the time of this writing further legal contests about a general settlement are on-going. In the United States – unlike the EU – 'food supplements' go almost entirely unregulated despite the harmful ingredients frequently found in them. These and many more observations, critiques, and criticisms of alienation and power in worlds of biomedicine or imitating them are more than enough for many to challenge the view that its activities make the world a better place.

To compound the question of benefit is the manner in which capital investment in hospitals and other institutions and organisations, along with payments for personnel, insurance, services, devices, and (of course) drugs, creates and reinforces all kinds of inequities among persons, most importantly racialised and gendered forms but also income, class, education, and so forth. It would clearly be an act of wilfulness to block out the economic interests that together with public and philanthropical ambitions support the contemporary biomedical establishment. For instance, the business section of a recent issue of the *Boston Globe* reports that in the wake of the success of investments in recent corona-virus vaccines, a local venture capital giant will be investing $3.4 billion 'on a generation of biotech companies focused on therapeutics, agriculture, and nutrition'.[60] A few days later the same section reported the announcement of the British Prime Minister about building the United Kingdom into a 'science superpower' in the wake of the success in producing a vaccine, 'harnessing 14.9 billion pounds a year for spending on research and development'.[61] Hopes for Britain's superpower status in biomedicine will certainly include continuing investments of about £1 billion per annum from one of the world's largest charities, the UK-based Wellcome Trust, with an estimated endowment of just over 29.1 billion pounds in 2020, an endowment originating in a Foundation set up as an offshoot of the

Figure 13.5 Chemist dispensing tablets, Sue Snell. Courtesy of Wellcome Collection.

Burroughs Wellcome pharmaceutical company, although the Trust sold its last holdings in that sector to a former rival, Glaxo, in 1995.[62]

Nevertheless, as we have seen, there remains a robust and understandable conviction among the people who produce the knowledge and substances of the biomedical sector, and among the patients and other masses of people who benefit from its successes, that it is a progressive field. Moreover, there is a long history of medical practitioners who have served other than privileged patients coming to understand the material conditions of life for the wider population, sometimes causing them to become reformers or even radicals. For most, the negative features mentioned earlier are commonly treated as side effects, or exceptions, or as problems that can be rectified with better regulation or ethical instruction – 'bugs' rather than features. The medical and scientific actors in it also take much personal satisfaction from working towards the advancement of their subject knowledge and sharing it with others, often achieving reward and prestige for doing so, as do the institutions – including universities – in which they work. (It is hard to imagine the current higher education sector without the biomedical aspects of it.) The competition for position and status can be fierce but it propels fields forward through considerable personal investments of time, attention, ingenuity, patience, and determination on the part of countless participants: researchers and practitioners are constantly developing new methods, making new findings, expressing new concepts, and developing new projects. While to an outsider some of the conceptual innovations may seem like tiny wrinkles in a larger fabric, when one is positioned next to a wrinkle, it can seem enormous. But in this limited sense, at least, an actor's category of progress in medicine is clear and active.

Commensurable materialism enables those kinds of creative innovations to move through the world. As the seventeenth-century writer on *bencao*, Zhang Lu, put it, a standard work like Shennong's 'is just like artisans having their own measuring tools. Once there is measure, then there are rules; with rules then comes the room for flexibility and innovation. Flexibility and innovation derive from intelligence and dexterity, yet are still grounded in the basic measures'.[63] In other words, commensurable measure establishes the condition for impersonal evaluation of material phenomena, encouraging certain kinds of innovation and exchange.

The history of how groups of humans shifted from qualitative evaluations of the powers of substances to commensurably materialistic ones, and how their standards grew into a global norm, point to the intertwining of biomedicine and larger systems of collective power that foster and support ever increasing population numbers. Whether those systems can sustain themselves in the long run is certainly an open question. But as the current pandemic reminds, it is difficult to imagine any hope for sustainability of materially aspirational societies without the inclusion of biomedical expertise. It is a very powerful tool, requiring care and discipline to be used for proper ends, which in turn require other kinds of collective decision-making. But in the meantime, it is not surprising that when people think of medicine they think of progress in the efficacy of medicines, too.

Notes

1. For a caution, see Arthur Imhof, 'The Implications of Increased Life Expectancy for Family and Social Life', in *Medicine in Society: Historical Essays*, ed. Andrew Wear (Cambridge: Cambridge University Press, 1992), 347–376.
2. My conviction about this point is based on much academic and administrative experience in universities, including service on various kinds of funding committees, promotion committees, and external assessment committees.
3. Yafeng Shan, 'A New Functional Approach to Scientific Progress', *Philosophy of Science* 86, (2019), for example, 747.
4. Thomas S. Kuhn, *The Structure of Scientific Revolutions*, 2nd ed. (Chicago: University of Chicago Press, 1970), xlii.
5. For instance, Roger Hart, 'Translating the Untranslatable: From Copula to Incommensurable Worlds', in *Tokens of Exchange*, ed. Lydia H. Liu (Durham: Duke University Press, 1999), 45–73; Ian Hacking, *Historical Ontology* (Cambridge, MA: Harvard University Press, 2002), 168–172; G. E. R. Lloyd, *Being, Humanity, and Understanding: Studies in Ancient and Modern Societies* (Oxford: Oxford University Press, 2012), Chapter 5.

6. Like so many I owe special acknowledgement to Steven Shapin and Simon Schaffer, *Leviathan and the Air Pump: Hobbes, Boyle, and the Experimental Life* (Princeton: Princeton University Press, 1986) and Bruno Latour and Steve Woolgar, *Laboratory Life: The Construction of Scientific Facts*, 2nd ed. (Princeton: Princeton University Press, 1986); also Barbara J. Shapiro, *Probability and Certainty in Seventeenth-Century England: A Study of the Relationship Between Natural Science, Religion, History, Law, and Literature* (Princeton: Princeton University Press, 1983) and Chandra Mukerji, *From Graven Images: Patterns of Modern Materialism* (New York: Columbia University Press, 1983).

7. Jacobus Bontius, *Tropische Geneeskunde/on Tropical Medicine*, Vol. 10 of Opuscula Selecta Neerlandicorum De Arte Medica (Amstelodami: Sumptibus Societatis, 1931), 396–397.

8. For more on Bontius and similar examples, Harold J. Cook, *Matters of Exchange: Commerce, Medicine and Science in the Dutch Golden Age* (New Haven: Yale University Press, 2007).

9. Alisha Rankin, *The Poison Trials: Wonder Drugs, Experiment, and the Battle for Authority in Renaissance Science* (Chicago: The University of Chicago Press, 2020).

10. The erasure of much of the formal expressions of medical explanation in the pre-contact Americas presents special difficulties, pushing this short essay to draw on Afro-Eurasian examples.

11. For 'medicamentum', Charlton Thomas Lewis and Charles Short, *A Latin Dictionary* (Oxford: Oxford University Press, 1975); for 'pharmakon', Hermann Herlinghaus, ed. *The Pharmakon: Concept Figure, Image of Transgression, Poetic Practice* (Heidelberg: Universitatsverlag Winter, 2018), and Ronny Spaans, *Dangerous Drugs: The Self-Presentation of the Merchant-Poet Jan Six Van Chandelier (1620–1695)*, transl. Ciarán Ó. Faoláin (Amsterdam: Amsterdam University Press, 2020).

12. Yan Liu, *Healing with Poisons: Potent Medicines in Medieval China* (Seattle: University of Washington Press, 2021).

13. But later commentators sometimes objected that a person as grand as Shennong would not have had to resort to taste, instead intuiting the properties of the drugs: He Bian, *Know Your Remedies: Pharmacy and Culture in Early Modern China* (Princeton: Princeton University Press, 2020), 81.

14. There is a fascinating philosophical literature on taste and truth, which arises more from Continental than Anglo-American outlooks, perhaps for linguistic reasons. For example, see Hans Gadamer, *Truth and Method* (New York: Crossroad, 1989), 35–42; Michel Serres, *The Five Senses: A Philosophy of Mingled Bodies (I)* (London: Continuum International, 2008), 152–197.

15. OED, 'proof', P 1462–1464.

16. Carla Nappi, *The Monkey and the Inkpot: Natural History and Its Transformations in Early Modern China* (Cambridge, MA: Harvard University Press, 2009).

17. Paul U. Unschuld, *What is Medicine? Western and Eastern Approaches to Healing*, transl. Karen Reimers (Berkeley: University of California Press, 2009), 110; C. Pierce Salguero, *Translating Buddhist Medicine in Medieval China* (Philadelphia: University of Pennsylvania Press, 2014); Ronit Yoeli-Tlalim, *Reorienting Histories of Medicine: Encounters Along the Silk Roads* (London: Bloomsbury Academic, 2021).

18. David Bulbeck, Anthony Reid, Lay Cheng Tan, and Yiqi Wu, eds. *Southeast Asian Exports Since the 14th Century: Cloves, Pepper, Coffee, and Sugar* (Leiden: Koninklijk Instituut voor Taal-, Land- en Volkenkunde Press, 1998), 60.

19. See Unschuld; also C. Pierce Salguero, *A Global History of Buddhism and Medicine* (New York: Columbia University Press, 2021).
20. Angela Ki Che Leung and Ming Chen, 'The Itinerary of Hing/*awei*/asafetida Across Eurasia, 400–1800', in *Entangled Itineraries: Materials, Practices, and Knowledges Across Eurasia*, ed. Pamela Smith (Pittsburgh: University of Pittsburgh Press, 2019), 141–164, 303.
21. For instance, Avner Greif, *Institutions and the Path to the Modern Economy: Lessons From Medieval Trade* (Cambridge and New York: Cambridge University Press, 2006); Ron Harris, *Going the Distance: Eurasian Trade and the Rise of the Business Corporation, 1400–1700* (Princeton: Princeton University Press, 2019).
22. Barbara J. Shapiro, *A Culture of Fact: England, 1550–1720* (Ithaca: Cornell University Press, 2000); William N. Goetzmann, *Money Changes Everything: How Finance Made Civilization Possible* (Princeton: Princeton University Press, 2016); Mitao Miyamoto and Yoshiaki Shikano, 'The Emergence of the Tokugawa Monetary System in East Asian International Perspective', in *Global Connections and Monetary History, 1470–1800*, ed. Dennis O. Flynn, Arturo Giráldez, and Richard von Glahn (Aldershot, Hants: Ashgate, 2003), 169–186.
23. Emanuele Lugli, *The Making of Measure and the Promise of Sameness* (Chicago: The University of Chicago Press, 2019).
24. *OED*, 'test', in T, pp. 219–221.
25. Joel Kaye, *Economy and Nature in the Fourteenth Century: Money, Market Exchange, and the Emergence of Scientific Thought* (Cambridge: Cambridge University Press, 1998), 11, 13; Joel Kaye, *A History of Balance 1250–1375: The Emergence of a New Model of Equilibrium and Its Impact on Thought* (Cambridge: Cambridge University Press, 2014).
26. Frank Perlin, 'The Other 'Species' World: Speciation of Commodities and Moneys, and the Knowledge-Base of Commerce, 1500–1900', in *Merchants, Companies and Trade: Europe and Asia in the Early Modern Era*, ed. Sushil Chaudury and Michel Morineau (Cambridge: Cambridge University Press, 1999), 145–172.
27. Michael McVaugh, 'Quantified Medical Theory and Practice At Fourteenth-Century Montpellier', *Bulletin of the History of Medicine* (1969): 397–413.
28. J. Williams, 'Mathematics and the Alloying of Coinage 1202–1700: Pts I and II', *Annals of Science* 52 (1995): 213–234, 235–263.
29. Garcia da Orta, *Colloquies on the Simples and Drugs of India*, transl. Clements Markham (London: Henry Sotheran and Co., 1913), 41st colloquy; Palmira Fontes da Costa, ed., *Medicine, Trade and Empire: Garcia De Orta's Colloquies on the Simples and Drugs of India (1563) in Context* (Farnham, Surrey: Ashgate, 2015).
30. Bian, *Know your Remedies*, pp. 87–92; more generally, Nicholas K. Menzies, *Ordering the Myriad Things: From Traditional Knowledge to Scientific Botany in China* (Seattle: University of Washington Press, 2021).
31. On Clusius dropping the qualities, Harold J. Cook, 'Trading in Medical Simples and Developing the New Science: De Orta and His Contemporaries', in *Medicine, Trade and Empire*, ed. Fontes Da Costa, 139–143.
32. See esp. the study of James E. Shaw and Evelyn Welch, *Making and Marketing Medicine in Renaissance Florence* (Amsterdam: Rodopi, 2011).
33. See James L. A. Webb, *Humanity's Burden: A Global History of Malaria* (Cambridge: Cambridge University Press, 2009).
34. Sabine Anagnostou, 'Jesuits in Spanish America: Contributions to the Exploration of the American Materia Medica'. *Pharmacy in History* 47 (2005): 3–17.

35. Saul Jarcho, *Quinine's Predecessor: Francesco Torti and the Early History of Cinchona* (Baltimore: The Johns Hopkins University Press, 1993); Fernando I. Ortiz Crespo, 'Fragoso, Monardes, and Pre-Chinchonian Knowledge of Cinchona', *Archives of Natural History* 22 (1995): 169–181; Matthew James Crawford, *The Andean Wonder Drug: Cinchona Bark and Imperial Science in the Spanish Atlantic, 1630–1800* (Pittsburgh: University of Pittsburgh Press, 2016); Samir Boumediene, *La Colonisation du Savoir: Une Histoire des Plantes Médicinales du 'Nouveau Monde' (1492–1750)* (Vaulx-en-Velin: Les Éditions des mondes à faire, 2016); Benjamin Breen, *The Age of Intoxication: Origins of the Global Drug Trade* (Philadelphia: University of Pennsylvania Press, 2019).
36. Beatriz Puente-Ballesteros, 'Jesuit Medicine in the Kangxi Court (1662–1722): Imperial Networks and Patronage', *East Asian Science, Technology, and Society: An International Journal* 34 (2011), 58–62.
37. Mary J. Dobson, *Contours of Death and Disease in Early Modern England* (Cambridge: Cambridge University Press, 1997), 316, referring to Morton's *Pyretologia*.
38. Harold J. Cook, 'Markets and Cultures: Medical Specifics and the Reconfiguration of the Body in Early Modern Europe', *Transactions of the Royal Historical Society* 21 (2011): 123–145.
39. For example, William Eamon, *The Professor of Secrets: Mystery Medicine and Alchemy in Renaissance Italy* (Washington, DC: National Geographic, 2010).
40. Crawford, *The Andean Wonder Drug*.
41. For recent approaches to the practice of chemistry in the period, see Lissa Roberts and Simon Werrett, eds. *Compound Histories: Materials, Governance, and Production, 1760–1840* (Leiden and Boston: Brill, 2018).
42. John E. Lesch, *Science and Medicine in France: The Emergence of Experimental Physiology, 1790–1855* (Cambridge, MA: Harvard University Pres, 1984).
43. Arjo Roersch van der Hoogte and Toine Pieters, 'Quinine, Malaria, and the Cinchona Bureau: Marketing Practices and Knowledge Circulation in a Dutch Transoceanic Cinchona-Quinine Enterprise', *History of Medicine and Allied Sciences* 71 (2016): 197–225; Wouter Klein and Toine Pieters, 'The Hidden History of a Famous Drug: Tracing the Medical and Public Acculturation of Peruvian Bark in Early Modern Western Europe (C. 1650–1720)', *Journal of the History of Medicine* 71 (2016): 400–421.
44. Rohan Deb Roy, *Malarial Subjects: Empire, Medicine and Nonhumans in British India, 1820–1909* (Cambridge: Cambridge University Press, 2017).
45. Roersch van der Hoogte and Pieters, 'Quinine, Malaria, and the Cinchona Bureau'.
46. Rohan Deb Roy, 'Quinine, Mosquitoes and Empire: Reassembling Malaria in British India, 1890–1910', *South Asian History and Culture* 4 (2013): 65–86.
47. Marcos Cueto and Steven Paul Palmer, *Medicine and Public Health in Latin America: A History* (New York: Cambridge University Press, 2015); Randall M. Packard, *A History of Global Health: Interventions into the Lives of Other Peoples* (Baltimore: Johns Hopkins University Press, 2016).
48. For instance, Londa L. Schiebinger, *Secret Cures of Slaves: People, Plants, and Medicine in the Eighteenth-Century Atlantic World* (Stanford: Stanford University Press, 2017); Jordan Goodman and Vivien Walsh, *The Story of Taxol: Nature and Politics in the Pursuit of an Anti-Cancer Drug* (Cambridge: Cambridge University Press, 2001).
49. Abena Dove Osseo-Asare, *Bitter Roots: The Search for Healing Plants in Africa* (Chicago: University of Chicago Press, 2014).

50. Bian, *Know Your Remedies*; Federico Marcon, *The Knowledge of Nature and the Nature of Knowledge in Early Modern Japan* (Chicago: University of Chicago Press, 2015); Dagmar Schäfer, *The Crafting of the 10,000 Things: Knowledge and Technology in Seventeenth-Century China* (Chicago: University of Chicago Press, 2011); Craig Clunas, *Superfluous Things: Material Culture and Social Status in Early Modern China* (Honolulu: University of Hawai'i Press, 2004).

51. The literature on iatrochemistry, patent medicines, and the history of chemistry is too large to review here, but see for example, Bruce T. Moran, *Distilling Knowledge: Alchemy, Chemistry, and the Scientific Revolution* (Cambridge, MA: Harvard University Press, 2005).

52. Harold J. Cook, 'Practical Medicine and the British Armed Forces After the 'Glorious Revolution', *Medical History* 34 (1990): 1–26.

53. Zack Dorner, *Merchants of Medicines: The Commerce and Coercion of Health in Britain's Long Eighteenth Century* (Chicago: University of Chicago Press, 2020).

54. Nancy Tomes, *Remaking the American Patient: How Madison Avenue and Modern Medicine Turned Patients Into Consumers* (Chapel Hill: University of North Carolina Press, 2016); Sherman Cochran, *Chinese Medicine Men: Consumer Culture in China and Southeast Asia* (Cambridge, MA: Harvard University Press, 2006).

55. Michael T. Taussig, 'Reification and the Consciousness of the Patient', *Social Science & Medicine* 14B (1980): 3–13.

56. For example, Michael Taussig, *The Devil and Commodity Fetishism in South America* (Chapel Hill, NC: The University of North Carolina Press, 1980); also see William Pietz, 'The Problem of the Fetish [Parts I, II, and IIIa]', *RES: Anthropology and Aesthetics* 9, 13, 16 (1985).

57. Michel Foucault, *The Birth of the Clinic: An Archaeology of Medical Perception*, trans. A. M. Sheridan Smith (New York: Vintage Books, 1973); Ivan Illich, *Medical Nemesis: The Expropriation of Health* (New York: Pantheon, 1976).

58. For examples of a huge literature see Warwick Anderson, 'Making Global Health History: The Postcolonial Worldliness of Biomedicine', *Social History of Medicine* (2014): 1–13; idem, *Colonial Pathologies: American Tropical Medicine, Race, and Hygiene in the Philippines* (Durham and London: Duke University Press, 2006); Pratik Chakrabarti, *Medicine and Empire, 1600–1960* (New York: Palgrave Macmillan, 2014); Schiebinger, *Secret Cures of Slaves*.

59. For example, Deborah Joy Neill, *Networks in Tropical Medicine: Internationalism, Colonialism, and the Rise of a Medical Specialty, 1890–1930* (Stanford: Stanford University Press, 2012); more generally, Helen Tilley, *Africa as a Living Laboratory: Empire, Development, and the Problem of Scientific Knowledge, 1870–1950* (Chicago: University of Chicago Press, 2011).

60. *Boston Globe*, 15 June 2021, D1.

61. *Boston Globe*, 21 June 2021, D3.

62. https://en.wikipedia.org/wiki/Wellcome_Trust, accessed 22 June 2021.

63. Quoted in Bian, *Know Your Remedies*, p. 112.

Part III
Related Issues

Part III
Related Issues

14 Scientific Progress and Scientific Realism

David Harker

Scientific realism has been defined, interpreted, and opposed in a variety of ways, but it is most centrally concerned with attitudes towards the unobservable entities that are posited by our best scientific theories, entities like photons, genes, proteins, and electrons. More specifically, scientific realism is commonly regarded as involving three key commitments, one metaphysical, one epistemological, and one semantic (Kukla 1998; Psillos 1999; Chakravartty 2017). First, scientific realists suppose that unobservable entities exist independently of our experiences of the world. Second, scientific realists suppose that we can acquire *knowledge* of such entities, thereby eschewing various forms of scepticism. Third, scientific realists suppose that the descriptions provided by our best scientific theories should be interpreted literally, even when those descriptions refer to entities that we can't observe directly. If scientists describe protons as having certain properties, for example, then we should regard these statements as being either true or false. Realists thereby distinguish themselves from various forms of instrumentalism, which suppose that statements about unobservable entities are better conceived as convenient shorthand for more complex statements that refer only to what's more directly observable.

Scientific realism doesn't presume that today's scientific theories are immune to revision, with respect to what kinds of things exist, or their properties and relations. Nowadays, at least, scientific realists are all fallibilists. Realists also recognise that scientific models and theories frequently involve idealisations and abstractions. A consequence of the fallibilism, idealisations, and abstractions is that realists describe scientific theories as *approximately* true, rather than strictly true. One arguably unfortunate connotation of the word 'approximately', in this context, is the implication that scientific realists suppose that science is *close to* the truth, or at least close to the truth in certain respects. If someone is described as being approximately six feet tall, we suppose that they're close to six feet tall. If we describe a given scientific law or theory as approximately true, we might similarly be understood as implying that the law or theory is close to the truth.

DOI: 10.4324/9781003165859-18

If the implication is unfortunate, it's because the epistemic commitment of scientific realism needn't be interpreted in such optimistic terms. If the sciences have achieved *some* knowledge of unobservables, then at least some interpretations of the realist position will be satisfied, even if those interpretations allow that we still have a great deal to learn, about fundamental particles, genomes, dark matter, and so on, and hence that there is something misleading about describing current theories as somehow 'close to' the truth. We can thus distinguish pessimistic scientific realists from optimistic ones. An optimistic realist supposes that the world is pretty much as our current theories describe. A pessimistic realist supposes that our theories may well undergo significant future change, but that science has nevertheless achieved some knowledge of unobservables. Pessimistic realism is conceptually and importantly distinct from scientific antirealist theses. Of particular significance to questions surrounding realism and scientific progress is that the pessimistic realist can allow that we might have more knowledge of unobservables now than we did in the past (or that our theories are now more truthlike). If scientific antirealism is understood as *denying* that we can justifiably claim knowledge of unobservables, then pessimistic realism is inconsistent with antirealism.

Scientific realism has also been defined in terms of scientific *aims*; on this approach, it is the thesis that science aims to provide true theories of the world (van Fraassen 1980; Lyons 2005). Questions about scientific progress and questions about the aims of science can appear very close (Dellsén 2016). However, it's unclear whether aiming for truth should be considered either sufficient or necessary for realism. First, the suggestion that science aims for truth entails neither that science attains or approaches truth (Kitcher 1993). Defining realism purely in terms of the aims of scientific inquiry can thus appear insufficient: most realists seek to defend the thesis that science doesn't merely aim at, but also achieves, to some extent, knowledge of unobservables (Chakravartty 2017). The suggestion that science aims for truth may not be a necessary condition of scientific realism either. It might be the case, for example, that much of scientific practice is best explained if we suppose that science aims to solve problems, or increase our understanding (Shan 2019; Dellsén 2016). But it might still be argued that in the pursuit of these goals, science does converge on the truth, or achieves knowledge of unobservables, as a by-product of its actual aims. Depending upon how one understands the concept of aiming, it's possible that scientific realism could be defended even without committing to the proposition that science aims for truth.

The most influential argument for scientific realism is commonly known as the no-miracles argument.[1] Those sympathetic to the argument draw attention to the remarkable, empirical successes of modern science, where empirical success is typically understood in terms of predictive scope and accuracy, is sometimes extended to include explanatory achievements, and often highlights verified novel predictions.[2] Advocates for the argument then ask how we're to explain the fact that science is empirically successful. If our best scientific theories are approximately true, so the argument goes, then their successes are unsurprising and hence, in some sense, explicable. If our scientific theories were radically false, however, then the successes of these theories would be highly surprising and perhaps even miraculous. It is a reasonable epistemic principle that we shouldn't invoke miracles to explain our observations when alternative explanations are available that don't invoke miracles. Given all this, we are urged to conclude that our best scientific theories must be approximately true, as the realist contends, since this provides the only reasonable explanation we have for scientific success. The argument has given rise to significant discussion surrounding the nature of the explanation offered by scientific realism, the significance of any apparent explanatory triumph, and the possibility of explaining scientific success in ways that don't presuppose, or appeal to, realist commitments.[3]

Two particularly influential arguments have been advanced in opposition to scientific realism.[4] One draws attention to the failure of available experimental and observational evidence to reliably gesture towards a unique, best, scientific theory. For any successful scientific theory, it is argued, there will always be alternatives that can account for all the same evidence, but in ways that appeal to a radically different set of theoretical assumptions. (The suggestion is not that such alternatives have already been proposed, articulated, or defended, necessarily, but that the existence of such alternatives is in some sense guaranteed, or at least highly probable.) If there exist radically different alternatives to our best scientific theories, which are consistent with available evidence, then it might seem to follow that available evidence can't justify a judgement that one theory is better confirmed than the alternatives, and hence it can appear that we're not justified in regarding our preferred theory as approximately true. Since the core insight is that available evidence doesn't *determine* a unique, best scientific theory, relative to that evidence, the challenge to realism is known as the underdetermination of theories by evidence or data.[5]

A second, widely discussed challenge to scientific realism appeals to historical considerations, drawing attention to scientific theories that were successful but which seemingly can't, by current lights, be described as even approximately true. The eighteenth and nineteenth century theories of phlogiston, aether, and caloric are the most frequently offered examples of scientific theories that enjoyed significant empirical success, but which appear so far removed from our modern understanding that we can't sensibly describe both the replaced and replacement theories as being close to the truth. Historically based arguments strike even more directly against the scientific realist's contention that scientific success implies approximate truth. If history contains enough examples of successful theories that can't be described as approximately true, then it no longer appears sensible to regard scientific success as a reliable indicator of the approximate truth of our successful theories.

The historical challenge has prompted several lines of response, but the most influential attempt to narrow the realist's commitments in some way. Some structural realists, for example, argue that we should not be realists about claims concerning the *nature* of unobservable entities, but only claims about their structure.[6] Entity realism is the view that we have good reason to accept that certain entities exist, if we acquire the ability to manipulate those entities in ways that interact with other phenomena and have observable effects, but we needn't suppose in addition that scientific theories are approximately true. Philip Kitcher (1993) argues that some constituents of a scientific theory receive evidentiary support from the theory's successes, while others don't, and that this inequality can help the realist explain why radically false theories are sometimes empirically successful. In each of these cases, it is hoped that by narrowing the realist's commitments, realists will be able to explain the success of science in a way that is consistent with instances of radical theoretical change that we observe within the history of science.[7]

Kitcher's general strategy will be important for arguments developed later in the chapter so is worth reviewing further. To illustrate the basic insight, consider any successful scientific theory and any outlandish proposition (that many species of dinosaur made highly ornate garments from biodegradable materials, for example). We can create a new 'scientific theory' by conjoining the original, successful theory and the outlandish proposition. *Prima facie*, our new theory can claim as support all those successes that our initial theory can adduce. Since our new theory includes the outlandish proposition, however, it might appear as if we have provided that proposition with empirical support, simply by conjoining it to a successful, scientific theory. Of course, no one would accept this conclusion, but how we should resist isn't entirely clear. One line of response begins with the idea that not all parts of a theory can justifiably claim equal levels of support from available evidence.

In much the same way, if we discover that replaced scientific theories were successful, but included assumptions that were redundant with respect to the success, it seems we should conclude that the redundant assumptions can gain no credit for the success. Even if success is indicative of approximate truth, as the realist contends, idle assumptions shouldn't receive realist's assent. Early twentieth-century geological sciences were successful, but it was assumed that the continents were all static, relative to one another. If this fixist commitment played no substantive role with respect to the explanatory and predictive successes of those disciplines, however, then the example need not cause realists any serious concern. Realist commitments, it would seem, should only ever have been extended to those parts of theories that stand in the right sort of relationship to the theories' successes.

Kitcher's strategy has given rise to what is often known as 'selective' scientific realism (e.g. Peters 2014; Carman and Díez 2015). If the strategy works for the purpose intended, then instances of genuine scientific success will typically be traceable to theoretical insights that have been largely retained within our modern scientific view.[8] Simultaneously, where past scientific frameworks depart radically from their modern replacements, the replaced presumptions will be seen largely to have played no substantive role in success. Many realists have embraced the strategy, but it remains to be seen how well selective realism can reconcile history with the conviction that success is a reliable indicator of approximate truth. New historical challenges have emerged, with several authors arguing that particular components of replaced scientific theories (i) contributed essentially to scientific success but (ii) cannot sensibly be regarded as approximately true, thereby challenging the historical cogency of even selective scientific realism (e.g. Chang 2003; Lyons 2006; Carman and Díez 2015; Tulodziecki 2017). A big part of the challenge for selective realists is articulating which components of a theory should be regarded as responsible for success, or idle with respect to it, and justifying why. This has proved far from straightforward and remains the subject of on-going debate (Vickers 2013; Harker 2013; Peters 2014). The strategy also inherits many of the same problems that plague non-selective forms of selective realism, concerning the explanatory potential and significance of scientific realism.

The concept of success is important for scientific realists, since it features prominently in the no-miracles argument. Typically, however, success is understood in non-comparative terms. Theories are either successful or they're not; either they've had novel predictions verified or they haven't. A non-comparative notion of success fits with the general tenor of the realist literature, which focuses on the question of whether we are justified in claiming any knowledge of unobservables, and which has thereby tended to marginalise questions about progress. Certainly one might suspect that most scientific realists would regard science both as achieving knowledge of unobservable entities and of converging on increasingly truthlike theories (or accumulating more knowledge).[9] It would be odd, for example, to suppose that at some point we acquired knowledge of electrons, but have subsequently failed to learn anything more about them. Nevertheless, the question of whether we can achieve knowledge of unobservable entities is clearly distinct from that of whether we have achieved progress with respect to our theories of unobservables.

The realist's focus on the former question might be explained in several ways. Establishing that we are accumulating more knowledge of unobservables (or that our theories are becoming more truthlike) presupposes that we have achieved at least some. Evaluating the case for scientific realism might therefore seem more fundamental than the case for a converging scientific realism. Second, defending the proposition that science is becoming more truthlike might seem to require some method for measuring proximity to the truth, for purposes then of arguing that more recent theories are typically more truthlike. While there exists a considerable literature surrounding the concepts of verisimilitude and truthlikeness, there is certainly no consensus about how we can measure proximity to truth and even whether it's meaningful. It might therefore seem prudent to approach questions about scientific realism in ways that don't require a formal notion of truthlikeness. Third, for optimistic scientific realists (who suppose that today's theories are largely correct in important respects), the question of convergence might hold less interest. Most importantly, I suspect, however, is that antirealist arguments are aimed at the very possibility of achieving knowledge of unobservables, and also provide profound challenges to realism. Working through the arguments, the variations, their consequences, the merits of various kinds of response, and relevant historical analysis, has involved significant amounts of work.

While success has typically been interpreted by realists non-comparatively, it is worth considering how scientific realism might fare if we adopt comparative notions of scientific success. Might scientific realism appear more plausible, if success is understood in terms of scientific progress?[10] Is it better to think of Thomson, Rutherford, Bohr, Schrödinger, and so on, each as having each discovered certain specific (approximate) truths about the nature of atoms? Or is there some benefit to regarding each as having instead increased our knowledge of atoms, or improved the truthlikeness of our atomic models? The difference between these two questions can seem slight: if Bohr improved our knowledge of atoms, for example, then it might seem to follow that he must have discovered some (approximately) true proposition that wasn't previously known. Nevertheless, the shift in emphasis – from understanding success in absolute terms to understanding success as progress – seems to me significant. In the remainder of this chapter I'll consider some consequences of understanding success in terms of progress, with respect to interpreting and defending scientific realism. The chapter will thereby also gesture towards ways that the scientific realism debate might profit from greater engagement with questions concerning scientific progress.

As several authors have noted, Bird (2007) inspired renewed interest in the issue of what it means to say that science makes progress: more specifically, what it means to say that science makes *cognitive* progress. Over time, science might achieve more influence within society, or increase diversity among researchers, or receive more funding, each of which might be regarded as a kind of progress. These should be distinguished, however, from cognitive progress, which – drawing inspiration from the four lead chapters in this volume – denotes something more like knowledge accumulation, increased truthlikeness, greater understanding, or increased usefulness of problem-solving capabilities. None of these approaches is inconsistent with scientific realism. Historically, understanding progress in terms of problem-solving has often been associated with the rejection of truth as an aim or as an achievement of science (Dellsén 2018). However, Shan's (2019) account of scientific progress emphasises problem-solving, but Shan describes his account as consistent with scientific realism.

My argument doesn't presume or privilege any particular conception of scientific progress, but does presume that cognitive progress can be profitably analysed in a variety of ways. One way to motivate this assumption is to recognise that analyses of cognitive progress can aim for greater or lesser generality. We might argue that most cases of scientific progress are instances of a particular kind of cognitive achievement, such as the accumulation of knowledge, hence that this achievement best captures what it means to say that science makes progress. We might alternatively argue that distinct cognitive achievements call for distinct notions of progress. This would give rise to questions about which types of progress are more fundamental, if any, which are easier to evaluate, which types we ought to prioritise, and so on.

A pluralist approach to questions of scientific progress is potentially important to scientific realists for two reasons. First, if realist attitudes towards scientific theories can become *more* reasonable over time, as a result of us gathering more evidence in support of those theories, for example, then this would require us to recognise epistemic justification as a kind of progress that's distinct from increased truthlikeness or growth in knowledge.[11] Admittedly, realists have not typically concerned themselves with the possibility that accumulation of empirical evidence might render realism more plausible, but it's a potential benefit of bringing the realism debate into closer contact with conversations about progress.

The second reason for realists to be pluralists about progress emerges when we return to the no-miracles argument, but now interpret success as some form of progress, rather than a property of theories that can be evaluated without regard for the historical context in which the theory is introduced. The no-miracles argument appeals to two kinds of successes: empirical success is taken to imply, perhaps via an inference to the best explanation, that our best scientific theories are successful in giving us (approximate) truths about unobservable entities and processes. Success in one sense is offered as reason to suppose that we've achieved success in a second sense. If we shift our attention and ask not whether theories (or parts thereof) are approximately true, but whether lineages of theories are becoming more truthlike (or providing more knowledge), then we can analogously distinguish between two kinds of progress. We can make scientific progress by explaining or predicting phenomena that were previously inexplicable or impossible to predict, solving problems that were previously unsolved, or rendering understandable what previously wasn't. We can also make progress, however, by increasing the truthlikeness of our theories or accumulating more knowledge.

In the same way that we might hope to evaluate whether theories' empirical successes reliably indicate the approximate truth of (parts of) successful theories, so we might hope to evaluate whether scientific progress (understood in terms of greater predictive scope and accuracy, greater problem-solving capabilities, etc.) is a reliable indicator of increased truthlikeness or knowledge. In what follows I'll thus distinguish between two kinds of progresses: for ease of exposition I'll refer to one as increased truthlikeness, although accumulation of knowledge would serve the argument just as well; the other I'll refer to as *empirical progress*, by which I mean some combination of increased explanatory and predictive scope and accuracy, increased understanding, and more useful problem-solving capacity. My concern is not to adjudicate between distinct accounts of progress, from a realist's perspective or any other, but to consider the relationship between empirical progress and increased truthlikeness, and how this might bear on the realism debate.

Empirical progress might be regarded as evidence of increased truthlikeness, but in the same way that selective realists emphasise how non-comparative successes can provide varying degrees of support for distinct components of the same theory, so instances of empirical progress gesture selectively to the significance of particular components of new scientific theories. The selective realist argues that if aspects of a theory are redundant, with respect to that theory's explanatory or predictive success, then we shouldn't regard those elements as being confirmed by the success. Similarly, if constituents of a theory are redundant, with respect to empirical progress, then those same constituents shouldn't be credited with having helped us achieve a *more* truthlike theory. Conversely, when theoretical novelty is introduced, and explanatory or predictive scope, depth or accuracy improves, a realist construal may regard the empirical progress as evidence of the increased truthlikeness of those particular novel constituents that are responsible.

The suggestion that empirical progress is attributable to only certain parts of a theory is most straightforward to illustrate when we recognise that progress is achieved only with the introduction of change. Some changes might be regarded more naturally as revisions to an existing framework; others might be interpreted as the introduction of a new theory, although this distinction won't matter for the present discussion. Any empirical progress that is achieved, however, is clearly not a product of conserved assumptions. Insofar as more recent models of the electron, for example, enjoy greater empirical adequacy than the models they replaced, such progress is not attributable to shared features of the models. Assumptions that are not shared across theoretical change may of course also play no role in apparent empirical progress.

In principle, we can distinguish the theoretical changes that improve the empirical strength and problem-solving capacity of our models and theories, from those that don't. If our reasons for embracing changes to existing frameworks are generally a result of us having achieved a more truthlike representation of the target domain, then we would expect such changes to be largely retained. The arguments that are offered in support of revising current scientific theory, in specific ways, might be interpreted by the selective, convergent realist as both evidence of empirical progress and evidence that we have inched towards a more truthlike account of the unobservable entities posited by science. This suggestion makes use of the selective realist's insight that a theory might be successful, despite some aspects of that theory playing no important role in the success. I've simply reoriented the argument to focus on empirical progress. As with other forms of selective realism, the thesis would help answer historically based concerns, if it turned out that empirical progress has typically been a product of theoretical modifications that have been largely retained across subsequent cases of theory change.

The historical challenge to scientific realism draws attention to radical discontinuities across instances of theory change, arguing that theories which straddle the divide are each successful, yet acutely unalike in terms of theoretical commitment. Selective realists concede that replacement theories can appear very different in some respects, but observe that there is often also significant continuity shared across theory change. The preservation of certain theoretical assumptions, alongside the revision and replacement, might be regarded as evidence for the approximate truth of those stable components. *Mere* continuity and retention have been criticised as providing a poor argument for even selective realism, however, since there are alternative, reasonable explanations for continuity that offer no comfort to realists (Chang 2003). Such alternatives may invoke considerations of human cognitive limitations or an epistemic strategy that recommends making as few changes to existing worldviews as possible.

What the selective realist ought to emphasise, in response, is that the historical narratives of interest concern more than mere retention. For selective realists who employ non-comparative notions of success, the concern is with the retention of those components of replaced theories that stand in the right relation to (i.e. were appropriately responsible for) that theory's successes. When selective realism interprets success in terms of empirical progress, then the narratives that concern us involve the retention of those particular insights that precipitate progress. How we understand this claim would clearly benefit from a better understanding of the differing kinds of empirical progress and, in particular, more detailed studies of the different ways that such progress has been achieved.

I have described empirical progress as involving some combination of increased predictive and explanatory strength, increased understanding, and greater problem-solving capacities, but these facets can each be analysed in greater levels of detail. Shan's functional approach to scientific progress focuses on problem-solving and problem-defining, for example, but Shan emphasises that each is multidimensional. Problem-solving requires appropriate vocabulary, methodology, and patterns of reasoning. On Shan's approach, empirical progress might be the result of new or improved patterns of reasoning, which in turn facilitate more useful problem-solving capabilities. Empirical progress might also be achieved as a consequence of learning how to gather new kinds of data, or more accurate data, or analyse larger quantities of data, or as we discover and overcome deficiencies and shortcomings in existing methods. These kinds of detail might not help settle questions about how scientific progress should be defined in the most general sense, but they are of interest to convergent scientific realists. Every instance of empirical progress is an opportunity to evaluate which changes in our theories of unobservables (if any) were responsible for the empirical progress and, furthermore, to observe whether those revisions that have brought about empirical progress in the past have persisted across subsequent cases of theory change.

Faced with historical evidence for the retention, across instances of theory change, of those parts of theories that had engendered *earlier* cases of empirical progress, and the explanatory burden for realists and antirealists is not simply one of accounting for their preservation. The explanatory burden is rather one of explaining why ideas that are introduced in order to *improve* the empirical adequacy of our theories, at one time, as well as our problem-solving abilities, are retained even as scientific disciplines continue to mushroom in terms of scope and accuracy, and as our methods become more rigorous. Of course this assumes that the relevant patterns of growth and retention are evident within the historical record. I don't mean to pretend that the evidence for such historical patterns is at all clear, or that such historical reconstructions are easy to produce. On the contrary, I suspect that the history of any scientific discipline is often sufficiently complicated, that using it to somehow test or support either antirealist or realist attitudes towards science is enormously ambitious. My point for now, however, is that if convergent realism is to be made plausible on historical grounds, or challenged, then a big part of the project could involve developing clearer notions of what constitutes different kinds of scientific progress and how these are achieved. Only then can we start to discern whether the harbingers of empirical progress tend to survive subsequent cases of theory change.

The challenges of utilising the historical record aren't limited to its complexity and size, however. Stanford (2006) argues that attempts to identify those parts of replaced theories that contributed to success must avoid inappropriate dependence on contemporary scientific theories, if they're to avoid begging the question against anti-realists. The selective realist would like to argue that it is plausible, on historical grounds, to regard as approximately true those components of replaced scientific theories which contributed in the right ways to the successes of those theories. This involves two judgements, about which components contributed to success, and about whether they are close enough to our modern understanding to plausibly be regarded as approximately true. Stanford's worry is that if scientific realists use the perspective of contemporary science both to evaluate which parts of past theories were approximately true, and to determine which parts were responsible for the successes of those theories, then these assessments are likely to agree, but as a result of the common stance we adopt when making the judgements. That's to say, there's a risk that we'll regard parts of past theories as responsible for success, because today's theories appeal to very similar principles to account for the same phenomena. Equally, our evidence that those parts of theories are approximately true is their resemblance to modern theories. Converging judgements may therefore result from the shared perspective from which the questions are addressed, and hence fail to provide any good reason for supposing that past theories were successful *because* some of their parts were approximately true.

The moral that Stanford draws is that any viable, selective realist strategy must adopt criteria that are '*prospectively applicable* and *historically reliable*' (ibid., 169). Some selective realists have accepted Stanford's diagnosis and have sought to defend their own criteria as meeting these standards. Convergent scientific realists might similarly hope to meet the challenge, along the lines described earlier. With a richer sense of how the sciences make empirical progress, we'll be better positioned to identify the kinds of changes to theory that induce progress. Perhaps these components of theories will prove prospectively identifiable. And if scientists' decisions to embrace these changes are generally indicative of more truthlike representations of a mind-independent world, then we would expect such modifications to be largely and generally preserved across subsequent theory change. History might thereby provide a test for our selective thesis.

The demand that scientific realists can prospectively identify those components that were responsible for success, however, has seemed unreasonable to others (e.g. Saatsi 2009). If realists were able to reliably distinguish those components of contemporary theories that are unlikely to undergo significant future revision, from those components that are idle and hence much better candidates for replacement, this would clearly be important news for science. The demand seems to require that scientific realists understand scientific theories and their supporting evidence better than the scientists. It's more sensible, perhaps, for selective realists to interpret the work of modern science as, in part, striving to identify which parts of our current theoretical frameworks are doing the work, and which parts can be modified to improve the fit between theory and data. An inability to provide a principled means of identifying the working parts of past or current theories, furthermore, wouldn't indicate that there isn't a coherent distinction to be made between working and idle posits. Despite these lines of response, if realists are not in a position to say which parts of theories they regard as worthy of realist commitment, then the thesis does appear somewhat attenuated.

But perhaps we shouldn't be too quick to concede Stanford's analysis of what selective realism requires. Is a complete embargo on current knowledge, for purposes of reassessing past theories, not unnecessarily restrictive? If it is, then perhaps there is nothing question-begging about using contemporary science to evaluate replaced theories. Declining to exploit the benefit of hindsight, for purposes of re-evaluating past scientific judgements, might be another instance where realists' tendency to ignore questions about progress has robbed them of important perspectives.

Realists and antirealists might disagree about the fundamental interpretation of what it means to make scientific progress, but often there is common ground insofar as most seem to agree that scientific progress does occur. Even those who have denied that we can measure progress in terms of increased truthlikeness, or knowledge accumulation, such as Kuhn (1962) and Laudan (1978) argue that there is a sense in which science makes cognitive progress. Progress entails that we are better situated now, in certain respects at least, to evaluate past theories, methods, and analyses. We need not suppose current practices and methods are perfect, or even close, to suppose that we have made progress. Denying ourselves access to such advantages, when evaluating past results, shouldn't be admitted too hastily. Availing ourselves of such advantages, evaluating past theories from the perspective of today's understanding, thus demurring from the challenge to identify prospectively applicable criteria, may not be compatible with the kind of realist thesis that Stanford is targeting, and that some realists perhaps hope to defend. We shouldn't conclude, however, that there aren't worthwhile and interesting realist theses that can be developed by explicitly utilising retrospective criteria for evaluating replaced theories.

With or without the benefit of hindsight, distinguishing those parts of replaced theories that were either idle or responsible, relative to a given theory's success, is a project that we can approach from either direction. We can try to identify what's idle, or we can try to identify what's responsible. These projects aren't equivalent: the properties of being either idle or responsible for success might be exhaustive, but having good reasons for supposing that certain parts of a theory were idle does not provide good reasons for supposing that the remaining components were responsible for any success (Vickers 2013). Nevertheless, we do often have sensible, independently justifiable, reasons for concluding that certain parts of old theories or models were defended on the basis of insufficient or even no data, or data that were gathered via methods that were unreliable or are no longer deemed sufficiently demanding. We might be aware of alternative explanations for certain phenomena, that weren't entertained by previous generations, couldn't have been reasonably discounted given the evidence that was available at the time, yet in light of more recent discoveries appear better explanations.[12] These replaced components were idle in the sense that they can be replaced by hypotheses that are better supported by currently available evidence, often without explanatory or predictive loss. Such arguments rely on the benefits of hindsight. What was idle, in this sense, was perhaps not prospectively recognisable as such.

Despite relying on hindsight, the kind of judgements described here are not based merely on the fact that certain ideas have been dispensed with. Such judgements do presume that we've made empirical progress and may help better reconcile realism with history. These are judgements, however, which suggest only that certain aspects of a theory were idle, without any accompanying argument that other parts of a theory played the kind of role that would justify realist assent. The question now is whether there is anything within these histories that might give selective realists a more positive argument. Does the benefit of hindsight enable us to gather evidence that our scientific theories are not just improving in ways that antirealists can admit, but improving in ways that are more suggestive of increased truthlikeness?

An older idea of some scientific realists was that replacement theories ought to explain the successes of the scientific theories they replaced (Sellars 1963; Post 1971). Laudan (1981) expressed some doubts about what exactly this means and whether empiricists couldn't endorse similar explanatory relations, but perhaps the focus should never have been about explaining past *successes*, but on explaining past *failures*. If science makes progress, then we are now better positioned to explain ways in which our predecessors were mistaken.

Suppose I have two maps. Each purportedly depicts the same geographical region and each is intended for the same kind of purpose, but suppose I have reasons to believe that one of these maps better represents its target than the other. The inferior map might be deficient in ways that prevents people successfully navigating the area, for example. Perhaps the scale is inconsistently applied, or perhaps the map contains landmarks and streets that have now disappeared or it fails to mark newer ones. These problems need not affect the utility of the map in all circumstances and hence some individuals will use the map successfully. We might not be sure in what *sense* we should say that the improved map *explains* the failures of the inferior map. It nevertheless seems clear that the better map can help us understand, or make sense of, or explain, why the inferior map was not always worthless and why it sometimes led people astray.

We needn't suppose that the superior map is completely accurate in all respects to make sense of its explanatory value, with respect to understanding why the poorer map performed less well. The former is a more useful, better map, as long as it serves its purpose better.[13] We can now imagine a sequence of maps, each purporting to represent the same target, and each more useful than its predecessor. Later maps in the sequence will often help us understand why earlier maps were less useful navigation tools. Assuming only that later maps represent the geographical region *better* will at least sometimes enable us to explain why earlier maps were inadequate. More useful maps will often be regarded as *correcting* the errors of those they replaced. Now, the important point for convergent realism is that at least in the case of maps, a sequence of representations that are increasingly useful, and produced by methods that we have reasons to suppose are increasingly reliable, is a sequence that we imagine would include significant continuity, where the amount of stable content would also increase. That continuity, furthermore, would give us confidence not only that future maps will retain these stable features, but also that the continuity is itself a consequence of us having better represented the target. The critical question is whether there is historical evidence within the sciences of such patterns of growth and preservation. When past theories were regarded as having corrected the errors of those that preceded them, are those corrections generally and largely endorsed by subsequent generations?

The unabashed reliance on hindsight is possible, because the realist thesis under consideration is concerned with increased truthlikeness. We don't suppose that parts of replaced scientific theories are approximately true, in virtue of those parts agreeing with contemporary scientific opinion. We are not privileging current scientific theory in a way that supposes that no future change is probable. Instead, we privilege our perspective only insofar as we regard our methods, models, laws, and theories as empirically *better* than those they replaced. From this vantage, does history looks like a growing number of increasingly stable theoretical claims, which help predict with greater accuracy, provide greater explanatory scope, and contribute to broader epistemic goals? As generations past have corrected and modified the ideas they inherited, do those changes generally look like improvements from our perspective?

Nothing guarantees that history must look like a steady march towards modern practices and conclusions. Historical analysis could reveal that ideas, methods, and research programmes become popular, then fade away as newer ideas, methods, and puzzles are judged more exciting or more interesting, rather than better. New ideas might emerge and become entrenched as a result of political influence, rather than evidence and reason. Ideas might become entrenched, because communities become increasingly intolerant of criticism or novelty. Such accounts have been offered, at least for the histories of some ideas, across some periods of time. Perhaps some, or even much, of the history of the sciences is best understood in terms of whims, social influence, and prejudice such that evidence-based reasoning plays only a nominal role. The point, however, is that if history does reveal significant evidence of retention and growth in theory and method, where corrections and modifications are made for reasons that we continue to endorse, then these patterns would deserve attention. An image of science which supposes that our practices are getting better at avoiding error and increasing truthlikeness would account for such patterns.

As should now be clear, the argument, even if it is well-supported by historical evidence, is not an overtly optimistic realist view. Even if such historical patterns are evident, they would suggest only that we are achieving more truthlike accounts, without taking a position on how close we are to the end of inquiry, a complete theory, or whatever other terminus we imagine. As such, despite some similarities with other selective realist theses, there are important differences. In some respects, the selective realist response to examples of historical discontinuity seems to me correct. We should insist that success amount to more than scientific popularity. We should balance evidence of radical discontinuity with careful attention to the continuities that persist across examples of theory change. We should insist that a realist attitude not be applied evenly to all scientific theories or parts of theories. However, perhaps the selective realist's desire to identify just those parts that deserve realist approval, as well as the exact scientific arguments that justify such approval, is not quite the right approach. Decisions at the level of scientific communities, to embrace certain changes to existing understanding, are important. If those changes are (largely) endorsed by later generations, then this might offer evidence both that our decision-making procedures are reliable and that our sciences are becoming more truthlike. Convergent realism might yet be defended on the grounds that it provides the best explanation for patterns of retention and revision in the history of science. Addressing this question not only requires significant historical investigation, but also clearly requires a better understanding of the ways in which science makes progress. Realists would be well-advised to pay more attention to matters of progress.

Notes

1. The earliest formulations of the argument are usually credited to Smart (1963) and Putnam (1975)
2. Verified, novel predictions are most commonly understood as the confirmed consequences of a scientific theory which that theory was not specifically designed to entail. See Barnes (2018).
3. Another objection to the argument is that it might commit the base-rate fallacy. See Magnus and Callender (2004).
4. Not everyone understands the arguments in the same way, so it is probably more accurate to think of these as families of similar arguments.
5. Influential discussions include Boyd (1973) and Laudan and Leplin (1991). For a helpful overview, see Stanford (2017).
6. Not all structural realists would describe matters this way. Worrall (1989) distinguished the structural content of scientific theories from content concerning the nature of unobservable entities. The structural realism of Ladyman (1998) and French and Ladyman (2003), however, has a different motivation and doesn't rely on distinguishing structural from non-structural parts of scientific theories.

7. Scientific realists have also suggested that we limit attention to 'mature' sciences. While it's not obvious how exactly we should define the notion of maturity, it seems likely that any realist's thesis should be limited to scientific work that has been widely tested, overseen by communities of investigators, developed along multiple lines of inquiry, and so on. The justification for this restriction is that if past failures are to give us good reason for doubts about current theories, then the replaced scientific work must be sufficiently similar in epistemically relevant respects.

8. The two qualifiers ('typically' and 'largely') are deliberate. Scientific realists might concede that success *sometimes* occurs as a consequence of theoretical error; hence they expect that what's responsible for success is only typically retained. Second, realists will likely concede that what's responsible for success can undergo *some* subsequent change, since their conviction is that the source of success is approximately true.

9. Some scientific realists have been explicit in their commitment to convergence on approximate truth and others have addressed issues of both scientific realism and scientific progress, Hardin and Rosenberg (1982), Kitcher (1993), Niiniluoto (1999), Kuipers (2009), Niiniluoto (2017).

10. Kuipers (2009) and Niiniluoto (2017) each think so, and offer arguments that overlap in several respects with my own.

11. This is related to Niiniluoto's distinction between real and estimated progress (Niiniluoto 1999).

12. This possibility is central to Stanford's argument from unconceived alternatives (see Stanford 2006).

13. Greater utility isn't the only reason to prefer one map over others. The standards we employ for purposes of gathering relevant data can become more demanding. Our techniques for producing maps can improve. We might thus have cause for greater confidence in a given map, as we learn more about how it was produced. Analogously, we might have cause for greater confidence in the deliverances of scientific investigation, concerning the properties of unobservables, if they have been held to higher epistemic standards. Such considerations may hold little sway with empiricists, but in combination with the kinds of continuity described, they may have significance for evaluating scientific realism.

References

Barnes, E. C. (2018). Prediction versus accommodation. In Edward N. Zalta (ed.), *The Stanford Encyclopedia of Philosophy* (Fall 2018 Edition), <https://plato.stanford.edu/archives/fall2018/entries/prediction-accommodation/>.

Bird, A. (2007). What is scientific progress? *Noûs, 41*(1), 64–89.

Boyd, R. N. (1973). Realism, underdetermination, and a causal theory of evidence. *Noûs, 7*(1), 1–12.

Carman, C., & Díez, J. (2015). Did Ptolemy make novel predictions? Launching Ptolemaic astronomy into the scientific realism debate. *Studies in History and Philosophy of Science Part A, 52,* 20–34.

Chakravartty, A. (2017). Scientific realism. In Edward N. Zalta (ed.), *The Stanford Encyclopedia of Philosophy* (Summer 2017 Edition), <https://plato.stanford.edu/archives/sum2017/entries/scientific-realism/>.

Chang, H. (2003). Preservative realism and its discontents: Revisiting caloric. *Philosophy of Science, 70,* 902–912.

Dellsén, F. (2016). Scientific progress: Knowledge versus understanding. *Studies in History and Philosophy of Science, Part A, 56,* 72–83.

Dellsén, F. (2018). Scientific progress: Four accounts. *Philosophy Compass*, *13*(11), e12525.

French, S. and Ladyman, J. (2003) Remodelling structural realism: Quantum physics and the metaphysics of structure, *Synthese*, *136*, 31–56.

Hardin, C. L. and Rosenberg, A. (1982). In defense of convergent realism. *Philosophy of Science*, *49*(4), 604–615.

Harker, D. (2013). How to split a theory: Defending selective realism and convergence without proximity. *The British Journal for the Philosophy of Science*, *64*(1), 79–106.

Kitcher, P. (1993). *The advancement of science: Science without legend, objectivity without illusions*. Oxford: Oxford University Press.

Kuhn, T. S. (1962). *The structure of scientific revolutions*. Chicago: University of Chicago Press.

Kuipers, A. F. T. (2009). Comparative realism as the best response to antirealism. In C. Glymour, W. Wei, and Dag Westerstahl (eds.), *Logic, Methodology and Philosophy of Science: Proceedings of the Thirteenth International Congress* (pp. 221–250). Rickmansworth: College Publications.

Kukla, A. (1998). *Studies in Scientific Realism*. Oxford: Oxford University Press.

Ladyman, J. (1998). What is structural realism? *Studies in History and Philosophy of Modern Science*, *29*, 409–424.

Laudan, L. (1978). *Progress and its Problems: Towards a Theory of Scientific Growth*. Berkeley, CA: University of California Press.

Laudan, L. (1981). A confutation of convergent realism. *Philosophy of science*, *48*(1), 19–49.

Laudan, L. and Leplin, J. (1991). Empirical equivalence and underdetermination. *Journal of Philosophy*, *88*, 449–472.

Lyons, T. D. (2005). Toward a purely axiological scientific realism. *Erkenntnis*, *63*(2), 167–204.

Lyons, T. D. (2006). Scientific realism and the *stratagema de divide et impera*. *The British Journal for the Philosophy of Science*, *57*(3), 537–560.

Magnus, P. D. and Callender, C. (2004). Realist ennui and the base rate fallacy. *Philosophy of Science*, *71*(3), 320–338.

Niiniluoto, I. (1999). *Critical Scientific Realism*. Oxford: Oxford University Press.

Niiniluoto, I. (2017). Optimistic realism about scientific progress. *Synthese*, *194*(9), 3291–3309.

Peters, D. (2014). What elements of successful scientific theories are the correct targets for "selective" scientific realism? *Philosophy of Science*, *81*(3), 377–397.

Post, H. R. (1971). Correspondence, invariance and heuristics: In praise of conservative induction. *Studies in History and Philosophy of Science Part A*, *2*(3), 213–255.

Psillos, S. (1999). *Scientific Realism: How Science Tracks Truth*. London and New York: Routledge.

Putnam, H. (1975). *Mathematics, Matter and Method*. Cambridge: Cambridge University Press.

Saatsi, J. (2009). Grasping at realist straws. Review of *Exceeding Our Grasp: Science, History, and the Problem of Unconceived Alternatives*, by P. Kyle Stanford. *Metascience*, *18*(3), 355–363.

Sellars, W. (1963), *Science, Perception and Reality*. New York: The Humanities Press.

Shan, Y. (2019). A new functional approach to scientific progress. *Philosophy of Science*, 86(4), 739–758.

Smart, J. J. C. (1963). *Philosophy and Scientific Realism*. London: Routledge & Kegan Paul.

Stanford, P. K. (2006). *Exceeding our Grasp: Science, History, and the Problem of Unconceived Alternatives*. Oxford: Oxford University Press.

Stanford, P. K. (2017). Underdetermination of scientific theory. In Edward N. Zalta (ed.), *The Stanford Encyclopedia of Philosophy* (Winter 2017 Edition), <https://plato.stanford.edu/archives/win2017/entries/scientific-underdetermination/>.

Tulodziecki, D. (2017). Against selective realism(s). *Philosophy of Science*, 84(5), 996–1007.

van Fraassen, B. C. (1980). *The Scientific Image*. Oxford: Oxford University Press.

Vickers, P. (2013). A confrontation of convergent realism. *Philosophy of Science*, 80(2), 189–211.

Worrall, J. (1989). Structural realism: The best of both worlds? *Dialectica*, 43(1–2), 99–124.

15 Scientific Progress and Incommensurability

Eric Oberheim

There has never been an unambiguous account of incommensurability in science. Neither of its two main proponents, Thomas Kuhn and Paul Feyerabend, managed to make themselves understood to the extent that a coherent consensus has formed about what incommensurability is, or what its implications for scientific progress are. Arguably, they were both trying to talk about the same thing, but they had different aims and approaches.

Feyerabend came from a normative, methodological background. He understood science as 'an attempt at a realistic interpretation of experience' and tried to explain 'how to be a good empiricist'.[1] He used 'incommensurable' while trying to describe and explain how progress in science is made.[2] He examined the methods scientists use to justify laws as explanations of observations, and how theories are used to explain such laws, and he speculated about what progress might be, given those methods of making it. He criticised the idea that new theories always explain why older theories were correct to the extent that they were by showing that the older theory can be deduced from its replacement. According to Feyerabend, this is incorrect. For example, classical mechanics cannot be deduced from relativity theory, and thereby be explained by it. But he did not just want to improve our understanding of science and progress. He was also trying to protect progress with a 'plea for pluralism'.[3] He argued that such views are not just wrong, they impede progress by stifling theory proliferation that might otherwise promote it.

Feyerabend was not the first to use the term 'incommensurable' or to develop the idea in this context.[4] He had read Albert Einstein's reflections on comparing competing universal theories and the underdetermination and incommensurability involved.[5] He had read Pierre Duhem on the aim and structure of physical theory (which had heavily influenced Einstein),[6] and he worked closely with Popper on the logic of scientific justification. Popper had argued that a theory (or any general statement) is never confirmed to be true. Instead, science is a game played by conjecture and refutation that provides the best available justification. While we cannot confirm theories to be true, we can refute them and replace

DOI: 10.4324/9781003165859-19

them with better theories that have not been refuted. According to Popper, when comparing theories quantitatively with respect to their deductive consequences, they can be equal, or one can be larger and one smaller. Otherwise, they are '*inkommensurable*', by which he meant that the relative sizes of the two sets of deductive consequences are quantitatively 'non-comparable'.[7] Popper suggested that this is the case whenever all the deductive consequences of one theory are not a proper subset of the other theory.

Feyerabend (citing Kant) was wondering how it is possible for there to be empirical evidence that refutes a universal theory, if a universal theory conditions all the evidence that can be used to test it by structuring all experience according to it. He was focused on universal theories because they apply to everything, and so are the best corroborated and most fundamental with respect to the ontology that they imply. He realised that Popper's logic of scientific justification is too simple.[8] It may explain how commensurable statements are tested empirically in the usual way, but it does not explain how incommensurable rival theories compete against each other through revolutions.

According to Feyerabend, incommensurability is a specific form of 'conceptual disparity'.[9] Two theories are incommensurable: 'if a new theory entails that all the concepts of the preceding theory have zero extension or if it introduces rules that change the system of classes itself'.[10] So that for there to be incommensurability: 'The situation must be rigged in such a way that the conditions of concept formation in one theory forbid the formation of the basic concepts of the other'.[11] Or put another way, 'When the meaning of their main descriptive terms depend on mutually inconsistent principles',[12] So that incommensurability means 'deductive disjointedness, and nothing else'.[13]

For Feyerabend, the main methodological implication incommensurability has for progress is that because progress is fallible and revisable, it is unpredictable. For example, even the best corroborated theoretical entity (once the ether) may turn out not to exist, while what may initially appear to be blatant incoherent nonsense, or even something impossible (such as space bending), might turn out to make sense as part of a better corroborated explanation. The best available theories, the assumptions, and methods used to make predictions with them, and the evidence used to test those predictions – any part, or even all of it together, may be fundamentally mistaken and in need of revision. They are all equally speculative[14] so that from a normative methodological perspective: anything goes, and whatever works, works.

Feyerabend argued that incommensurability implies that pluralism and tolerance best promote progress. The example he used to illustrate his point was Einstein's prediction of the statistical behaviour of Brownian motion, which, he claimed, served as a crucial experiment between incommensurable rivals.[15] Without a better corroborated alternative, such anomalies can be explained away with *ad hoc* hypotheses, or simply ignored. The trick is not to show definitively that these recalcitrant experiences refute the existing point of view by conclusively deducing them from it. It is to show that the recalcitrant experiences are exactly what to expect, as deduced from a better corroborated rival. When a new conceptually incompatible theory is used to explain measurement results used as evidence for that theory, the results are not neutral with respect to it. They are interpreted according to that new theory. Pre-existing measurement results will count as two different kinds of evidence when they are deduced from incommensurable rival theories. This shows, according to Feyerabend, why pluralism best promotes progress: Even if a theory predicts all known facts, judging its practical success or factual adequacy still requires comparing it to incommensurable rivals that might explain those facts better.[16]

Feyerabend thought that understanding a theory is not just the ability to use it to explain laws by deducing predictions from them (and thereby to solve problems with it), but also to be able to understand its predecessors and how they had provided a different realistic interpretation of experience. Only then can we fully understand how and why the current theory became the best available theory. Although we can only experience one reality at a time coherently, there were previous, coherent worldviews. A worldview is the whole set (theories, the auxiliary assumptions, including any methods, and corroborating evidence) taken as a unit, realistically interpreted when used to understand and explain. From this Feyerabendian perspective, scientific progress is not a process that converges towards a single, abstract ideal. It is an ever-expanding set of alternative worldviews, each forcing the others into greater articulation, all potentially contributing to improving our understanding, as they compete to be the best.[17]

Feyerabend never satisfactorily explained his views on meaning change, nor exactly how incommensurable theories preclude one another. How can incommensurable theories be based on inconsistent principles and be deductively disjoint at the same time? His suggestion that incommensurable rivals are compared by 'indirect refutation' was dismissed as incoherent as an indirect refutation is no refutation at all. Similarly, his suggestion that incommensurable rival theories are part of each other's empirical content seems muddled and was found unconvincing, especially given that they are deductively disjoint. Much confusion resulted, with Hilary Putnam wondering how Feyerabend could claim that something is unintelligible and then go on to explain it, leaving Feyerabend with the task of reiterating that theories are only unintelligible until they are learned, which may require patience and tolerance, because suspension of otherwise well-corroborated beliefs may be needed to understand proposed improvements.[18]

Kuhn also wanted to understand science and progress, but unlike Feyerabend, he examined how science is learned and what can be learned from looking historically at science as a community-based practice. He had noticed that discoveries and revolutions that may look like single events from a present-centred perspective were, upon closer inspection, actually extended phases (e.g. the discovery of Uranus and the Copernican revolution) that required communities gradually revising consensus about how best to describe and explain something.[19] Kuhn tried to explain scientific progress by suggesting historical laws that dictate how and why scientific communities make the different kinds of progress (normal and revolutionary) that they do.[20]

Kuhn's views on comparative historiography (methods of writing history) suggest that while present-centred (or 'developmental') history is typically (perhaps even necessarily) used to teach science, it takes 'hermeneutic' historiography to understand it. Present-centred history, such as Einstein and Infeld's *The Evolution of Science* (1938), explains current theories and how they developed out of earlier ideas by abstracting from what actually happened. Hermeneutic historiography reveals what actually happened by abstracting from contemporary theories and the ideas used to state them. Hermeneutic historiography reveals two kinds of progress in science (normal and revolutionary), separated by a 'crisis' phase (during which theories compete),[21] in an evolutionary process driven by an essential tension between tradition and innovation.[22] He realised that use of incommensurable theories as explanations of facts changes how members of competing communities experience what is real. He concluded that communities with incommensurable worldviews compete to explain the experiences that they each make possible.

Kuhn's views about incommensurability also caused quite a confusion. After his early vague and sometimes inconsistent remarks about paradigms and incommensurability,[23] he spent the rest of his career trying to come to terms with it. His views became more specific with the introduction of the 'no-overlap principle'.[24] He was still trying to explain incommensurability and its implications in a book about 'The Plurality of Worlds' that he was writing at the time of his death, and his final and most detailed views on incommensurability remain unpublished.[25]

According to Kuhn, incommensurable theories reclassify some of the same things into incompatible sets of kinds that break the no-overlap principle with respect to the reigning point of view. For example, the theory that all planets travel around the Earth is conceptually incompatible (or 'incommensurable') with the Copernican theory that all planets travel around the sun, because the two theories cross-classify some of the same things (the Sun, Earth, and moon) into mutually exclusive sets of non-overlapping kinds of those things (the Earth, planets, and stars versus planets, stars, and moons). This invalidates inferential relations between statements that use these kinds and can lead to apparent nonsensical statements. For example, today, anyone who claims the sun is a planet does not understand 'planet' (as it is understood today) and appears to be making a category mistake from the past. For a time, the sun and the moon were 'planets'. That turned out to be wrong. Then at some point there were thousands of 'planets', while recently it was discovered that there are only eight, down from nine. What counts as a 'planet' has changed and is always open to revision.

Kuhn emphasised that incommensurable theories are mutually exclusive both conceptually and ontologically.[26] Use of one theory precludes coherent use of the other in the same statement or argument, and both cannot be true: either all planets orbit the sun, or they orbit the Earth (assuming one is stationary with respect to the other). The reason that they both cannot be true is that they are part of ontologically incompatible descriptions of some of the same things, not because they logically contradict each other by making inconsistent claims about the same things (as is the case with rival commensurable theories). Although two sentences may appear to make inconsistent claims about planets, because they are not talking about the same set of things with 'planets', they do not. Instead, they make two incompatible statements as interpreted by two theories about some of the same things described as mutually exclusive kinds of things. Ptolemaic and Copernican planets are incompatible kinds of 'planets' as described by incommensurable theories.

Kuhn emphasised that progress was not made by replacing incoherent nonsense with a better theory. Both theories made sense, and for a time were even equally empirically supported by the available evidence. For a time, they were both taken to be true descriptions of what really happens. Progress was made by *replacing* established theories with incommensurable rivals that *revise* what is experienced as real by the new community (revolutionary scientific progress); not always merely by adding new commensurable theories to what was already known (normal scientific progress). For Kuhn, scientific progress is 'a process whose successive stages are characterized by an increasingly detailed and refined understanding of nature'.[27] However, it is not progress towards a goal set by nature in advance. It is evolution away from what we thought we knew, not evolution towards what we think we want to know. Truth is not the aim of science. Solving puzzles is. Normal science makes progress by solving more puzzles. Revolutionary science resolves an outstanding anomaly by recasting the puzzles and indicating how to re-solve those puzzles better.

In many respects, Kuhn and Feyerabend's views inform each other. They arrived at similar conclusions, starting from different kinds of assumptions belonging to different approaches. By building on aspects of both of their accounts, we can better understand incommensurability as part a description of progress as the selection of the most fruitful explanation as manifest as a consilience of corroborations. To this end, we will focus on the following questions: What is incommensurability? What causes incommensurability? What are the consequences of incommensurability for intelligibility? What does incommensurability imply about theory comparison? And finally, what does it imply about truth, reality, and progress?

1.   What Is Incommensurability?

Incommensurability is a relation between a theory and a revolutionary rival. They have no common measure. The evidence used to test the two theories does not provide a set of neutral facts against which they compete as the best (most fruitful) explanation of what scientists who use that theory both take to be real.[28]

Scientific revolutions are conceptually and ontologically revisionary. They reclassify things into new sets of kinds of things as beliefs are revised about what kinds of things are real. Revolutionary incommensurable theories are incompatible because they offer conceptually incompatible explanations of the same measurement results explained as ontologically incompatible processes. This 'taxonomic' aspect of incommensurability can be distinguished from a 'methodological' aspect of' incommensurability, according to which incommensurable theories allow rational disagreement due to a lack of common measure in the form of neutral, epistemic values (such as fruitfulness, simplicity and accuracy) used to assess their relative merits. Just as homology of taxonomic lexical structure allows for individual variation within scientific communities conceptually without resulting in incoherence, epistemic values may also vary within and between communities without undermining rationality. Such differences make rational disagreement within and between communities possible, as new facts may lead to better theories and better methods, new theories may lead to better methods that explain new facts, and better methods may lead to new facts explained by better theories.

These are purportedly facts about the logic of scientific justification; or more specifically, about how scientific theories compete to be the best explanations of reality through revisionary scientific progress called scientific revolutions.

2.   What Causes Incommensurability?

Incommensurability is caused by reclassification that breaks the no-overlap principle. Revolutionary progress happens when scientists propose new incommensurable theories that use a 'lexical taxonomy' (a taxonomically structured set of kinds) that cross-classifies some of the same things into a mutually exclusive set of kinds, thereby breaking the no-overlap principle with respect to the kinds used to state the reigning theory. As the new theory gradually becomes better corroborated than its predecessor, the scientific community gradually forms a new consensus about which theory is best corroborated by the available empirical evidence. The once reigning theory and the lexical taxonomy used to state it, together with how it uses experience to understand what is real, are replaced by the new theory and its lexical taxonomy, which states new relations between new kinds that turn out to offer better, but incompatible, descriptions and explanations of some of the same things. That is what causes incommensurability.

Abiding by the no-overlap principle is necessary for using the results of measurements to test theories based on empirical predictions that logically follow from them. Otherwise, deductions from the theory to the evidence will not be valid, nor will predictions made by it be reliable. The no-overlap principle is purportedly a necessary condition for testing any logically coherent set of general descriptions and explanations empirically. It does not matter if they are theories, laws, models, or methods, as long as they can be represented as a set of general statements with the logical form 'all x are y'. Then, they can be tested against their predictions understood as deductive consequences that can be measured,[29] and thereby used to explain those measurements.[30]

A lexical taxonomy is incompatible and mutually exclusive with another lexical taxonomy, if the kinds of one lexical taxonomy cannot be introduced into the other lexical taxonomy without breaking the no-overlap principle (i.e. without cross-classifying some of the same things into what are supposed to be non-overlapping kinds). If that were allowed to happen (breaking the no-overlap principle), some of the things to which the kinds of the two incommensurable theories refer would be subject to different sets of laws, resulting in conflicting expectations about the same things, loss of logical relations between statements made about those things and kinds of things, and ultimately nonsensical incoherence.[31]

In this way, the no-overlap principle limits the scope of logical implications based on referential relations of kind terms to particulars within a single hierarchically structured lexical taxonomy, keeping conceptually incompatible theories logically disjoint. Homology of taxonomic lexical structure suffices for coherence, while allowing for individual variation within communities as to how they identify members of those kinds. Breaks of the no-overlap principle can be used to identify and differentiate language that must be kept distinct when learned and understood for it to remain coherent (e.g. during periods in which two incommensurable theories compete, or when engaged in hermeneutic historiographical investigation of antiquated theories).

In sum, incommensurability is conceptual incompatibility caused by reclassification that breaks the no-overlap principle resulting in revisions with respect to the kinds of things that are experienced as real.

3. What Are the Consequences of Incommensurability for Intelligibility?

Incommensurability causes mutual ineffability. With just the conceptual resources of either theory, the other point of view cannot be coherently formulated. To be intelligible, only one can be used at a time when describing or explaining. In other words, a direct consequence of introducing a rival theory whose taxonomic lexical structure breaks the no-overlap principle with respect to that of the reigning theory's is mutual ineffability. Moreover, it will remain unintelligible until the new set of kind terms used to make it has been learned. This goes both ways. For example, without an alternative, such as Ptolemy or Brahe's theory, 'all planets orbit the Earth' appears unintelligible, as the Earth is a planet, so how could it orbit itself? Conversely, the new theory was literally ineffable with the taxonomic lexical structure imposed by the preceding theory. They are mutually ineffable due to their conceptual incompatibility, which results because they cross-classify some of the same things into incompatible kinds of things.

Overcoming ineffability requires two abilities: learning and unlearning. New incommensurable kinds must be learned together as sets of statements about new kinds and their newly conceived relations. Moreover, the reigning theory and its lexical structure must be suspended (at least temporarily) so that the ability to coherently state and understand the new theory (according to its incompatible taxonomic lexical structure) can be learned. Otherwise, how the new theory describes and explains what happens remains nonsensical and incoherent (e.g. space bending from a Newtonian perspective). Similarly, learning to understand antiquated theories requires suspending (at least temporarily) contemporary theories, which otherwise distort older theories making them appear nonsensical and incoherent. When incommensurable theories compete as the best available explanation, there can also be much talk at cross-purposes as a new consensus about how best to describe and explain what happens forms. Avoiding miscommunication requires those theories, the methods used to test them, and facts used to corroborate them are kept separate, when making statements (for them to be coherent) and arguments (for them to be valid).[32]

A single (uninterpreted) sentence may be used to test two incommensurable theories interpreted as two incompatible statements. For example, the same table of measurement results of the positions of celestial bodies can be used to corroborate three incommensurable rivals, just as measuring how deep a falling ball penetrates a surface could be used to measure its impetus or its momentum, and thus be used to corroborate statements about Aristotelian impetus or Newtonian mass.[33] In this example, Aristotle and Newton cross-classify the stone, the Earth (and the other planets) breaking the no-overlap principle. For Aristotle, a falling stone is a different kind of thing that behaves differently (obeys different laws) than the Earth (which does not fall) towards which it falls, which is a different kind of thing than the planets (which are not falling). For Newton, they are both the same kind of things, and both are falling (the Earth towards the sun and the ball towards the Earth). Things that were different kinds (stones and planets) were the same kind of things (falling masses), and things that were the same kind of thing, the moon, Mercury, and the sun, are now different kinds of things. Each theory implies that a different ontological process is taking place as described by conceptually incompatible kinds. They are incompatible because they cross-classify some of the same things into mutually exclusive sets of kinds.

If a sentence stating a measurement result is interpreted by two incommensurable theories, it makes two incompatible statements that cannot logically contradict one another. Each of the statements can corroborate the theory from which it can be deduced allowing that theory to explain those measurements.

4. What Does Incommensurability Imply About Comparing Theories?

Empirical evidence is often erroneously understood as a definitive, neutral arbiter, used to settle scientific disputes about how best to describe and explain something objectively. However, neither refutations, nor corroborations, are definitive, nor are measurement results neutral when used as evidence to compare incommensurable rivals. To be evidence for a theory, measurement results must be interpreted by that theory. Theories state relations between the kinds that it uses to describe and explain what happens. To do this, they must classify particular 'things' as belonging to (non-overlapping) kinds of those things.[34] The assumption is that particular things behave the way they do because they are the kinds of things that they are, and because of the regular relations to other kinds of things that those kinds of things have.

To connect a theory to measurement results used to test it requires deducing those measurement results from that theory, by inserting the relevant law and initial conditions, and deducing a particular prediction from it that can be measured. This prediction is then compared to the measurements. That is how a theory explains them. The reason for believing that the theory explains the measurement results includes all the reasons for believing in all the assumptions used to connect the two. This may include the methods, models, or any other tools, used to connect the measurement to the regularity that explains it. This is how empirical measurements serve as evidence for a theory (or any generalisation). The prediction cannot be compared to the measurements by itself, but only together with all the assumptions used to connect that theory to the measurements by deduction *as a unit*.[35]

Normally, competing commensurable conjectures can be compared to neutral evidence. All the competing hypotheses can be refuted by the same experiments according to the same interpretation of the evidence, but one. Popper had already described how to compare commensurable hypothesis based on falsifications or 'refutations' in great detail, and he explained that while general statements cannot be confirmed by particular evidence to be true, particular evidence can still refute general statements.[36] Otherwise there could be no cumulative progress in science. The question is, does this procedure also work when comparing incommensurable rivals?

The answer, of course, is no. Commensurable theories can compete against neutral evidence produced by the same measurements by refutation – by finding evidence that refutes all available hypothesis but one. Incommensurable theories compete as conceptually incompatible interpretations of some of the same measurements by corroboration – by finding evidence that corroborates only one of the available alternatives. For example, Ptolemy, Brahe, and Copernicus proposed incommensurable theories that all equally explained the same measurement results. They used incompatible concepts to describe ontologically incompatible processes that explain some of the same measurements. Moreover, with respect to the best measurement procedures available, for a time, they were all equally corroborated. Theory choice was underdetermined by the evidence, until new tools provided new kinds of evidence that corroborated only one (telescopic observations of the phases of Venus), thereby empirically undermining its ontologically incompatible rivals. The reigning theory was rejected and replaced because it was ontologically incompatible with a better corroborated conceptually incompatible theory, not because evidence that was deduced from the reigning theory had finally definitively refuted it.

As we have seen, Feyerabend's favourite example of a crucial experiment between incommensurable rivals was the corroboration of Einstein's prediction of the statistical behaviour of Brownian motion. It shares the same structure. The measurement results that settled the dispute in favour of atomism were made by deducing predictions from the kinetic theory of heat (which posits atoms). The new theory was needed to make the calculations that corroborated the decisive measurement results in its favour. The evidence was not neutral with respect to both theories. It was 'directed' or 'mediated' by the theory that it corroborates.[37] An anomaly (potential refutation in the form of some recalcitrant experience) may *cause* crisis and theory proliferation, but the crisis ends as a revolution through a consilience of corroborations of the new theory, not by deducing more results from the old theory that refute it.[38] Recalcitrant evidence can be explained away with *ad hoc* hypotheses, or ignored, which is what happens when there are no known alternative explanations of it. But mounting corroboration (by different kinds of evidence that all point in the same direction) of an unrefuted incommensurable rival is not as easily ignored.[39]

The anomaly causing the crisis does not have to be new measurement results. The problem causing the crisis in the Copernican revolution was not some new empirical evidence that the reigning theory could not explain (epicycles could be used to explain retrograde motion). The problem was that three conceptually incompatible rivals were all equally well corroborated by the same measurement results interpreted as ontologically incompatible processes, and there was not enough evidence available to settle the dispute. In Kuhnian terms, a crisis, not an anomaly, caused the crisis that led to progress.

The evidence corroborating a theory (it's empirical content) is not a fixed quantity. It changes when new evidence is discovered by existing methods, or when different assumptions are used to connect it to existing measurements, or when new methods of measuring produce new measurement results that it predicts, or when incommensurable rivals replace them as the best available explanations. The empirical content of a theory is in flux.[40] Revolutions can revise what counts as evidence for what. For example, the failed attempts to measure ether were not useless. Although they did not corroborate or refute theories about ether, they were shown to corroborate a more fruitful incommensurable alternative about the speed of light, leaving theories about ether with nothing left to explain. Moreover, because theories, methods, and facts are all corroborated as units, improvements in any of them can improve of all of them together.

Each statement of the theory can be tested if some of the others are taken as definitions so that what functions as a definition in one context can be used as an empirical claim that can be tested and is open to revision in another context. In these ways, the theory, as a set of statements about kinds and their relations, can act like a set of synthetic *a priori* principles that make a correspondence between experience and reality possible so that what counts as facts is relativised to which theory is used to interpret them (see the next section).

Underdetermination may be overcome by a new theory fulfilling its promise to be more fruitful as manifest, in retrospect, as a consilience of corroborations as it is increasingly used to explain more of what happens, or the reigning theory may remain in place in a state of crisis, or the new theory could just be ignored as nonsense, just as older theories have been forgotten. The community gradually decides over time as individuals make decisions about what to believe. The reigning theory has the chance to try to explain the new evidence, and if it does, choice between them remains underdetermined. But it is sufficient that the reigning theory is not corroborated by some evidence that does corroborate an incommensurable rival. That is enough for it to count as part of the evidence for one and thereby against its ontologically incompatible rival, potentially tipping the balance as evidence mounts as a consilience of corroborations. Because incommensurable theories use incompatible sets of kinds to describe ontologically incompatible processes, if evidence is deduced that corroborates one of two incommensurable rivals, it cannot be the same evidence that refutes the other, because it can only be evidence for or against a theory, once it is deduced from that theory, so that even the same measurement will be interpreted as the results of incompatible processes, and thus used as different kinds of evidence if used to test each theory.

Universal theories are not refuted by consequences deduced from them. They are superseded by a better corroborated incommensurable alternative, according to *its* deductive consequences. Because in retrospect, communities eventually form a consensus with respect to the best empirically corroborated theory, scientific progress can erroneously appear to be an algorithmic process in which shared, independent empirical evidence always explains how and why consensus forms about how things are. But as we have seen, this is not always the case. Decisive empirical grounds to settle scientific disputes empirically are not always available (transient underdetermination) allowing for rational disagreements between the proponents of incommensurable theories. This seems to be a reoccurring historical phenomenon as consensus form for periods of normal (commensurable) scientific progress, only later to be replaced, because an incommensurable rival eventually turned out to be more fruitful.

Methodologically, no matter how closely a theory seems to reflect the facts, no matter how universal its use, nor however necessary it may seem to be to those speaking the corresponding idiom, its factual adequacy can be assessed only after it has been confronted with alternatives, whose invention and detailed development must therefore precede making such judgements.[41] That is why incommensurability implies pluralism. New incommensurable theories that make different kinds of predictions can provide a more rigorous test of a reigning theory than is possible merely by testing that theory by deducing predictions from it. A consilience of corroborations of an incommensurable rival may undermine the justification of the reigning theory as it is gradually replaced.

5. What Does Incommensurability Imply About Truth, Reality, and Progress?

Both Kuhn and Feyerabend used incommensurability to explain revolutionary progress as 'world change' and to argue that incommensurability undermines scientific realist conceptions of truth and progress as approximation to it. Instead, the truth of a statement is relative to the theory used to interpret it.[42] If what is taken to be truth is relative to a particular lexical taxonomy, scientific progress is not progress towards truth, or any kind of series of better approximations to truth, as science tracks truth.[43] Instead, revolutions happen when the selection of the most fruitful available theory revises what is taken to be real, leading to new knowledge about some of the same things reconceptualised as members of incompatible sets of kinds of things as used to describe ontologically incompatible processes.

Incommensurability implies that when there is a change that breaks the no-overlap principle during scientific revolutions, there is also world change. Put another way, world change accompanies kind change during the process of revisionary reclassifications called 'scientific revolutions'.[44]

World change happens when established facts (established based on the taxonomic lexical structure of the reigning theory) are reclassified into new facts (established based on the taxonomic lexical structure of the new incommensurable theory), which corroborate the new theory and can thereby undermine its ontologically incompatible rival. As the reigning rival gradually loses its status as the most fruitful available explanation, and is eventually replaced, it loses the role it plays in organising experience into reality, which is increasingly played by the new theory. With hindsight, we can see a repeating pattern. When new theories replace established theories, mistaken or erroneous facts are replaced by new facts according to the new lexical structure. These new facts can be deduced from the new theory and thereby used to corroborate it so that the older theory was not just wrong, it was also wrong about the facts that were used to corroborate it as reformulated by the newer theory.

Incommensurability implies that world change accompanies kind change that breaks the no-overlap principle, given the additional assumption that using theories to describe and explain has ontological consequences.[45] They contribute to our understanding of what happens. Theories have ontological consequences because they imply the kinds of things that we take to be real when we use them to explain experience. They act as synthetic *a priori* principles for individuals in communities that share a homologous lexical structure so that when the facts follow logically from our theories, our theories help us understand those facts by providing a conception of reality that is explained by them.[46]

Sociologically, community consensus of scientists determines what counts as the facts of empirical reality (as there is no higher authority). Psychologically, how an individual within that community understands their experiences of empirical reality is empirical reality for that individual within that community. World change happens when our experience of some 'things' as kinds of 'things' change, altering what we experience to be real as individuals in communities. Collective reclassification, by a community, that breaks the no-overlap principal results in new experiences of reality that are justified by explanations that use conceptually incompatible sets of kinds to describe ontologically incompatible processes.

When we observe communities describing and explaining, we see people living in worlds of kinds that they understand according to the most fruitful theories available to them. When science improves our understanding by revising what we thought we knew, there is kind change that breaks the no-overlap principle resulting in world change. When world change happens, established facts (established using the taxonomic lexical structure of the reigning theory) are reclassified into new incompatible facts (established using the lexical taxonomy of the new incommensurable theory). These 'alternative facts' corroborate the new theory and can thereby undermine rivals that are ontologically incompatible with it. Not because the two incommensurable theories logically contradict each other, and not because statements deduced from both about measurement results contradict each other (which can only be the case when testing commensurable hypothesis), but because they cannot both be right due to their ontological incompatibility. In short, revisionary progress leads to established facts no longer being real from the perspective of better theory, which through a consilience of corroborations gradually replaces its predecessor as the new theory spreads throughout the community forming new empirical realities and new truths. Progress involves world change

As we have seen, incommensurability implies there are at least two kinds of progress (commensurable and incommensurable), which becomes clear when examining two methods of writing history used to understand science. Hermeneutic historiography allows access to antiquated theories and the worldviews to which they contributed allowing us to understand science as it developed, in the hope that a more accurate view of the history of science will help us better understand what science is and how it makes progress. Present-centred history helps us understand contemporary science by showing how previous conceptual insights and advances led to the development of our current best theories.

These two methods of writing history support incompatible accounts of scientific progress. Present-centred historiography explains progress as a series of better approximations to current theories assuming they are true (or at least approximately true and the closest to the truth available). It is used to support scientific realist views of progress as progress towards truth, and it explains current theories and how they were developed by providing a history of ideas and their justifications that abstracts from history as it really happened. Hermeneutic historiography, by contrast, explains progress as it happened. Communities used taxonomically structured kinds to describe and explain their experiences of what is real according to relations between those kinds. Truth and reality are relative to the taxonomic structure shared by individuals within a community. When theories with new taxonomic structures that break the no-overlap principle with respect to the reigning theory are used by a community to describe and explain some of the same things, they can replace what counts as truth and reality about those things for that community.

Progress with respect to the best corroborated theories used to explain what is real is not the only kind of scientific progress. Theories that are no longer accepted as part of the best (most fruitful) available explanations may make progress, even without a realistic interpretation. For example, progress is still made with and within classical mechanics, even after it no longer provides the best available realistic interpretation of experience by being the most fruitful explanation of it according to the scientific community.[47] Just because it has been superseded in its role in organising experience into reality by a community, it does not mean that it stops being a powerful tool for solving problems and can no longer lead to improvement or be itself improved. The next best explanation of everything might even revive some of its ideas in some form.

In his late philosophy, Kuhn revised his phase model, giving up the idea that there can be only one normal (cumulative, commensurable) scientific tradition per discipline at a time. He represented normal and revolutionary progress with a cyclical *branching* phase model (like Darwin's tree of life) suggesting that incommensurability may play a function analogous to reproductive barriers and isolating mechanisms involved in speciation:

> To anyone who values the unity of knowledge, this aspect of special-ization – lexical or taxonomic divergence, with consequent limitations on communication – is a condition to be deplored. But such unity may be in principle an unattainable goal, and its energetic pursuit might well place the growth of knowledge at risk. Lexical diversity and the principled limit it imposes on communication may be the isolating mechanism required for the development of knowledge.[48]

Feyerabend looked closely at methodology, abstracting from actual his-tory. Kuhn looked closely at how communities actually made progress. They both tried to explain what progress is from different perspectives, and they both used incommensurability to describe theories separated by a revolution that changed how experience was used to understand what happens. Incommensurability can make sense if science is understood as a process of improving our collective understanding, as fallible speculations compete, as we try to learn from *all* our mistakes. Seen in this way, science is a collective activity of improving our understanding through conjecture, corroboration, and selection of the best explanation.

As we have seen, in his early papers, Feyerabend argued that incom-mensurability implies pluralism so that theory proliferation best promotes progress. Kuhn, in his late philosophy, appears to have come to a similar conclusion. He claimed that when we look at progress, we see specialties proliferate by transitioning to a new, usually much narrower, taxonomi-cally structured vocabulary, but:

> What lies outside of it becomes the domain of another scientific spe-cialty, a specialty in which an evolving form of the old kinds remains in use. Proliferation of structures, practices, and worlds is what preserves the breadth of scientific knowledge, intense practice at the horizons of individual worlds is what increases its depth.[49]

So in the end, pluralism and theory proliferation are the best strategies for improving our understanding, according to both Kuhn and Feyerabend.

Incommensurability implies that science is a game where you can break the rules and there is not always a neutral or impartial arbiter that decides who wins. Epistemic values that guide progress vary within communities and may have to be suspended or re-envisaged to make progress, pending further developments or revisions. Scientific progress can make the impossible possible through reclassification that breaks the no-overlap principle. Revisionary reclassification allows new incommensurable theories to act as new synthetic *a priori* principles so that facts and impossibilities are only facts and impossibilities relative to the theories that they corroborate. What counts as the facts may change as better explanations of what is happening become available. Because everything (theories, methods, facts) is explained and tested together, and revisions may be needed anywhere, it seems impossible to predict what progress will look like in the future (nor exactly when or where it will happen), but with enough hindsight, we seem easily to know it when we see it. It is recognised in retrospect as a consilience of corroborations of a new emerging worldview.

To understand a theory is not just to have the ability to apply it to solve the right kind of problems (by finding evidence that corroborates it, and thus is explained by it). It also includes learning to understand why it is currently the best available explanation *in contrast to its rivals*. Kuhn and Feyerabend both emphasised that understanding antiquated theories helps understand what progress is and even promotes progress in science. Whether from a logical perspective and from a historical/sociological perspective, revolutionary progress is best explained merely neither as a series of conjectures and confirmations, nor even a series of conjectures and refutations. It appears better described as a competition to provide the most fruitful realistic explanation of experience as manifest historically in a consilience of corroborations that changed the set of kinds taken to be real.

Understanding of incommensurability and how pluralism promotes progress may improve, as we improve our understanding of how theories, facts, and methods are used to improve each other, and as conceptually incompatible communities with ontologically incompatible beliefs continue to compete to provide the most fruitful explanations of what is real. We should try to learn from *all* our mistakes. After all, mistakes *are* progress, when we learn from them.

Acknowledgements

Thanks to Paul Hoyningen-Huene and Yafeng Shan for helpful comments and suggestions.

Notes

1. See the titles of Feyerabend (1958) and Feyerabend (1963).
2. See Feyerabend (1962).
3. See Feyerabend (1963).
4. See Oberheim (2005)

5. Einstein used 'incommensurable' in Einstein (1949). For discussion, see Oberheim (2016).
6. See Duhem (1954 [1906/1914]) and Howard (1990).
7. Compare Popper (1959 [2005]), p. 98 and Popper (1935 [1989]), p. 70.
8. Feyerabend had worked for Popper and was widely seen as a Popperian at the time. See Collodel (2016). With his discussion of incommensurability, he was trying to distance himself from Popper. A footnote specifically identifying Popper as the target of his criticism was removed from a draft of Feyerabend (1962). See Collodel and Oberheim (forthcoming), *Vol. 3*.
9. See Feyerabend (1970a), p. 222.
10. Feyerabend (1965a), p. 268.
11. Feyerabend (1970a), p. 222. Compare Feyerabend (1978), p. 68, and Feyerabend (1975), p. 269.
12. Feyerabend (1965b), p. 227. Compare Feyerabend (1975), pp. 269–270 and p. 276.
13. Feyerabend (1977), p. 365.
14. Feyerabend decided there was no problem of the existence of theoretical entities because all entities are theoretical when used to explain what is real.
15. See Feyerabend (1962).
16. See Feyerabend (1965c), p. 150.
17. Compare, for example, Feyerabend (1965b), pp. 224–225 and Feyerabend (1975), p. 30.
18. See Putnam (1981), p. 114 and Feyerabend (1987).
19. See Kuhn (1962/1970).
20. Kuhn's views became increasingly less naturalistic: 'many of the most central conclusions we drew from the historical record can be derived instead from first principles'. Kuhn (1992) in Kuhn (2000) p. 112. For discussion, see Bird (2008). In stark contrast, Feyerabend always adamantly rejected any form of historical necessity. See, for example, Hoyningen-Huene (1995).
21. Kuhn argued that without hermeneutic history to counterbalance present-centered history, we cannot understand its actual historical development, as scientific revolutions would be invisible. See Kuhn (1962/1970), Ch. XI. Others have argued that although there may appear to be revolutions, they are illusory:

> The principles which constituted the triumph of the preceding stages of the science, may appear to be subverted and ejected by the later discoveries, but in fact they are (so far as they were true) taken up into the subsequent doctrines and included in them. They continue to be an essential part of the science. The earlier truths are not expelled but absorbed, not contradicted but extended; and the history of each science, which may thus appear like a succession of revolutions, is in reality, a series of developments.
>
> Whewell (1967 [1847]), p. 8

22. See Kuhn (1977a).
23. In the postscript added to Kuhn (1962), he suggests that members of incommensurable language communities 'become translators'" which later, he realised is something that incommensurability precludes, see Kuhn (1962/1970), p. 175.
24. See especially Kuhn (1991) in Kuhn (2000), pp. 92–96.
25. See Kuhn's *The Plurality of Worlds* (unpublished/1994).
26. They may also be mutually exclusive cognitively. For discussion, see Bird (2008).
27. See Kuhn (1962/1970), p. 171.

28. See Sankey and Hoyningen-Huene (2001). For Kuhn's views on 'methodological incommensurability', see especially Kuhn (1977b) in Kuhn (1977a), pp. 320–339.
29. See Popper on the 'hypothetico-deductive model' of testing theories in Popper (2005 [1959]). Compare Duhem (1954 [1906/1914]), who had a more refined account of the logic of scientific justification, according to which theory comparison is a process of attempting to refute all the available alternatives except one, which always leaves open the possibility that there may be as yet better unconceived alternatives. Compare Stanford (2006).
30. See Hempel and Oppenheim (1948).
31. See Kuhn (1993) in Kuhn (2000), pp. 232–238.
32. Nonsense may still have a positive role to play pedagogically. See, for example, Baltas (2008).
33. This was one of Feyerabend's examples. See Feyerabend (1962) and Oberheim (2006), p. 54, fn. 353.
34. The 'things' could be any kind of 'things', for example, objects, processes, properties, relations, or whatever.
35. Duhem stresses this point in his (1954 [1906/14], which Einstein repeated. For detailed references and discussion, see Howard (1990) and Oberheim (2016).
36. See Popper (2005 [1959]) and Popper (1963).
37. For the details of how 'theory-mediated measurement' was used to produce observations of Brownian motion that count as *evidence*, see Smith and Seth (2020), For details of how Newton theory had used theory-directed measurements, see Harper (2011).
38. That is why Feyerabend thought incommensurability undermines Popper's falsificationist account of scientific justification as conjectures and *refutations*.
39. Whewell calls a similar idea a 'consilience of inductions'. See Whewell (1967 [1847]), p. 65. While Whewell's "consilience of inductions" has been interpreted in several different ways, his basic intuition seems to be similar: the more ways there are to test a theory by its predictions, the greater the severity of those tests, and therefore the greater the corroborative support conferred. See Fischer (1985), p. 254.
40. See Laudan and Leplin (1991).
41. Compare Feyerabend (1965c), p. 150.
42. According to Feyerabend

> Even the stability of a testable hypothesis cannot be regarded as a sign of its truth because it may be due to the fact that, owing to some particular astigmatism on our part, we have has yet overlooked some very decisive tests. There is no sign by which factual truth may be recognized.
>
> Feyerabend (1962), p. 70, fn. 86

43. See Psillos (1999).
44. For more on world change, see Hoyningen-Huene (1993) and Horwich (1993).
45. This is the assumption that is incompatible with scientific realism. Ontological commitments, even when mistaken, can lead to progress. See De Regt (1999).
46. Compare:

> We want the observed facts to follow logically from our conception of reality. Without the belief that it is possible to grasp reality with our theoretical constructions, without the belief in the inner harmony of our world, there could be no science.
>
> Einstein and Infeld (1938), p. 312

47. See Feyerabend (1970b).
48. Kuhn (1991), pp. 98–99.
49. Kuhn (1993), in Kuhn (2000), p. 250.

References

Baltas, A. (2008) "Nonsense and Paradigm Change". In L. Soler, H. Sankey, and P. Hoyningen-Huene (eds.) *Rethinking Scientific Change and Theory Comparison: Stabilities, Ruptures, Incommensurabilities*. Dordrecht: Springer, pp. 49–70.

Bird, A. (2008) "Incommensurability Naturalized". In L. Soler, H. Sankey, and P. Hoyningen-Huene (eds.) *Rethinking Scientific Change and Theory Comparison: Stabilities, Ruptures, Incommensurabilities*. Dordrecht: Springer, pp. 21–39.

Collodel, M. (2016) "Was Feyerabend a Popperian? Methodological issues in the History of the Philosophy of Science". *Studies in the History and Philosophy of Science A* 57: 27–56.

Collodel, M. and Oberheim, E., eds. (forthcoming) *Feyerabend's Formative Years. Volume 3. Feyerabend among Popperians*. Chicago: University of Chicago Press.

De Regt, H. (1999) "Pauli versus Heisenberg: A case study of the heuristic role of philosophy". *Foundations of Science* 4: 405–426.

Duhem, P. (1954 [1906/1914]) *The Aim and Structure of Physical Theory*. Princeton: Princeton University Press.

Einstein, A. (1949) "Autobiographical notes". In P. Schilpp (ed.) *Albert Einstein: Philosopher-Scientist*. La Salle: Open Court, pp. 3–95.

Einstein, A. and Infeld, L. (1938) *The Evolution of Physics: From Early Concepts to Relativity and Quanta*. Cambridge: Cambridge University Press.

Feyerabend, P. (1958) "An attempt at a realistic interpretation of experience". *Proceedings of the Aristotelian Society. New Series* 58: 143–170.

Feyerabend, P. (1962) "Explanation, Reduction and Empiricism". In H. Feigl and G. Maxwell (eds.) *Scientific Explanation, Space and Time. Minnesota Studies in the Philosophy of Science, Volume III*. Minneapolis: University of Minneapolis Press, pp. 28–97.

Feyerabend, P. (1963) "How to Be a Good Empiricist: A Plea for Tolerance in Matters Epistemological". In B. Baumrin (ed.) *Philosophy of Science: The Delaware Seminar, Vol. II*. New York: Interscience Publishers, pp. 3–39.

Feyerabend, P. (1965a) "On the 'Meaning' of Scientific Terms". *The Journal of Philosophy* 62: 266–274.

Feyeraben, P. (1965b) "Reply to Criticism: Comments on Smart, Sellars and Putnam". In R. Cohen and M. Wartofsky (eds.) *Proceedings of the Boston Colloquium for the Philosophy of Science 1962–64: In Honor of Philipp Frank* (Boston Studies in the Philosophy of Science, Volume II). New York: Humanities Press, pp. 223–261.

Feyerabend, P. (1965c) "Problems of Empiricism". In R. Colodny (ed.) *Beyond the Edge of Certainty. Essays in Contemporary Science and Philosophy*. Pittsburgh: CPS Publications in the Philosophy of Science, pp. 145–260.

Feyerabend, P. (1970a) "Consolations for the Specialist". In I. Lakatos and A. Musgrave (eds.) *Criticism and the Growth of Knowledge. Proceedings of the International Colloquium in the Philosophy of Science, London, 1965*. Cambridge: Cambridge University Press, pp. 197–230.

312   Eric Oberheim

Feyerabend, P. (1970b) "In Defense of Classical Physics". *Studies in History and Philosophy of Science* 1: 59–85.

Feyerabend, P. (1975) *Against Method. Outline of an Anarchistic Theory of Knowledge*. London: New Left Books.

Feyerabend, P. (1977) "Changing Patterns of Reconstruction." *British Journal for the Philosophy of Science* 28: 351–382.

Feyerabend, P. (1978) *Science in a Free Society*. London and New York: New Left Books.

Feyerabend, P. (1987) "Putnam on Incommensurability". *The British Journal for the Philosophy of Science* 38: 75–81.

Fischer, M. (1985) "Whewell's Consilience of Inductions – An Evaluation". *Philosophy of Science* 52: 239–255.

Harper, W. (2011) *Isaac Newton's Scientific Method. Turning Data into Evidence about Gravity and Cosmology*. Oxford: Oxford University Press.

Hempel, C. and Oppenheim, P. (1948) "Studies in the Logic of Explanation." *Philosophy of Science* 15: 135–145.

Horwich, P., ed. (1993) *World Changes: Thomas Kuhn and the Nature of Science*. Cambridge, MA: MIT Press.

Howard, D. (1990) "Einstein and Duhem". *Synthese* 83: 363–384.

Hoyningen-Huene, P. (1993) *Reconstructing Scientific Revolutions. The Philosophy of Science of Thomas S. Kuhn*. Chicago: Chicago University Press.

Hoyningen-Huene, P., ed. (1995) "Two Letters by Paul Feyerabend to Thomas S. Kuhn on a Draft of 'The Structure of Scientific Revolutions'". *Studies in History and Philosophy of Science A* 26: 353–387.

Kuhn, T. (1962/1970) *The Structure of Scientific Revolutions*. Chicago: University of Chicago Press. Page references are to the second enlarged edition with new "Postscript – 1969" published 1970, and unchanged in subsequent editions.

Kuhn, T. (1977a) *The Essential Tension*. Chicago. Selected Studies in Scientific Tradition and Change. Chicago: University of Chicago Press.

Kuhn, T. (1977b) "Objectivity, Value Judgement, and Theory Choice". In Kuhn (1977a), pp. 320–339.

Kuhn, T. (1991) "The Road since Structure". In Kuhn (2000), pp. 90–104.

Kuhn, T. (1992) "The Trouble with the Historical Philosophy of Science". In Kuhn (2000), pp. 105–120.

Kuhn, T. (1993) "Afterwards". In P. Horwich (1993), pp. 311–341. Reprinted in Kuhn (2000), pp. 224–252.

Kuhn, T. (2000) *The Road since Structure. Philosophical Essays, 1970–1993, with an Autobiographical Interview*. Chicago: University of Chicago Press.

Kuhn, T. (unpublished/1994) *The Plurality of Worlds. An Evolutionary Theory of Scientific Development*.

Laudan, L. and Leplin, J. (1991) "Empirical Equivalence and Underdetermination". *Journal of Philosophy* 88: 449–472.

Oberheim, E. (2005) "On the Historical Origins of the Contemporary Notion of Incommensurability: Paul Feyerabend's Assault on Conceptual Conservativism". *Studies in the History and Philosophy of Science A* 36: 363–390.

Oberheim, E. (2006) *Feyerabend's Philosophy*. Berlin and New York: De Gruyter.

Oberheim, E. (2016) "Rediscovering Einstein's Legacy. How Einstein Anticipates Kuhn and Feyerabend on the Nature of Science". *Studies in the History and Philosophy of Science A* 57: 17–26.

Popper, K. (1935 [1989]) *Logik der Forschung. Zur Erkenntnistheorie der modernen Naturwissenschaft*. Vienna: Julius Springer. Ninth edition 1989. Tübingen: Mohr Siebeck.

Popper, K. (1959 [2005]) *The Logic of Scientific Discovery*. London: Hutchinson and Co. Published 2005. London and New York: Routledge. Translation of Popper (1935 [1989]).

Popper, K. (1963) *Conjectures and Refutations*. London: Routledge & Paul Kegan.

Psillos, S. (1999) *Scientific Realism. How Science Tracks Truth*. London: Routledge.

Putnam, H. (1981) *Reason, Truth and History*. Cambridge: Cambridge University Press.

Sankey, H. and Hoyningen-Huene, P. (2001) "Introduction". In P. Hoyningen-Huene and H. Sankey (eds.) *Incommensurability and Related Matters*. Dordrecht: Kluwer, pp. vii–xxxiv.

Smith, G. and Seth, R. (2020) *Brownian Motion and Molecular Reality*. Oxford: Oxford University Press.

Stanford, P. (2006) *Exceeding Our Grasp: Science, History, and the Problem of Unconceived Alternatives*. New York: Oxford University Press.

Whewell, W. (1847 [1967]) *History of the Inductive Sciences, from the Earliest to the Present Times*. London: J. W. Parker and Son.

16 Scientific Progress and Aesthetic Values

Milena Ivanova

1. Introduction

Aesthetic values have featured in scientific practice for centuries, shaping what theories and experiments are pursued, what explanations are considered satisfactory, and whether theories are trusted. How do such values enter in the different levels of scientific practice and should they influence our epistemic attitudes? In this chapter I explore these questions and how throughout scientific progress the questions we ask about the role of aesthetic values might change. I start this chapter with an overview of the traditional philosophical distinction between context of discovery and context of justification, showing how aesthetic values were taken to be relevant to scientific discovery and not scientific evaluation, which was regarded value-proof. I then proceed with an exploration of different levels of scientific activities, from designing experiments and reconstructing fossils, to evaluating data. In this discussion we will see that the traditional context of discovery and justification seems to break down, as aesthetic values shape all levels of scientific activity. I then turn our attention to the epistemological question: can beauty play an epistemic role, is it to be trusted, or is it a suspect value that might bias scientific inquiry? I explore how we could justify the epistemic import of aesthetic values and present some concerns as well. In the last section I ask whether we should expect the questions surrounding aesthetic values in scientific practice to change with scientific progress, as we enter the era of post-empirical physics, big data science, and make more and more discoveries using AI.

2. Discovery and Justification

Stories of scientific discovery are filled with intriguing personal reflections of sudden illuminations, random and chaotic processes, and personal guiding values. It is no surprise that for so long the philosophical approach to discovery was one of pure mystery; romanticist approaches to creativity emphasise the role of genius and mysticism surrounding the discovery process, and scientists in their turn described their discoveries

DOI: 10.4324/9781003165859-20

dressed in inspirationalist cloaks. From Keluke's dream to Poincaré's stroll, discoveries in science and mathematics have been usually presented to us without reference to the context within which the problematisation happens, without much discussion of the community effort and ground work required for the work to take place, and much emphasis has traditionally been given to the sudden moment of illumination without reflection on what happens before and after the light bolt moment.

Personal values have often been at the forefront of the stories told about discoveries. Take for instance Poincaré's account of creativity, which he develops in his chapter 'Mathematical Discovery' from *Science and Method*. Similarly to inspirationalism and the combinational account of creativity, Poincaré takes creativity to be the identification of unsuspected relations among known facts. But what is rather original in his account is the emphasis on the aesthetic sensibility. For Poincaré the creative process takes several stages: preparation, incubation, insight and revision. During preparation the subject consciously studies the problem at hand, while in incubation the mind freely explores possibilities without being conscious or constrained, then the 'the unconscious machine' comes up with 'sudden illuminations'. Fortunately Poincaré tells us a little more about what happens next, as discovery is not just that one moment of illumination, which would have problematic repercussions for our credit attribution practises, given that it is an unconscious process. He tells us that the sudden illumination is followed by a critical conscious reflection in which the produced ideas are assessed, but this assessment is delivered by the scientist's aesthetic sensibility, which acts as a 'delicate sieve', selecting the theories or proofs that best suit our aesthetic requirements.

Poincaré made discovery a little less mysterious and random compared to traditional inspirationalist accounts,[1] by giving it clearly defined stages, and opened the door to more productive explorations of scientific creativity. It nevertheless continues to place a central role to the aesthetic sensibility, implying that our aesthetic values shape what ideas we come up with. It is no surprise, then, that for so many decades philosophers of science refused to engage systematically with discovery. Hans Reichenbach (1938) drew a sharp distinction between context of discovery and context of justification, which was for decades adopted by the logical positivists' school and also formed Karl Popper's (1963) views on the scientific method. The idea behind this distinction is that aesthetic values are deemed psychological and subjective, and while they may be operating in the context of discovery, they have no place in our theory of knowledge. For Reichenbach, in constructing our theory of knowledge we should focus only on how our evidence relates to our hypotheses and ignore subjective or psychological factors that are operating in the discovery context. As he argues, 'It would be a vain attempt to construct a theory of knowledge which is at the same time logically complete and in strict correspondence with the psychological process of thought' (Reichenbach 1938, 5).

Thus, from the time of the Vienna Circle up to even recent literature, aesthetic considerations were not the focus of systematic philosophical attention. If their presence in scientific practice was acknowledged, they were only given relevance to the context of discovery, rendering them not part of our rational acquisition of knowledge.

More recent reflections on scientific practice, however, have called into question the sharp delineation between the context of discovery and justification. It seems that value is pervasive in all levels of scientific theorising, and even selecting, processing, and shaping the nature of the evidence is not without the involvement of value. From feminists critiques of science, exposing the heavy gender bias operating in how the evidence is selected and interpreted (Mann 2020; Martin 1991; Criado-Perez 2019), to showing the operation of value in the selection and interpretation of theories (Duhem 1954; Kuhn 1962), it seems like values are everywhere in the practice of science and the meaning of our observations and experimental results are not immune to our values. Whether this is bad news, a highly problematic aspect of science that needs to be amended for, or simply a fact about how we as agents operate in this world and shape our enquiry, has been disputed. For those who follow in the logical positivists' path, science should be value free in order to deliver the epistemic goods we aim for: objective knowledge about the world. On the other hand, feminist philosophers such as Helen Longino (1990) have argued that values do not threaten the objectivity of science, but simply shift our understanding of objectivity not as an aspect of individual agents but of communities. Longino and Philip Kitcher (1993) have argued that the values operating in science need to reflect those of the communities that science serves, making values subject to critical democratic evaluation.[2]

In addition to the dissolution of the context of discovery and justification, certain philosophical developments in philosophy of science over the last decades have opened the door to studying the relationship between art and science. The popularity of the semantic approach to scientific theories put at the centre of attention the notion of representation, which led philosophers of science to draw analogies between scientific models and artworks as representational vehicles (van Fraassen 2008; Frigg and Hunter 2010). Additional momentum to the interaction between philosophy of science and aesthetics was given by Bueno et al. (2017), and Ivanova and French (2020), opening further questions for investigation, including what aesthetic responses are elicited by scientific products, how aesthetic values shape our understanding, pictorial and other forms of representation and their aesthetic values, and the aesthetic nature of thought experiments. Instrumental in drawing connections between art and science has also been Catherine Elgin's (1991) work, which compared literary works to thought experiments, showing how our understanding can be advanced through notions such as exemplification. We will return to the normative question whether aesthetic values undermine epistemic goals in the later sections of this chapter. For now, let us have a look at how aesthetic considerations shape scientific practice.

3. How Aesthetic Values Shape Scientific Practice

Aesthetics values seem to feature rather prominently in all types of scientific activities, and we can identify at least three levels of scientific practise at which we can find aesthetic judgements at play. First, the subject of our investigations, nature herself, is often regarded to afford aesthetic experiences, to be beautiful. From sunsets and beehive honeycombs, to rainbows and snowflakes, nature is regularly regarded as aesthetically pleasing, generating is us feelings of awe and wonder.[3] Second, the very products of scientific activities are also aesthetically appraised. The images scientists produce are often the subject of aesthetic appreciation, from pictures of oscillating particles, to Leonardo da Vinci's careful depiction of the human body, to Robert Hooke's drawings of the flee, scientific images have for centuries been compared to, or considered to be, artworks and claimed to afford aesthetic experiences in the viewer. Likewise, scientific models like the double helix structure of DNA molecules or the different models of the atom, mathematical proofs like Euclid's Elements, experiments like Rutherford's explorations of uranium radiation, and many scientific theories, from Newton's and Einstein's theories of gravity, to Darwin's theory of evolution, the standard model in particle physics and string theory, all are claimed to be aesthetically valuable, beautiful, elegant, and simple, and to elicit in us aesthetic responses (Ivanova 2017a). Last, the very process by which scientists arrive at a product, whether constructing a proof or an experiment or arriving at a theory, can be subject to aesthetic judgements. The French physicist Pierre Duhem argues that:

> [I]t is impossible to follow the march of one of the great theories of physics, to see it unroll majestically its regular deductions starting from initial hypotheses, to see its consequences represent a multitude of experimental laws down to the small detail, without being charmed by the beauty of such a construction, without feeling keenly that such a creation of the human mind is truly a work of art.
>
> (Duhem 1954, 24)

Ernest Rutherford similarly reflects that theories can be seen as artistic productions: 'a strong claim can be made that the process of scientific discovery may be regarded as a form of art. . . . A well constructed theory is in some respects undoubtedly an artistic production' (quoted in McAllister 1996, 14). In the context of designing and performing an experiment, scientists are also very often compared to artists. Reflecting on the Michelson-Morley experiment, which aimed to detect ether drift, the method employed' (quoted in Holton 1969, 157). The comparisons between scientific and artistic production can be traced back to the natural philosophers, during a time where generating feelings of amusement was one of the goals of the experimenter and experimental practice was often seen as a public spectacle (Ivanova 2021b; Parsons and Reuger 2000; Wragge-Morley 2020).

Let us explore some specific examples that shed light on how aesthetic values operate in the different levels of scientific practice. Caitlin Wylie (2015), and more recently Derek Turner (2019) and Adrian Currie (2020), have offered an illuminating discussion of how aesthetic factors function in the preparation of fossils in palaeontology. Wiley observed that when palaeontologists reconstruct a specimen, they employ their knowledge as well as aesthetic sensibility and enhance the aesthetic features of the specimen. This means that our aesthetic values have already entered the scientific activities before we have even started theorising, in the very preparation of the evidence we have. Glenn Parsons (2012) has also reflected on how aesthetic factors feature in chemistry, where it is not only that molecules are found to display aesthetic value, by displaying certain formal properties such as symmetry and elegance, but also the very process of synthesis employed in producing new molecules is a source of aesthetic appreciation (2012, 578). When it comes to medicine, Brendan Clarke and Chiara Ambrosio (2018) have discussed how visual aesthetic factors have influenced studies in anatomy. Alexander Waggle-Morley (2020) has further discussed how drawings of specimen by natural philosophers, such as Robert Hooke, embody aesthetic qualities, showing that the aim of the natural philosophers was not simply to depict the specimen, but to enhance nature's aesthetic qualities and provoke an aesthetic response in the viewer.

Aesthetic considerations are also part and parcel of designing experiments. Scientists care for the creative process behind a well-designed experiment, and praise a good experiment for its beauty and its designers for their creativity and ingenuity (Ivanova 2021b; Parsons and Reuger 2000). And interestingly, even the reception of an experiment can be subject to aesthetic values. Experimental results are often evaluated on their beauty, regarding how clear they are. For instance, the famous Meselson-Stahl experiment in molecular biology, which confirmed the method by which DNA replicates, has been regarded as beautiful not only because of its design but also because of the clear and significant results it produced (Ivanova 2021a, 2021b). And even thought experiments have received aesthetic praise for their elegance and simplicity, with Brown (2004) arguing that simplicity and original set-up makes Galileo's thought experiment on falling bodies the most beautiful thought experiment in science.

One could claim that the aforementioned activities and practices are all part of the context of discovery; ultimately evidence settles which hypotheses or theories withstand testing. We have already seen that scientific practice does not seem to allow us to make this distinction so clearly, and the extraction and evaluation of data is subject to values too, but it is worth exploring next how even our epistemic stance towards a theory can be effected by aesthetic values. In the following, I explore how scientists use beauty to justify their belief in scientific theories. Before I proceed, let me just mention that beauty can be understood as an irreducible aesthetic quality, and some scientists have certainly taken this stance, but in the following discussion, it will become clear that often scientists take a reductionist understanding of beauty, analysing it in terms of aesthetic values such as elegance, simplicity, and symmetry.

4. Beauty: Systematic Bias or Guide to the Truth?

Beauty is often not only seen as a motivator in scientific enquiry, but is also given significant epistemic weight by being regarded as an indicator of the theory's truth. When it comes to its motivational role, scientists have claimed that they study nature in order to find the beauty within it. The French mathematician and scientist Henri Poincaré was instrumental in the development of this idea, that beyond the practical benefits of scientific progress, scientists are invested in studying the world because this activity delivers an aesthetic pleasure. Uncovering nature's beauty, Poincaré argues, is what motivates scientists' work. The beauty scientific practice uncovers is not one that is easily perceivable with our senses, he argues, but rather is uncovered by our intellect's engagement with nature shaped by our understanding. In his own words:

> [t]he scientist does not study nature because it is useful to do so. He studies it because he takes pleasure in it, and he takes pleasure in it because it is beautiful . . . I am not speaking, of course, of the beauty which strikes the senses, of the beauty of qualities and appearances. I am far from despising this, but it has nothing to do with science. What I mean is that more intimate beauty which comes from the harmonious order of its parts, and which pure intelligence can grasp.
> (Poincaré 2001, 368)

The Nobel laureate Subrahmanyan Chandrasekhar, who dedicated much of his time thinking about the role of beauty in science, claims that scientists and artists ultimately have the same aims and that is to discover beauty in all its forms (Chandrasekhar 1987, 52).

Being a motivator is one thing, but it hardly tells us that beauty can play a role outside of the context of discovery. However, in addition to being a motivator, beauty is often also seen as a pragmatic element in theory development as well as in theory evaluation. It is argued that it is convenient for us to develop theories that satisfy our aesthetic requirements and if we need to make choices between hypotheses that are equally supported by the evidence, it is most convenient for us to choose simplicity over complexity, beauty over ugliness. As the Austrian physicist Ernst Mach (1984) argues, we should always aim to explain the phenomena in the most conceptually economical manner, because simple theories are easier to use. He argues that science should be seen as an economical description of our observations, with simplicity being a guiding principle in the construction and evaluation of scientific hypotheses. In a similar manner, Poincaré argues that there is a relationship between utility and simplicity, in that a simpler theory is easier to use, making simplicity a regulative principle in the construction and choice of theories. According to him, 'care for the beautiful leads us to the same selection [of theories] as care for the useful' and an economical theory 'is a source of beauty as well as a practical advantage' (Poincaré 2001, 369). In the more recent literature on scientific realism, Bas van Fraassen has also acknowledged the need to appeal to non-empirical factors in scientific practice. However, these aesthetic considerations are regarded as purely 'human concerns, a function of our interests and pleasures, which make some theories more valuable or appealing to us than others. These values provide reasons for using a theory but cannot rationally guide our epistemic attitudes and decisions' (van Fraassen 1980, 87). The aesthetic values are thus informing our activities, but should not guide our beliefs in the truth of a theory.

Contrary to this empiricist take on the role of aesthetic factors in science, many scientists take beauty to do much more than just act as a heuristic tool for convenience; they believe beauty stands in a special relationship to truth. Many contemporary and past scientists have defended the idea that a beautiful theory is more likely to be true, and if faced with a choice between two theories, the simplest or more beautiful theory should be epistemically privileged. A known defender of this idea is Paul Dirac, whose defence of the general theory of relativity, prior to its empirical confirmation, was motivated by his inherent believe in the epistemic import of beauty. He claimed that 'one has a great confidence in [a] theory arising from its great beauty, quite independently of its detailed successes', continuing that '[o]ne has an overpowering belief that [the theory's] foundations must be correct quite independently of its agreement with observation' (Dirac 1980, 40). Such convictions were shared by many physicists at the time, including Eddington himself as well as Werner Heisenberg, who claimed that '[i]f nature leads us to mathematical forms of great simplicity and beauty we cannot help thinking that they are 'true', that they reveal a genuine feature of nature' (Heisenberg 1971, 68). In a similar manner, in his 1979 lecture 'Beauty and the Quest for Beauty in Science' presented at the Fermi National Accelerator Laboratory, Chandrasekhar claims that we should trust scientists with well-trained aesthetic sensibilities, because their aesthetic preferences have epistemic import: 'we have evidence that a theory developed by a scientist, with an exceptionally well-developed aesthetic sensibility, can turn out to be true even if, at the time of its formulation, it appeared not to be so' (Chandrasekhar 1987, 64). With Einstein's general theory of relativity in mind, he claims that the theory's beauty – its achieving the unification of our fundamental concepts of space and time, and the concepts of matter and motion with 'unerring sense for mathematical elegance and simplicity' – speaks in favour of its truth (ibid., 71). He argues that the discovery of beauty in nature is the most significant of achievements, that it is an 'incredible fact that a discovery motivated by a search after the beautiful in mathematics should find its exact replica in Nature' (ibid., 54).

The debate on the role of beauty in science has taken a new turn recently with two domains in contemporary physics where the question of whether and how beauty should feature in the progress of fundamental physics is receiving a lot of attention. There are two ways to think about this discussion. First, in the last decades, fundamental physics has aimed to advance the standard model and a lot of the work particle accelerators were after was to confirm the standard model, by detecting the Higgs boson which the theory had predicted, and to further advance our understanding of fundamental physics beyond the standard model. Part of the quest in the last decades has been to discover 'SUSY' particles: these are super symmetric particles that are entailed by the theory. Sabine Hossenfelder observes that the symmetry principle has become something of an imperative in the particle physics community, noting that the Nobel laureate Marry Gell-Mann, whose use of symmetry principles led to the advancement of the standard model and the discovery of the previously unknown particle now called omega minus, has defended the deep connection between beauty and truth (Hossenfelder 2018, 37). But while many remain enthusiastic about the fertility of the symmetry principle and believe we need to invest efforts in building more powerful colliders to discover SUSY particles, others are questioning whether trust in the symmetry principle is well placed. Hossenfelder challenges the contemporary trust in beauty among the community of particle physicists, asking whether beauty might in fact be a systematic bias in contemporary physics, leading physicists to pursue research programmes that are not fruitful (Hossenfelder 2018). Second, again in contemporary high-energy physics, our theories are becoming concerned with energy ranges that will be impossible in principle to subject to test. The discussions surrounding the status of string theory concern exactly whether we should trust this theory given its non-empirical status, with Richard Dawid (2013) calling us to redefine the scientific method to accommodate for what modern physics looks like, claiming that non-empirical constraints should be introduced to guide trust and acceptance of theories in the 'post empirical' stage of science. This discussion has led us to consider the demarcation question anew, with Ellis and Silk (2014) asking whether string theory, as a theory that will not be empirically confirmed, should even be taken to be genuinely scientific. They express grieve concerns with the contemporary use of aesthetic ideals and other non-empirical considerations among the physicist community, arguing that such 'breaking with centuries of philosophical tradition of defending scientific knowledge as empirical' is dangerous and calls into question the integrity of science (Ellis and Silk 2014, 321).

It is easy to see why the debate on the role of aesthetic ideals in science has become so prominent in contemporary high-energy physics, with so much at stake. This leads us to the question of how we can justify the epistemic significance of aesthetic values, should we trust beauty to be something beyond a heuristic and motivational tool? One way to respond to this question is to argue that beauty is a learned ideal, one that scientists acquire through practice by working with successful theories. We can argue that our aesthetic ideals are ultimately shaped by our education and practice. James McAllister (1996) calls the formation of aesthetic values that a community collectively establishes an 'aesthetic canon'. The idea is that scientists learn to operate with such aesthetic canons and while those canons are based upon reflections on the successful theories of the past and present, the canons help to mould and shape decisions and confidence in new hypotheses and theories. Let's explore these ideas with an example. We could claim that a reflection on our past successful theories that exhibited empirical success can justify confidence in trusting unity, since our best theories in physics achieved high levels of unification. Copernicus unified terrestrial and celestial phenomena in one planetary framework, Maxwell unified electric and magnetic phenomena, Einstein unified gravity and inertial mass, and the concepts of space and time into a space-time continuum. This reflection serves to boost our confidence in the unifying power of theories. As Richard Boyd (1984) has argued, such reflection on the aesthetic features of past successful theories helps us evaluate the plausibility of new hypothesis, by establishing whether they fit our aesthetic canon (to use McAllister's term). Thus, according to this account, one could be confident about the plausibility of new theories even before they are confirmed by the evidence, if these theories fit our aesthetic canon.[4]

In *Lost in Math*, however, Hossenfelder questions such inferences. She recognises that beauty can play a diversity of roles in science, from a motivator and guide to research, to generating feelings of reward in a scientist, but she also argues that often beauty can be a very systematic bias in a scientific community that needs to be identified, evaluated, and amended for if found to go against community goals. Hossenfelder reflects on the track record argument, often implied by contemporary physics, suggesting that their 'faith in beauty's guidance is built on its use in the development of the standard model and general relativity; it is commonly rationalized as an experience value: it worked so it seems reasonable to keep using it' (2018, 26). But against such track record argument, she draws our attention to the fact that some of our most successful theories have failed to fit our aesthetic ideals, while other much beloved theories that fitted those ideals have failed to gain empirical support. Hossenfelder argues that 'not only does the history of science thrive with beautiful theories that turned out to be wrong, but on the flipside we have the ugly ides that turned out to be correct' (2018, 31).

We can certainly reflect on the history of science and see beautiful theories that we do not regard to be true. Copernicus' heliocentric system was abandoned in favour of Kepler's despite the initial resistance to replace perfect cycles for epicycles. Newton's theory of gravitation is not only considered unifying and beautiful but also is at best an approximation. Aristotelian biology has simplicity and scope, but is false. Darwinian evolution lacks the clear boundaries between species and adds chance in their evolution, which is much messier than Aristotelian appeal to telos and essence and proper function. Unificationist projects in high-energy physics have been abandoned due to difficulties despite their aesthetic appeal, with the Kaluza-Klein theory providing a beautiful, elegant way to unify gravity and all the other gauge fields but abandoned. There is also the problem with some contemporary theories that are very successful, but seem to fail our aesthetic requirements. Our most successful theories, the standard model and quantum mechanics, are often regarded as inelegant and not of aesthetic appeal. Kaku and Thompson observe that the standard model might be our most empirically successful theory, but it is ultimately also one of our ugliest theories, claiming that the 'reason why the Standard Model is so ugly is that it is obtained by gluing, by brute force, the current theories of the electromagnetic force, the weak force, and the strong force into one theory' (Kaku and Thompson 1997, 75). It seems that replying on such track record arguments can go both ways, we can look at the history of science to find beauty failures and ugly successes, but such arguments are ultimately inconclusive (Ivanova 2020).

Earlier I also noted that contemporary discussions on the status of string theory are also a place where non-empirical evaluation has received a lot of attention. Dawid advocates that theories like string theory need to be evaluated on non-empirical grounds, but has been cautious when it comes to employing aesthetic factors, on the grounds of being subjective and contingent. Dawid takes aesthetic values to be psychological and contingent in nature, and for this reason aims to offer more formal arguments that can boost our confidence in a theory. Specifically, Dawid (2013) argues that our trust in string theory is justified on the following (philosophical) grounds: (1) No Alternatives Argument. despite systematic attempts, there are no viable alternatives to string theory; (2) The Argument of Unexpected Explanatory Interconnections: string theory has explanatory success beyond its originally intended domain of application; and (3) The Meta-Inductive Argument: string theory is part of a research programme that has a long record of success. All of these arguments, however, can be disputed. First, is our failure to conceive of alternative theories grounds for justification? Kyle Stanford's (2006) argument from unconceived alternatives might suggest otherwise. The argument questions whether our collective failure to conceive of an alternative explanation at a particular time should guide our trust in its (approximate) truth. The argument appeals to a long list of past cases where the scientific community had not been in a position to entertain an alternative (and better) explanation, for instance Aristotelians failed to conceive of the concept of action at a distance, but this failure had nothing to do with the truthlikeness of the agreed upon explanation. Second, is it the case there are no viable current alternatives to string theory? Some think there are indeed alternatives – loop quantum gravity – but there are important social factors affecting their development Smolin (2006). Ellis and Silk (2014) further argue that '[w]e cannot know that there are no alternative theories. We may not have found them yet. Or the premise might be wrong. There may be no need for an overarching theory' (2014, 321). Third, is unexpected explanatory success to be taken as more significant to pure accommodation? The epistemic asymmetry presupposed here between novel prediction and accommodation has been called into question too, identifying the often-contingent factors that lead to a prediction being considered novel (Barnes 2008). But what is perhaps most interesting in this discussion is whether Dawid succeeds in avoiding the problem he levied at those who defend string theory on aesthetic grounds. It is not clear that our judgements of explanatory success, for example, are free from aesthetic and other subjective values, so it can be argued that perhaps the aesthetic values come into play within Dawid's framework too, just at another level of analysis.

It is clear that today, perhaps more than ever, scientists are faced with choices that inevitably involve aesthetic considerations and these choices go way beyond the choice of what theory to pursue and develop, the aesthetic values seem to be playing a much more substantial role in shaping the community's trust in theories and their confidence in research programmes. This section explored one way to justify such trust, by exploiting the track record of beauty in the success of science, but we also saw that such arguments suffer from selection bias and can well be made to defend the very opposite conclusion. In what follows, I will reflect on how scientific progress, and the way science evolves, might change the kinds of questions we focus on when it comes to aesthetic values in science.[5]

5. Science, Progress, and Aesthetic Values

Before we begin our reflection on the new questions about aesthetic values that emerge in science today, I want to clarify how the accounts we explored earlier on whether aesthetic values are truth indicative, heuristic, or motivational relate to accounts of scientific progress. The question is whether our commitments about the role of aesthetic values in science also commit us to a particular construal on what scientific progress consists in. Scientific progress can been understood in several ways: (1) in terms of increasing knowledge (epistemic approach), developed originally by Bury (1920) and more recently by Bird (2007); or in terms of increasing verisimilitude (semantic approach), developed by and Niiniluoto (1980) and Rowbottom (2008); (2) as an increase in problem-solving abilities, developed originally in Kuhn 1962; Lakatos 1968; Laudan 1981 and more recently by Shan (2019); and (3) as increase in understanding, developed by Dellsén (2021). We can see an obvious parallel in how progress is construed with the role of aesthetic values in science. Taking aesthetic values to be truth indicative is continuous with taking progress to be the increase of verisimilitude or knowledge. Taking aesthetic values to be convenient heuristic or motivational tools is continuous with the functional construal of scientific progress, and taking progress to be the increase of our understanding is continuous with taking aesthetic values to be conditions of our understanding. Despite the clear parallel, our stance on these two questions does not necessarily have go hand in hand. That is, our stance on the epistemic significance of aesthetic values does not impose a particular commitment to scientific progress; one could easily commit to any other combination. For instance, one could commit to taking scientific progress to be construed as increase of knowledge but when it comes to aesthetic values, deem them purely instrumental or motivational, rather than truth-tracking. And while these combinations have not explicitly been explored in the literature yet, it is particularly interesting to consider whether construing aesthetic values as constraints in our understanding, as developed by Elgin (2020) and Ivanova (2020), blends in interesting ways with Dellsén's (2021) account of scientific progress.

With this caveat in mind, let us explore some new questions on the role of aesthetic values that emerge currently in science. Has the role of aesthetic values become more prominent in post empirical physics? Are such factors going to be features of future science or are they going to be eliminated from future reasoning, as the positivist envisioned? Science is a human activity, so it seems plausible to assume this activity will continue to reflect our goals, values, and aims. But with the emergence of computer-assisted discoveries, and AI now being part of scientific practice, should we expect the debate surrounding aesthetic values to change and how?

We have seen earlier that as science evolves, its methods have also changed. Traditionally empirical domains are now entering a post-empirical phase. On the other hand we are also living in the age of big data. We now produce data in abundance and by a variety of sources, from agriculture and weather forecast, to transportation and the devices we use in our daily life. A variety of interesting questions with regards to the epistemology of big data have recently received systematically attention (Leonelli and Tempini 2020; Pietsch 2021). It has become clear that the way in which scientists use big data is very much dependent on their values and aims, with data travelling from different domains and applications to serve plurality of goals. It remains to be studied how aesthetic considerations affect the selection, evaluation, and utilisation of big data, but some initial insights are worth mentioning. Leonelli (2020) for instance is apprehensive about the significance of elegance in biology, given that complexity is often more appropriate for capturing facts about biological systems. This is an important point, since it questions whether we can have a universal set of aesthetic values, rather than domain specific ones, that can very much be found in conflict with each other (Ivanova 2017a). On the other hand, Dykes and Wood (2009) explore how beauty can emerge from big data by uncovering underlying patterns and unexpected connections, resonating with how other thinkers, such as Poincare, have previously construed the role of beauty in science (Ivanova 2017b).

As science evolves we also see that much of the practice of science is now delegated to intelligent machines. AI is now successfully used in data extraction and data mining, in the design and implementation of certain experiments, and it has even led to some surprising discoveries. One such discovery was the much-celebrated recent prediction of protein folding by Deep Mind's Alpha Fold. Many interesting questions emerge. For one, we need to re-evaluate the role of AI in our discoveries – should we attribute creativity and thus credit AI for scientific discoveries? We could say that currently AI lacks the level of independence required for credit in discoveries; perhaps its role can be paralleled to those of technicians, which in itself opens an interesting door for exploration of our accreditation practices. Claire Anscomb (2021) has recently explored the notion of discovery as a collective practice, asking whether technicians and assistants who enable the production of an artwork or a scientific discovery deserve credit and recognition, arguing that creative autonomy should be at the centre of our credit attribution in science and art. This leads us to consider whether AI will require creative credit in computer assisted discoveries and whether AI can be creative[6] in designing scientific experiments. Another interesting consideration is also whether machine-learning algorithms will exhibit similar aesthetic preferences to those seen in human reasoning (Lombrozo 2016), and whether computer-assisted discoveries will be valued in a similar way to human discoveries. If we construe of aesthetic values as constraints imposed by human agents of particular capacities, it is perhaps to be celebrated that AI will not exhibit these constraints. But one reason for which one could be sceptical about the aesthetic value of computer-assisted discoveries comes from the reaction of mathematicians to the computer-assisted proof of the four colour theorem by Appel and Haken, an example of the method of proof by cases. While the discovery has certainly been of great epistemic significance, mathematicians have argued that the proof is 'ugly' and lacks the aesthetic merit of human discoveries (Montano 2014). Why is that? The computer-assisted proof operates with a 'bottom-up' approach: it scans around 2,000 cases in an important step in the proof. This approach is considered aesthetically and cognitively inelegant. As Montano argues, mathematicians consider this proof to go against the simplicity of traditional mathematical proofs, and this aesthetic constraint has a cognitive dimension 'simplicity facilities understanding the proof, while complexity hinders it' (Montano 2014, 36).

This leads us to consider the role of aesthetic value from a different perspective, aesthetic values play a cognitive role not in the traditional way – seen as linked to truth – but as cognitive constraints agents impose on their reasoning. Recently, attention has been given to the traditional question of what constitutes the aim of science: acquisition of truth, or in gaining understanding, which can be possible even in the absence of truth. This in its turn has also given rise to reconsidering what constitutes scientific progress (discussed in the first chapters of this book). Elgin (2020) and Potochnik (2017) have argued that we cannot accommodate for the use of idealisations and modelling in science if we construe the aim of science to be truth, given that models and idealisations often make successful predictions by making knowingly false assumptions about the target system. What these models allow us to do, however, is gain understanding and manipulate the system, which they see as the ultimate goal of science. Within this framework, we can construe of the role of aesthetic values as constrains on our reasoning, which we impose on our explanations, models, and theories, in order to gain understanding of the world. Aesthetic values function as constraints on the kind of reasoning we engage in, what theories and hypotheses we develop and ultimately operate as 'gatekeepers', reflecting human facts about human capacities rather than the world (Elgin 2020; Ivanova 2020). This does not mean that they need to be seen as rigid. We can accept hypotheses and explanations that conflict with our aesthetic values (Lombrozo 2016; Elgin 2020). Values can be updated and re-evaluated in light of their performance and success in practice, as argued originally by McAllister (1996), and more recently by Turner (2019) and Currie (2020) who see the use of aesthetic values to forms an 'epistemic feedback' effect, which ultimately grounds and justifies the use of such values. But whether the aesthetic values operating in current science will continue to form prevalent constraints in our sciences, whether machine algorithms are going to break away from such constraints and offer epistemic goods in spite of this, and what this will imply for the role of aesthetic values in science, remains to be seen.

Notes

1. Inspirationalist accounts go back to antiquity, where the production of artwork was credited to inspiration of the Muses, during the romantic period creativity is understood in light of divine inspiration. For an overview, see Gaut (2010).
2. For a recent overview of the debate on values in science and objectivity, see John (2021).

3. Different accounts have been proposed regarding the aesthetic appreciation of nature. For a detailed discussion, see Carlson (2011), Parsons (2002) and Turner (2019). Brady (2011) also defends that we can aesthetically engage with objects in nature that are not beautiful. Wraggle-Morley (2020) explores how natural philosophers, were committed to the idea that nature is beautiful due to being a product of divine creation, reconciled their belief with the existence of ugly or displeasing natural objects.
4. While James McAllister aims to show that it is rational to use aesthetic values in science, Richard Boyd is concerned with the claim that the virtues of a theory can be truth indicative and justifying our believe in a theory's truth. The analogies of their views are explored in (Ivanova forthcoming).
5. It is worth noting that in the aforementioned discussion we made a certain assumption: that the aim of science is epistemic, construed either as achievement of empirical adequacy, truth, or understanding. One could, however, claim that the aim of science is to deepen and enhance our aesthetic engagement with nature and our epistemic advances in science are supplementary to this primary aesthetic goal.
6. Whether AI can be creative has been explored recently after Deep Mind's Alpha Go beat the Go world champion player. Halina (2021) analyses whether Alpha Go was creative, arguing that whereas the programme did not exhibit domain generality, the Monte Carlo tree exploration and the ability to make a highly unexpected and surprising move, makes Alpha Go meet some minimal requirements for creativity.

References

Anscomb, C. (2021) Visibility, Creativity, and Collective Working Practices in Art and Science. *European Journal for Philosophy of Science*, 11, 5. https://doi.org/10.1007/s13194-020-00310-z

Barnes, E. C. (2008) *The Paradox of Predictivism*. Cambridge University Press.

Bird, A. (2007) What is Scientific Progress? *Nous*, 41, 64–89.

Boyd, R. N. (1984) The Current Status of Scientific Realism. In J. Leplin (ed.), *Scientific Realism* (pp. 41–82). Berkeley and Los Angeles: University of California Press.

Brady, E. (2011) The Ugly Truth: Negative Aesthetics and Environment. *Royal Institute of Philosophy Supplement*, 69, 83–99.

Brown, J. R. (2004) Why Thought Experiments Transcend Experience. In Christopher Hitchcock (ed.), *Contemporary Debates in Philosophy of Science* (pp. 23–44). Maiden, USA: Blackwell.

Bueno, O., Darby, G., French S., and Rickles, D., eds. (2017) *Thinking about Science, Reflecting on Art*. London and New York: Routledge.

Bury, J. B. (1920) *The Idea of Progress*. New York: Macmillan.

Carlson, A. (2011) Aesthetic Appreciation of Nature and Environmentalism. *Royal Institute of Philosophy Supplement*, 69, 137–155.

Chandrasekhar, S. (1987) *Truth and Beauty: Aesthetics and Motivation in Science*. The University of Chicago Press.

Clarke, B., and Ambrosio, C. (2018) The Nervous System and the Anatomy of Expression: Sir Charles Bell's Anatomical Watercolours. In C. Ambrosio and W. MacLehose (eds.), *Imagining the Brain: Episodes in the History of Brain Research* (pp. 109–138). Elsevier.

Criado-Perez, C. (2019) *Invisible Women: Exposing Data Bias in a World Designed for Men*. Vintage.

Currie, A. (2020) Epistemic Engagement, Aesthetic Value & Scientific Practice. *British Journal for Philosophy of Science* (online first).

Dawid, R. (2013) *String Theory and the Scientific Method*. Cambridge University Press.

Dellsén, F. (2021) Understanding Scientific Progress: The Noetic Account. *Synthese*, 199, 11249–11278.

Dirac, P. A. M. (1980) The Excellence of Einstein's Theory of Gravitation. In M. Goldsmith, A. Mackay, and J. Woudhuysen (eds.), *Einstein: The First Hundred Years* (pp. 41–46). Oxford: Pergamon Press.

Duhem, P. (1954 [1906]) *The Aim and Structure of Physical Theory*. Princeton University Press.

Dykes, J., and Wood, J. (2009) The Geographic Beauty of a Photographic Archive. In Toby Segaran and Jeff Hammerbacher (eds.), *Beautiful Data* (pp. 85–105). O'Reilly.

Elgin, C. Z. (1991) Understanding: Art and Science. *Midwest Studies in Philosophy*, 16(1), 196–208.

Elgin, C. Z. (2020) Epistemic Gatekeepers. In Milena Ivanova and Steven French (eds.), *The Aesthetics of Science: Beauty, Imagination, Understanding* (pp. 12–36). New York and London: Routledge.

Ellis, G., and Silk, J. (2014) Scientific Method: Defend the Integrity of Physics. *Nature News*, 516(7531), 321.

Frigg, R., and Hunter, M., eds. (2010) *Beyond Mimesis and Convention*. Dordrecht: Springer.

Gaut, B. (2010) The Philosophy of Creativity. *Philosophy Compass*, 5(12), 1034–1046.

Halina, M. (2021) Insightful Artificial Intelligence, *Mind and Language*, 36(2), 315–329. http://doi.org/10.1111/mila.12321?af=R

Heisenberg, W. (1971) *Physics and Beyond*. New York: Harper & Row Publishers.

Holton, G. (1969) Einstein, Michelson, and the 'Crucial' Experiment. *Isis*, 60, 132–197.

Hossenfelder, S. (2018) *Lost in Math: How Beauty Lead Physicists Astray*. Basic Books.

Ivanova, M. (2017a) Aesthetic Values in Science. *Philosophy Compass*, 12(10). http://doi.org/10.1111/phc3.12433

Ivanova, M. (2017b) Poincaré's Aesthetics of Science. *Synthese*, 194(7), 2581–2594.

Ivanova, M. (2020) Beauty, Truth and Understanding. In *The Aesthetics of Science: Beauty, Imagination and Understanding* (pp. 86–104). Routledge.

Ivanova, M. (2021a) *Duhem and Holism*. Cambridge Elements, Cambridge University Press.

Ivanova, M. (2021b) The Aesthetics of Scientific Experiments. *Philosophy Compass*, 16(3), e12730.

Ivanova, M. (forthcoming) Theory Virtues and Theory Acceptance. *Lauener Series in Philosophy: Bas van Fraassen's Contribution to Philosophy of Science*.

Ivanova, M., and French, S. (2020) *The Aesthetics of Science: Beauty, Imagination and Understanding*. Routledge.

John, S. (2021) *Objectivity in Science (Elements in the Philosophy of Science)*. Cambridge University Press. http://doi.org/10.1017/9781009063647

Kaku, M., and Thompson, J. T. (1997) *Beyond Einstein*. Oxford: Oxford University Press.

Kitcher, P. (1993) *The Advancement of Science: Science Without Legend, Objectivity without Illusions*. Oxford: Oxford University Press.

Kuhn, T. (1962) *The Structure of Scientific Revolutions*. University of Chicago Press.

Lakatos, I. (1968) Criticism and the Methodology of Scientific Research Programmes. *Proceedings of the Aristotelian Society*, 69, 149–186.

Laudan, L. (1981) A Problem-Solving Approach to Scientific Progress. In Ian Hacking (ed.), *Scientific Revolutions* (pp. 144–155). Oxford: Oxford University Press.

Leonelli, S., and Tempini, N. (2020) *Data Journeys in the Sciences*. Springer Open. https://doi.org/10.1007/978-3-030-37177-7

Lombrozo T. (2016) Explanatory Preferences Shape Learning and Inference. *Trends in Cognitive Sciences*, 20(10), 748–759. http://doi.org/10.1016/j.tics.2016.08.001

Longino, H. (1990) *Science as Social Knowledge: Value and Objectivity in Scientific Inquiry*. Princeton University Press.

Mann, K. (2020) *Entitled: How Male Privilege Hurts Women*. Allen Lane.

Martin, E. (1991) The Egg and the Sperm: How Science Has Constructed a Romance Based on Stereotypical Male Female Roles. *Signs*, 16(3), 485–501.

McAllister, J. W. (1996) *Beauty and Revolution in Science*. Ithaca, NY: Cornell University Press.

Montano, U. (2014) *Explaining Beauty in Mathematics: An Aesthetic Theory of Mathematics*. Cham: Springer.

Niiniluoto, I. (1980) Scientific Progress. *Synthese*, 45, 427–462.

Parsons, G. (2002) Nature Appreciation, Science, and Positive Aesthetics. *The British Journal of Aesthetics*, 42(3), 279–295. https://doi.org/10.1093/bjaesthetics/42.3.279

Parsons, G. (2012) The Aesthetics of Chemical Biology. *Current Opinion in Chemical Biology*, 16, 576–580.

Parsons, G., and Reuger, A. (2000) The Epistemic Significance of Appreciating Experiments Aesthetically. *British Journal of Aesthetics*, 40, 407–423.

Pietsch, W. (2021) *Big Data* (Elements in the Philosophy of Science). Cambridge University Press. http://doi.org/10.1017/9781108588676

Poincaré, H. (2001) Science and Hypothesis. In Stephen Gould (ed.), *The Value of Science: Essential Writings of Henri Poincaré* (pp. 3–181). New York: Modern Library.

Popper, Karl (1963) *Conjectures and Refutations: The Growth of Scientific Knowledge*. Routledge.

Potochnik, A. (2017) *Idealization and the Aims of Science*. Chicago and London: The University of Chicago Press.

Reichenbach, H. (1938) *Experience and Prediction. An Analysis of the Foundations and the Structure of Knowledge*. Chicago, IL: The University of Chicago Press.

Rowbottom, D. P. (2008) N-Rays and the Semantic View of Scientific Progress. *Studies in History and Philosophy of Science Part A*, 39(2), 277–278.

Shan, Y. (2019) A New Functional Approach to Scientific Progress. *Philosophy of Science*, 86, 739–758.

Smolin, L. (2006) *The Trouble with Physics: The Rise of String Theory, the Fall of a Science, and What Comes Next*. Houghton Mifflin Harcourt.

Stanford, P. K. (2006) *Exceeding Our Grasp*. Oxford: Oxford University Press.

Turner, D. (2019) *Paleoaesthetics and the Practice of Paleontology*. Cambridge University Press.

van Fraassen, B. (2008) *Scientific Representation: Paradoxes and Perspectives*, Oxford: Oxford University Press.

Waggle-Morley, A. (2020) *Aesthetic Science: Representing Nature in the Royal Society of London, 1650–1720*. Chicago University Press.

Wylie, C. (2015) 'The Artist's Piece is Already in the Stone': Constructing Creativity in Paleontology Laboratories. *Social Studies of Science*, 45(1), 31–55.

17 Scientific Progress and Idealisation

Insa Lawler

1. Introduction

Intuitively, science progresses from truth to truth. We learn more facts based on previously learnt facts. A glance at history quickly reveals that this idea is mistaken. We often learn from theories that turned out to be false. Take theories about cancer causes. According to the blastema theory (developed in the nineteenth century), cancer cells are developed from budding elements (blastema) that are scattered among normal tissue. We now know this to be incorrect, but we learned from this theory that cancer is made up of cells (and not of lymph, as previously believed). This discovery remains unchallenged. Replacing the false lymph theory with the false blastema theory thus looks to be progressive.

We could propose that progress is possible whenever we have good reasons to think that our theories are correct. However, such a view is challenged by *deliberately* used *idealisations* in science. Especially when scientists use models or simulations to analyse phenomena, they work with assumptions that are known to be false, like the stipulations that there is no intergenerational overlap, or that humans are perfectly rational.

Some idealisations are 'crutches' for scientific reasoning. We hope to eventually replace them with more accurate assumptions. Other idealisations are considered to be *indispensable*; we need them for realising scientific achievements. Yet surprisingly, both kinds of idealisations do not seem – in principle – to impede scientific progress. There are numerous examples of *empirically successful* scientific inquiries that involve such idealisations, for example, some idealised models can more or less accurately predict the behaviours of interest. A convincing account of scientific progress must do justice to this fact.

This chapter analyses the challenge from idealisation for accounts of scientific progress, with an eye on indispensable idealisations. I first describe idealisations in more detail (Section 2). I do not defend a particular theory of scientific progress. Instead, I proceed by analysing how well the four dominant accounts can cope with the challenge: the problem-solving account (Section 3.1), the noetic account (Section 3.2), the

DOI: 10.4324/9781003165859-21

epistemic account (Section 3.3), and the truthlikeness account (Section 3.4). One upshot is that, on all accounts, idealisations can promote progress. Only some accounts allow them to constitute progress.

2. The Challenge From Idealisation

Scientific inquiry – perhaps inevitably – involves falsehoods. Theories turn out to be false and mistakes happen – from miscalculations to mislabelled substances in experiments. Mistakes typically impede scientific progress. (But think of examples like the discovery of penicillin.)[1] The case is less obvious for what Strevens (2017, 37) calls 'deliberate falsehoods'. He is not interested in *illegitimate* ones, such as fabricating results to get a grant. He is concerned with deliberate falsehoods that are accepted although they are known to be false, paradigmatically idealisations. These are the falsehoods I focus on.

Some terminology before I characterise idealisations in more detail: I use 'scientific account' as an umbrella term for theories, models, and other forms of theorising, and 'idealisation' for deliberately used *false assumptions*. The word 'idealisation' is also used for the *practice* of idealising, or for models, theories, or laws that involve such falsehoods. I follow Potochnik (2017) in labelling the latter *idealized representations* or accounts.

Idealisations can be distinguished from *abstractions* and *approximations*.[2] Following Jones (2005), an abstraction is an *omission* that does not lead to a misrepresentation.[3] Take a description of a falling object that omits its colour. For Jones (2005, 175), an important difference is that an idealisation is a (clearly) *false* assumption, for example the assumption that gas molecules do not exert any long range forces on each other. An abstraction involves the omission of something *true*, such as omitting that the falling object is tangerine.

Following Norton (2012), idealisations should also be distinguished from *approximations*. An approximation merely describes a target system inexactly, such as describing roundish objects as round or streamlining collected data, like a smoothed curve of data points gained by ignoring outliers. What counts as an approximation or inexact description is context-sensitive. One has to specify a proximity range for the value of interest (perhaps relative to a given context). A description is no longer an approximation if the value is not within this specified proximity (but it can be an idealisation).[4]

Neither abstractions nor approximations seem to – in principle – impede progress. If a property does not affect a phenomenon – say, an object's colour when analysing its speed – abstracting from away it seems harmless. Indeed, such abstractions can *promote* progress by enabling scientists to focus on the (potentially) relevant properties. Approximations can be beneficial because we often must trade off precision against other values like applicability across domains (see e.g. Levins, 1966). They can also simplify (mathematical) reasoning. The case of idealisations is less clear. Adding falsehoods to scientific reasoning seems to be a form of *regress*. On the other hand, using idealisations can be progressive. As I illustrate in the following, idealisations can be a (perhaps essential) part of *empirically successful* investigations. I describe this puzzle about idealisations in more detail (Section 2.1), before I focus on two kinds of idealisations (Sections 2.2–2.4).

2.1.   A Puzzle About Idealisations

As Cartwright (1983) demonstrated for the case of physics, idealisations are used in scientific theories or laws. However, they are most prominently discussed in the case of scientific models and simulations. The majority of them involve at least some idealisations. For instance, according to the aorta model developed by Caruso et al. (2015), the walls of the aortas are *rigid* (although they are elastic).

Scientists even use different idealised models for analysing the same target phenomenon *at the same time*. Weisberg (2007) calls this 'multiple-models idealisation'. Even more puzzling, these models can be *incompatible*. One example in physics are the *optical* Glauber model and the *Monte Carlo* Glauber model for analysing collisions of atomic nuclei. According to the former, nuclei are perfect spheres of energy. According to latter, nuclei are not such spheres; the protons and neutrons constituting the nucleons are distributed at random. These assumptions are contradictory; the models are not compatible. Another example is that water is construed as a continuous fluid in some models and as being composed of discrete particles in others (Teller, 2001). These examples are not considered to be bad scientific practice (among other things because the models focus on different aspects of the phenomena), and it does seem that our theories about nuclei collisions and water are progressing.[5]

Relatedly, scientists often keep using idealised models even if they have de-idealised or less idealised versions, as Elgin (2017, ch. 12), Potochnik (2017, ch. 2), and Dellsén (2016) highlight. For instance, the ideal gas law is widely used (e.g. for teaching purposes) although we have less idealised options, such as the van de Waals equation.

That scientists frequently use idealisations is puzzling. What is attractive about working with assumptions known to be false? Why is this not considered to be improper scientific practice? The answer leads to another puzzle.

As Potochnik (2017, ch. 2) emphasises, one reason to work with idealisations is to *reduce complexity*. When exploring complex phenomena, it can help to start with a simplified analysis. More accurate models can be built when the phenomena are better understood. Idealised models can support *explorative* purposes, such as the simulation of various climate scenarios (see e.g. Gelfert, 2016, ch. 4; Rohwer & Rice, 2016). Some models are employed to *formulate causal hypotheses*, which can be empirically tested (see e.g. Alexandrova, 2008; Pincock, 2014). And models are used to *evaluate competing hypotheses*. For instance, Schmid-Hempel and his team explored why honeybees sometimes leave food sources even if their honey sacks are only partially filled. Their model was not designed to adequately capture this behaviour, but to '. . . investigate how much of the bees' behaviour can be accounted for by purely energetic models of nectar collecting' (Schmid-Hempel et al., 1985, 61).[6] That honeybees maximise their energy *efficiency* rather than their energy *intake* turned out to be more plausible.

These reasons are not exhaustive but suffice for our purposes (for more suggestions see Potochnik, 2017, 48). There is another aspect that justifies the use of idealisations, and often governs it, namely, their *empirical success*: working with idealisations often leads to the desired results. For example, many idealised models can more or less accurately reproduce or predict the behaviours of interest, or match observed data. For instance, the behaviours predicted by the aforementioned honeybee model match the data surprisingly well. Some simulated possible climate scenarios turn out to be realistic. And so forth. This success is important. Scientists typically employ an idealisation based on educated guesses about which aspects of a phenomenon they can idealise. When the scientific account involving it is *not* empirically successful, the idealisation is often revised or removed. Take the Glauber models mentioned earlier. The 'perfect sphere' idealisation used in the optical Glauber model led to false predictions. According to it, the collision of two nuclei should result in the formation of an *ellipsoid*. Measurement showed that other shapes can result. The Monte Carlo Glauber model (which was subsequently invented) does not use this idealisation and fares better with the shape predictions. (The optical Glauber model is still used for cases where the shapes are not crucial, such as explaining how much energy there is in the intersection of two colliding nuclei.)

The frequent success of idealisations is challenging for accounts of scientific progress. Arguably, empirically successful scientific accounts are progressive. But how we can progress by working with clearly false assumptions? In what follows, I introduce two kinds of idealisations: Galilean idealisations (Section 2.2) and *indispensable* idealisations (Section 2.3). This is not an exhaustive distinction, but these kinds constitute two extremes for the challenge from idealisation for scientific progress. Galilean idealisations can often be replaced with empirically accurate counterparts. While we clearly progress with such replacements, we still need to account for empirically successful intermediate stages. Indispensable idealisations are considered to be *irreplaceable* and are thus the most puzzling.

2.2.   Galilean Idealisations

Galilean idealisation is the use of idealisations to mathematically simplify a scientific account of a phenomenon to render it (more) computationally *tractable*. This idealisation strategy is named *Galilean* because Galileo used it in his scientific investigations (see McMullin, 1985; Weisberg, 2007). Whereas Galilean *idealisation* is a strategy, I dub the falsehoods it employs Galilean *idealisations*.

As Weisberg (2007) emphasises, this idealisation practice is *pragmatically* justified: we want to gain tractability. When the phenomenon is better understood and the mathematical methods are more advanced, we can de-idealise the model and replace it with increasingly accurate ones. He illustrates this with an example of models used in computational chemistry to compute wave functions for molecules of interest. The early very simplified models were gradually replaced with more accurate models. Galilean idealisations are thus not meant to be permanent. We hope to replace them with what Strevens (2008, 300) calls a *veridical counterpart*, which correctly captures the respective properties. A veridical counterpart of the idealisation that, say, gas molecules exert no long range forces on each other would accurately capture how these forces interact.

Empirical success is important. A simplified model that does not produce empirically adequate results (such as successful predictions) is often not further used. On the other hand, a successful de-idealisation of an empirically successful model shows that 'the [idealized] model does give a relatively good fit to the real structure of the explanandum object' (McMullin, 1985, 264). The idea is that the model approximately captures the target phenomenon because its idealisations turn out not to be essential.

2.3.   Indispensable Idealisations

Whereas Galilean idealisations can often eventually be replaced with veridical counterparts, *indispensable* idealisations are considered to be not replaceable.

Following Rohwer and Rice (2016, 1134), idealisations are indispensable in the case of idealised models '. . . when removing the idealisations has the effect of "destroying" or 'dismantling' the model . . .'. A model is dismantled when it stops working, for example when it does not compute meaningful results. An idealisation can be essential to the model's mathematical representation of the target phenomenon. A model is also dismantled when it is no longer empirically successful once the idealisations are removed or replaced. For instance, while the idealised model gives us predictions that match the empirical data, no de-idealised model delivers such a match. An example is given by Rice (2018): idealisations can be *entrenched* with the accurate assumptions such that once they are removed or replaced, the models are no longer empirically successful. His examples are optimality models in biology. Such models are highly idealised and used to explain behaviours or phenotypic traits (such as the honeybees' foraging behaviour) by determining evolutionary optimal strategies for them. In other cases, we can no longer obtain the desired *scientific explanations* if we remove the idealisations. As Rice (2018, 2809) puts it:

> these modeling techniques are often essential and ineliminable because they allow scientific modelers to extract the desired explanatory information that would otherwise be inaccessible.

Several scholars gave examples of indispensable idealisations (e.g. Wayne, 2011; Bokulich, 2011; Kennedy, 2012; Batterman, 2002; Batterman & Rice, 2014; Rice, 2018, 2019). Batterman (2002) suggested *asymptotic* idealisations as one key example. Such idealisations take some sort of limit, such as the *infinite population idealisation* according to which the population of interest is *infinitely large*. It is used in population genetics to create deterministic models of changes in a population's gene frequency (for an analysis, see e.g. Strevens, 2019). Another example is the *thermodynamic limit*, which states, roughly speaking, that a system's number of particles and its volume are arbitrary large. It is used, for instance, by phase-transition models. Phase transitions are abrupt changes of a system's or substance's qualitative macroscopic properties, such as water's freezing into ice or the magnetisation of iron. According to some scholars, this limit is an indispensable idealisation because such transitions cannot be reproduced with a model that assumes finite particles (rather than infinite ones). By employing finite systems, say, systems based on statistical mechanics, we cannot model phase transitions (see e.g. Batterman, 2002; Batterman & Rice, 2014; Rice, 2018). Others (e.g. Butterfield, 2011; Norton, 2012) dispute this claim and consider the thermodynamic limit and other asymptotic idealisations to be dispensable. I do not take a stance here (for discussion, see e.g. Shech, 2018), but I take for granted that there are some indispensable idealisations.

2.4. *Two Approaches to Idealisations*

A key issue debated is whether the understanding, explanations, or scientific accounts that can be obtained from working with indispensable idealisations are non-factive (i.e. contain falsehoods). Many general idealisation accounts have been proposed. I focus on two opposing approaches. According to what I call the *rationalisation approach*, scientific explanations or accounts can involve (indispensable) idealisations and thus be non-factive as long as the idealisations are *justified*, for example they fulfil a certain purpose. For Bokulich (2011), an idealised model can explain a phenomenon if the latter falls in its *application domain* (among other things). Rice (2018) proposes that being in the same *universality class* can justify exploring a phenomenon with an idealised model. Elgin (2017, chs. 11 and 12) argues that we can be justified in using idealisations when they *exemplify* properties of interest, especially when those are difficult to detect. Potochnik (2017) makes the case for accepting idealisations if they help us to *identify causal patterns*. And so forth.

In this chapter, I cannot go into any details. Instead, I contrast this approach with an approach according to which the explanations or accounts in question are *factive* despite being obtained using (indispensable) idealisations. Lawler (2021) describes an *extraction approach* to idealisation that serves to capture insights from Alexandrova (2008), Pincock (2014), Bokulich (2016), and Rice (2016, 2018, 2019). According to this approach, idealisations merely *enable* explanations or accounts (and the understanding they can generate), but are not an element of their content. When working with such idealisations only the truths we extract are elements of the content of our explanations, accounts, or our understanding. For instance, the empirical success of the honeybee model suggests that it is true that bees prioritise energy efficiency over energy intake. This information is part of our explanation of their behaviour and not the idealised model's content. Pincock (2021) advocates a similar view according to which idealised models can be explanatory when truths are underlying each falsehood relevant for the explanation in question (see also Pincock, 2014).

These approaches to idealisations are not exhaustive and I cannot discuss their plausibility. But it is useful to have them in mind when we turn to examining how well the dominant accounts of scientific progress can accommodate the fact that (indispensable) idealisations persist in many of our empirically successful scientific inquiries.

3.   Accounts of Scientific Progress and the Challenge From Idealisation

'Scientific progress' is a normative notion. In the sense relevant here, it is a form of *cognitive improvement* (Niiniluoto, 2019, Section 2.2).[7] Advancing from the lymph theory of cancer to the blastema theory was such an improvement. But what precisely constitutes cognitive improvement in science? Following Niiniluoto (2019) and Dellsén (2018a), I focus on four dominant accounts. They respectively explicate scientific progress in terms of an increase or decrease: a decrease in *unsolved problems* (the problem-solving account), an increase in the *truthlikeness* of scientific theories (the truthlikeness account), an increase in *knowledge* (the epistemic account), and an increase in *understanding* (the noetic account).

In what follows, I briefly characterise these accounts and examine how they can cope with the challenge from idealisation. I start with the account that *prima facie* can do this most easily because it lacks a factivity requirement, that is a requirement that every progressive claim, theory, or solution must be true: the problem-solving account (Section 3.1). Then, I turn to the accounts that explicate progress in terms of epistemic achievements: the noetic account (Section 3.2) and the epistemic account (Section 3.3). I end with the truthlikeness account (Section 3.4); it might struggle the most because it focuses on the truth of theories.

My analysis has an eye on whether idealisations or idealised accounts can *constitute* or merely *promote* progress on a given account of progress (if at all). This distinction has been emphasised by Bird (2008, 280) and Dellsén (2018b, 73). Something *constitutes* progress when the relevant achievement is thereby fully or partially realised, such as an increase in, say, knowledge or understanding. By contrast, something *promotes* progress when it merely renders it more likely that this increase can be achieved. For instance, buying better lab equipment does not constitute cognitive progress. We do not automatically gain more, say, understanding by having better equipment. But it arguably promotes cognitive progress by raising the probability that we will increase our understanding in the long run. Likewise, idealisations might merely increase such a probability.

3.1. The Problem-Solving Account and the Challenge From Idealisation

The problem-solving account of scientific progress was suggested by Kuhn (1970) and has been developed by Laudan (1977, 1981). Shan (2019) offered a related account.

3.1.1. The Kuhn-Laudan Problem-Solving Account

The Kuhn-Laudan account identifies *scientific problems*[8] as the key currency for scientific progress. Laudan suggests defining progress in terms of *problem-solving effectiveness*, considering the *amount* of the solved and unsolved problems and their *significance* relative to a research tradition (see later). We progress by decreasing the amount of significant unsolved problems – by solving or dismissing problems, or downgrading their significance.

Importantly, scientific problems are determined relative to a *research tradition*,[9] which involves various methodological, conceptual, and ontological commitments, for example assumptions about how to test theories or what kinds of entities exist. The research tradition that is dominant in a given discipline at a given time determines its problems and their importance. There are no objective factivity requirements for identifying a problem. If a problem presupposes what later turns out to be incorrect or if its target phenomenon is not real, it is still a problem for that research tradition. As Laudan (1977, 16, emphasis in original) puts it: 'A problem need not accurately describe a real state of affairs to be a problem: all that is required is that it be *thought to be* an actual state of affairs'. For instance, according to the caloric theory, 'caloric' (a self-repelling fluid) is the substance of heat. It is now known that caloric does not exist. To explain how caloric accounts for temperature phenomena was nonetheless a problem for this theory's research tradition.

This relativity of the problem-solving account is inspired by Kuhn's famous observation that theoretical frameworks (or 'paradigms') can radically change in the event of 'scientific revolutions' – think of the change from Ptolemaic cosmology to a Copernican one. Kuhn argued that such frameworks are *incommensurable*. Accordingly, whether a scientific problem is solved, can be dismissed, or downgraded is also determined by the scientists in a research tradition. Again, there are no objective factivity requirements. If a solution to a problem turns out to be incorrect but fulfilled the tradition's solution criteria, the problem is solved in that tradition.[10] Accordingly, 'progress' and 'regress' are relative notions. For instance, a problem concerning cancer causes can be solved in one research tradition but unsolved in a subsequent one, say, because it rejects the solution's assumptions.

This brief characterisation shall suffice for our purposes (for details and discussion, see e.g. Bird, 2016; Dellsén, 2018a; Niiniluoto, 2019; Shan, 2019).

3.1.2.   The Kuhn-Laudan Problem-Solving Account and the Challenge From Idealisation

Prima facie, the problem-solving account can cope well with idealisations. Because no objective factivity requirements are imposed on problems or their solutions, idealised accounts can be progressive if they solve problems (or leave fewer problems unsolved).

However, the mere fact that inaccuracy is not considered an obstacle to progress is not enough to account for idealisations. For a start, the problem-solving account has factivity constraints albeit *subjective* ones. The phenomenon a problem targets must be thought to be actual, and solutions to a scientific problem should be recognised as correct or adequate by the respective scientists. Such an 'apparent correctness' condition is *prima facie* in conflict with the fact that idealisations are *deliberate* falsehoods. Accordingly, the problem-solving account faces a similar challenge as factive accounts of progress (i.e. ones that have objective factivity requirements): it must show how accounts known to be idealised can provide (seemingly) correct solutions (or be part of them).

The problem-solving account, however, has more flexibility in meeting this challenge. Factive accounts of progress must demonstrate that even indispensable idealisations can generate de facto correct scientific accounts, for instance, by adapting a version of the extraction approach described earlier according to which idealisations function as a (perhaps indispensable) tool to generate such accounts. The problem-solving account can follow suit, but it can also adopt a version of the rationalisation approach according to which idealisations can be part of justified non-factive explanations or accounts – as long as these are (part of) solutions that the relevant scientists accept as correct or adequate.

Note that idealisations only *constitute* progress if they are part of the problem solutions. If the problem-solving account adopts an extraction approach, idealisations merely promote progress by increasing the probability that such solutions can be found.[11]

The problem-solving account can also explain why some idealised models continue to be used even when de-idealised versions are available. If they still solve relevant problems or have another function (e.g. a pedagogical one) and do not hinder solving problems, there is no need to abandon them.

Although the challenge from indispensable idealisations can be met by adopting a suitable idealisation approach, the problem-solving account struggles with an aspect of *Galilean* idealisations. Consider a case where a mathematically simplified model is ultimately replaced with a de-idealised version, with there being intermediate stages with increasingly more accurate models. Let us suppose that all involved models are empirically successful. The predictions made by the initial model roughly match the data, the predictions made by its first revision match them better, and so forth, until we have a de-idealised model that accurately matches all data. One issue is that it is not clear how many problems we are concerned with.[12] Either way, these models solve at least a problem of the form 'Construct a model that matches the data to the proximity degree x'. However, this is not enough for progress. An improved model needs to solve *more* problems. We could say that it additionally solves the problem 'Replace the idealisations used in its predecessor model with more accurate idealisations', or 'Construct a model that matches the data to a higher proximity degree than the previous model'. But this does not fully solve the issue. Such problems are solved exactly once (at least if scientific problems are genuine challenges), and we can have more than one intermediate stage between the initial model and the fully de-idealised one. One might think that the more accurate models solve all problems of their predecessors and thus more problems, but that is not straightforwardly true. If we are concerned with constructing a model that matches the data to a specific proximity degree, then a fine-grained model does not solve the problem of a more coarse-grained one. It is the other way around. A *less precise* model can capture a wider proximity range.[13] If so, the more idealised models solve *more* problems than the less idealised ones. Perhaps the more advanced models progress by solving more *significant* problems. Constructing a less idealised model might be more valuable than constructing a more idealised one. However, it is not clear how to weigh the number of problems solved against their significance – a general challenge for the problem-solving account.

To sum up: Although the problem-solving account has great flexibility in accounting for idealisations, it might struggle with cases where a more de-idealised representation does not solve more problems.

3.2.   The Noetic Account and the Challenge From Idealisation

The *noetic* account defines progress in terms of an epistemic achievement, namely, *understanding*. Scientific progress consists in an increase of scientific understanding. Making progress researching cancer causes consists in obtaining a better understanding of them.[14] This account has been developed by Dellsén (2016, 2018b, 2021) (see also Bangu, 2015).

3.2.1.   Dellsén's Noetic Account

Dellsén's account proceeds from the assumption that understanding is not reducible to corresponding knowledge.[15] Understanding requires an epistemic commitment to a systematic account of the target phenomenon, for example to an explanation of it. The nature of this commitment is contested (for details, see Hannon, 2021). For instance, it is debated whether understanding is *factive* and whether the commitment must be *epistemically justified* (i.e. roughly speaking, whether one needs to possess good reasons or evidence for it). Accordingly, there could be vastly different noetic accounts.

Dellsén (2021) couples his view with the following analysis of understanding: someone understands a target phenomenon if and only if they grasp an adequately accurate and comprehensive *dependency model* of it. Such a model primarily specifies the dependence relations the phenomenon is involved in, such as causal relations to other things, but it can additionally specify what the phenomenon does *not* depend on. Realising this can be important, for example discovering that a disease was not genetic can help to better understand its origins. Dependence relations encode information about how the phenomenon would have been different if other things had been different. So, someone who grasps a dependency model acquires the ability to *explain* the phenomenon and to *predict* how its features would change in different circumstances. The degree of someone's understanding can be determined based on how accurate and comprehensive the model is that they grasp. An understanding subject does not have to believe the dependency model or be epistemically justified in committing to it. It suffices that the subject accepts the model (where accepting roughly means treating it as given for a certain purpose). Dellsén (2016) tries to show that an unjustified theory (or model) can foster understanding and progress. This brief characterisation shall suffice for our purposes (for details see Dellsén, 2021).

3.2.2.   Dellsén's Noetic Account and the Challenge From Idealisation

How the understanding component is specified matters for a noetic account's ability to cope with the challenge from idealisation. For instance, an account that assumes that understanding involves *believing* the dependency model in question cannot straightforwardly explain how scientists progress with deliberate falsehoods. Dellsén's account does not face this issue; he only requires that the scientists accept the models. However, his resources to cope with idealisations are limited; his account is broadly *factive*. That a dependency model is adequately *accurate* is necessary for understanding and thus for progress. Dellsén (2021, Section 3.2) allows for inaccuracies when they increase the *comprehensiveness* of the model, but many idealisations are substantial inaccuracies. Dellsén is aware of this issue. To solve it, he adopts the non-difference maker view of idealisations developed by Strevens (2008, 2013, 2017). According to it, idealisations are compatible with factive reasoning as long as they '. . . indicate that certain factors make no difference to the phenomenon to be explained . . .' (Strevens, 2017, 37) – or as Dellsén (2021) puts it, point to the absence of a dependence relation. For instance, assuming that a population can be arbitrarily large indicates that the precise population size is not explanatorily relevant. An idealised model's content has then two parts (Strevens, 2008, 318):

> The first part contains the difference-makers for the explanatory target. . . . The second part is all idealisation; its overt claims are false but its role is to point to parts of the actual world that do not make a difference to the explanatory target.

An explanation involving idealisations '. . . is correct if the propositions expressing its explanatory content, as opposed to its literal content, are true' (Strevens, 2013, 512). This content contains veridical counterparts of the idealisations (Strevens, 2017, 45):

> [The idealized model] derives [the representation of?] the phenomenon to be explained from a mix of real difference-makers and fictional non-difference-makers. Take the fictional non-difference-makers away (substituting veridical statements such as "Long-range forces are small") and you can still derive the phenomenon.

Dellsén (2016, 2021) follows Strevens (2008, sect. 8.2) in using the ideal gas law to illustrate this analysis. Indeed, it seems possible to explain the gas regularities in question with the veridical parts of the ideal gas law and veridical counterparts of its idealisations. Let us stipulate that this example and related cases work well. Then, replaceable *Galilean* idealisations can be accommodated similarly. Because we can ultimately de-idealise the model, it seems that the idealised aspects do not make a difference to the analysed phenomenon and that veridical counterparts can be used for the desired explanations. Moreover, Dellsén could classify models of the intermediate stages as progressive because he does not demand that the dependency models be fully accurate.

Indispensable idealisations pose a greater challenge. Although Strevens' account is meant to apply to all idealisations (Strevens, 2008, 316; Strevens, 2017, 38), it is not clear that it fully works for such idealisations. Recall what makes them special: they might be necessary for the model's mathematical representation or be entrenched with the nonidealised assumptions such that the models are no longer empirically successful once they are removed or replaced. Strevens' proposal to explain phenomena using the idealisations' veridical counterparts and the correct parts of the model thus does not seem to work, as Rice (2018) highlights. Perhaps indispensable idealisations nonetheless point to nondifference makers, as Strevens (2019) argues for the case of asymptotic idealisations.[16] But Rice (2018) gives examples of such idealisations that *distort* difference makers.

While Dellsén adopts the non-difference maker concept, he neither endorses nor rejects the derivation part of Strevens' account. But since he does not address the issues with indispensable idealisations, the noetic account lacks a solution for them. Dellsén could adopt a version of the extraction approach. However, this would come with a cost. On Dellsén's view, idealisations *constitute* progress, as he (2018b) highlights. Re-interpreted as non-difference makers they are part of the dependency models; they highlight what the phenomenon does not depend on. Dellsén (2021, Section 3.2) emphasises that this explains why we sometimes keep idealised models alongside their de-idealised counterparts; they contain additional information about non-dependence relations. On the extraction approach, idealisations are not part of dependency models, but only *promote* their construction, for example by providing information about non-difference makers.

To summarise: Dellsén's noetic account can accommodate replaceable Galilean idealisations and other idealisations that can be analysed with Strevens' non-difference maker view. Its struggle with indispensable idealisations can be solved by adopting the extraction approach (at least for indispensable idealisations) and sacrificing the proposal that all idealisations can constitute progress.

While Dellsén's account is broadly factive, a noetic account coupled with a *non-factive* understanding view (and thus a non-factive progress view) would have more options to deal with the challenge from idealisation, such as adopting a rationalisation approach.

3.3. The Epistemic Account and the Challenge From Idealisation

The epistemic account also offers a definition of progress in terms of an epistemic achievement: Science progresses just in case it increases scientific *knowledge*. The more knowledge we acquire, the more we progress. This account has been suggested by Barnes (1991) and Cohen (1980), but was only recently developed by Bird (2007, 2008, 2016).

3.3.1. Bird's Epistemic Account

As in the case of understanding, there is no universally accepted theory of knowledge, but it is generally agreed that knowledge implies a true belief that is epistemically justified. The nature of epistemic justification is contested (for details, see Ichikawa & Steup, 2018), and there might be other necessary conditions for knowledge. Accordingly, there could be different variants of the epistemic account. Bird (2007, 72) does not commit to a specific analysis of knowledge, but he emphasises that the beliefs should not be *accidentally* true, as in the case of Gettier (1963)'s famous examples.[17]

On the epistemic account, progress demands justification. Discovering truths about, say, cancer causes is only progressive when these are justified, for example by having sufficient scientific evidence. Lucky guesses are not progressive (Bird, 2007, 2016).[18]

While the problem-solving account requires solutions to problems for progress and the noetic account dependency models, Bird (2007, 76) offers a broad concept of progressive achievements: 'Scientific knowledge will (locally) grow when any scientific proposition becomes known . . . .' For instance, we progress by gaining knowledge of unreliable methods. Assuming that understanding reduces to knowledge, Bird (2007, 84) suggests supplementing his account with an analysis of the *significance* of pieces of knowledge for progress, for example a gain in knowledge that is also a gain in understanding might be more significant and thus more progressive than a collection of irrelevant data.

This brief characterisation shall suffice for our purposes (for details and discussion, see e.g. Rowbottom, 2008; Cevolani & Tambolo, 2013; Dellsén, 2018a; Niiniluoto, 2019; Saatsi, 2019).

3.3.2. Bird's Epistemic Account and the Challenge From Idealisation

Any variant of the epistemic account is straightforwardly factive because knowledge is taken to be factive. Only true beliefs can constitute progress. It is thus more restricted than the problem-solving or the noetic account.

To address progress from false theories to false theories, Bird (2007, Section 3.3) initially focused on the concept of *approximate truth*.

Approximately true claims can be interpreted as truths by using an 'approximately' operator. For instance, the description of a nearly round object as round is approximately true, and the claim 'It is approximately true that the object is round' is true. This proposal is subject to serious objections (see e.g. Niiniluoto, 2014, 2019, Section 3,6; Saatsi, 2019). But Bird (2016, Section 5) broadened his proposal: False theories can also involve other truths, such as claims like 'The theory is highly truthlike', and false but progressive claims typically have *logical implications* that are true. For instance, the false claim that today's humidity is 58% implies the true claim that today's humidity is less than 60%. He thinks it is likely (but not guaranteed) that someone who believes a claim (dispositionally) believes such implications, and that such implications can account for the progress made in at least many cases featuring false theories.

Bird's approximation proposal could be applied to idealisations that are approximately true. Some *Galilean* idealisations of intermediate stages with increasingly accurate models would be examples. Even if this works, the proposal cannot accommodate numerous other idealisations. As Niini-luoto (2014) emphasises, it does not work for idealisations that are vastly incorrect. For instance, that there is no intergenerational overlap is far from being approximately true. This issue can affect initial Galilean ide-alisations, and it affects many *indispensable* idealisations. Take the infinite population idealisation: It is not approximately true that the population size is arbitrarily large. Moreover, idealisations that are *entrenched* with the model's true assumptions such that the model cannot produce (empiri-cally successful) results if they are removed or replaced with veridical coun-terparts cannot be substituted with 'It is approximately true . . .' statements.

While the approximation proposal is of limited use, Bird also broadens his implication proposal. When working with successful idealised models, scientists (Bird, 2016, 558–559)

> will typically have some idea regarding some of the implications of theory that these are supported by the evidence and reasoning whereas others are not. For example, the simple kinetic theory of gases is clearly false . . . but scientists do believe nontrivial implica-tions of the theory: that gases are constituted by particles; that the temperature of a gas is in large part a function of the kinetic energy of the particles; and that the ideal gas equation holds with a high degree of approximation for gases at moderate pressure and temperature. So false theories, even those known to be false, can contribute to progress on the epistemic view because they often have significant true content or true implications that are believed by scientists on the basis of good evidence and reasoning.

Bird focuses on the knowledge that we can gain from the veridical parts of the models (i.e. the true assumptions) and from the models' empirical success (e.g. that the ideal gas law's predictions approximately match the data from real gases). His (not fully worked-out) proposal does not render it clear whether only the implications of the models' true assumptions are meant. If so, it could not accommodate indispensable idealisation and initial Galilean idealisations. However, Bird could adopt a version of the extraction approach. Then, the various pieces of true information we can extract from working with idealised models would be additional forms of model implications (in a broad sense). The true information extracted could be the basis for the knowledge gained from working with the idealised models. Since on the epistemic account any kind of scientific knowledge constitutes progress (to some degree), it could additionally adopt other proposals for how to gain knowledge from idealisations, such as Greco (2014)'s proposal that we can know how the idealisations relate to the target phenomenon.

Seemingly unaware of Bird (2016)'s implication proposal, Park (2017, 577) offers a suggestion in the vicinity of the extraction approach to support the epistemic account:

> the generation of idealized theories counts as progress, given that they facilitate inferences about observables, and that those inferences are accompanied by an accumulation of observational knowledge.

As Dellsén (2018b) highlights, Park claims that such idealised theories would *constitute progress*, although they only facilitate inferences that lead to knowledge. This is because Park (2017) suggests that achieving the means to increase knowledge constitutes progress. However, Dellsén (2018b) correctly notes that for Bird (2008) facilitating knowledge acquisition promotes progress but does not constitute it. It only renders it likely that we gain knowledge. Bird (2016)'s own conclusion is that false theories can promote progress, and his approximation and implication views license that idealisations only promote it too. The re-interpreted or implied truths can constitute progress though.

Bird does not address how to account for the fact that we sometimes keep (or create) an idealised model although we have a de-idealised version. Dellsén (2016, 81) argues that the epistemic account cannot accommodate such cases because '. . . there would be no point in having [such] idealised theories from a purely epistemic point of view'. However, Park (2017, 577) seems to be right that Bird

> would reply that a non-idealized theory might be useless to generate observational knowledge. We should consider the increase and the decrease in observational knowledge when we determine whether idealisations are beneficial or detrimental to the accumulation of knowledge.

An idealised account might be more beneficial to knowledge gain than its de-idealised counterpart, and in any case, if it leads to new scientific knowledge, it is worth keeping. Take science education. Successfully teaching students, say, how gas regularities work using the ideal gas law leads to a potential gain in scientific knowledge (e.g. because these students might become the new gas experts).

To sum up: Because the epistemic account treats any piece of knowledge as constituting progress, it can adopt various proposals to accommodate idealisations, such as the approximation proposal for approximately true idealisations, and the extraction approach for indispensable ones. Because idealised models can be used to extract truths, they can have epistemic value even in light of de-idealised counterparts.

3.4. The Truthlikeness Account and the Challenge From Idealisation

The truthlikeness account builds on Popper (1963)'s work. Its basic idea is that progress in the case of one scientific theory replacing another one occurs when the former is more *truthlike* than the latter. The account was mainly developed by Niiniluoto (1984, 1987, 2014) (see also Kuipers, 2009; Cevolani & Tambolo, 2013). I focus on Niiniluoto's view.

3.4.1. *Niiniluoto's Truthlikeness Account*

The truthlikeness account does not require theories to be fully true for science to progress – among other things because most past theories turned out to be false. It suffices that they are 'truthlike'. Truthlikeness is not the same as *approximate* truth. Truthlikeness is also measured in how *informative* a theory or claim is, as Cevolani and Tambolo (2013) emphasise. For instance, the true claim that cancer cells are not made up from lymph (a negative claim) is less truthlike than the false but informative claim that cancer cells develop from blastema.

Niiniluoto (1987)'s view implies that truthlikeness is *language dependent*. Theories are formulated in a language and truthlikeness is measured in how similar a theory's content is to the maximally specific claim formulated in that language that fully captures the truth about the subject of inquiry. (The details do not matter for our purposes.) Scientific progress is thus language dependent and tied to theories on this account. Gathering new information, conducting new kinds of experiments, etc. is only progressive when it goes hand in hand with theory change, such as developing a more truthlike theory, endorsing more truthlike claims, or correcting false claims. No such theory change has to be recognised as more truthlike to be progressive. Progress does not require its recognition.

This brief characterisation shall suffice for our purposes (for details and discussion, see e.g. Bird, 2016; Cevolani & Tambolo, 2013; Dellsén, 2018a; Niiniluoto, 2019).

3.4.2. Niiniluoto's Truthlikeness Account and the Challenge From Idealisation

The truthlikeness account can straightforwardly analyse replaceable *Galilean* idealisations or their intermediate stages. Replacing simplified models with more accurate ones is clearly an increase in truthlikeness. Niiniluoto (2014)'s example is Galileo's model of free fall. While it ignored air resistance, this resistance could be modelled using Newton's idealised account of mechanics. He also mentions the progress from the ideal gas law to the van der Waals equation (which was later refined in statistical thermodynamics). The truthlikeness account can also accommodate approximately true idealisations. If the scientific accounts featuring them are more truthlike than relevant competitors, we progress. In all these cases, idealisations constitute progress because they are part of the more truthlike accounts.

The truthlikeness account struggles with *indispensable* idealisations. Since these are considered to be irreplaceable, we do not get a more truthlike theory. Scientific accounts featuring them are also not always the result of being more truthlike than previous accounts. Take the honeybee model. It did not replace a previous model, but it looks to be progressive. We learned that honeybees presumably increase their energy efficiency rather than energy intake. The truthlikeness account could adopt the extraction approach to accommodate indispensable idealisations. This would come at the cost of not considering all idealisations to constitute progress. The truthlikeness account focuses on *theories* (or accounts) and not on the accumulation of various scientific truths (in contrast to the epistemic account). Adding single truthlike claims thus does not constitute progress. It does *promote* progress insofar as it increases the likelihood of new truthlike theories.

Dellsén (2021) argues that the truthlikeness account cannot explain cases of idealised accounts where a de-idealised or less idealised counterpart is available. We should focus on scientific accounts that are most truthlike. In contrast to the epistemic account, the truthlikeness account cannot use the extraction approach to rebut this objection. While indispensable idealisations might promote more truthlike accounts, we already have these accounts. However, the truthlikeness account can adopt the suggestion that *pedagogical purposes* can justify keeping the idealised account. Using the ideal gas law to successfully teach students how gas regularities work can lead to a gain in truthlike theories (e.g. because these students might develop them).

To sum up, the truthlikeness account can easily accommodate Galilean idealisations and other idealisations. It can account for indispensable idealisations if it adopts the extraction approach and accepts that such idealisations only promote progress.

4. Concluding Remarks

The ubiquitous use of (indispensable) idealisations in empirically successful scientific theorising poses a challenge to any account of scientific progress. I have argued that all four dominant accounts of progress can accommodate some idealisations and struggle with others. The *problem-solving* account can employ any suitable idealisation approach, but it has difficulties accommodating cases where a more de-idealised account does not solve more problems. Adopting the non-difference maker account of idealisations, the *noetic* account can accommodate many idealisations but not indispensable ones. This issue can be solved by using the extraction approach. The *epistemic* view can account for cases of Galilean idealisations but needs something like the extraction approach to account for all other kinds of idealisations. The *truthlikeness* account straightforwardly meets the challenge for all idealisations other than indispensable ones, but it can utilise the extraction approach to fill this explanatory gap.

The accounts differ regarding whether idealisations can constitute progress rather than merely promoting it. On the problem-solving account, all idealisations other than the Galilean ones it struggles with can constitute progress. On the noetic and the truthlikeness account, all idealisations other than indispensable ones can constitute progress, whereas these can promote it. On the epistemic account, all idealisations merely promote progress.

All four accounts could adopt alternative views of idealisation or might find ways to fully accommodate idealisations without relying on approaches like the extraction approach. However, these and other issues must be explored on another occasion.

Acknowledgements

I'm grateful to Finnur Dellsén, Yafeng Shan, and an anonymous reviewer for helpful comments and suggestions. Research for this paper was funded by the Icelandic Centre for Research (grant number: 195617-051).

Notes

1. Fleming saw that bacteria on an agar plate had been killed close to where a mold was accidentally growing. The analysis of this mould led to the discovery of penicillin.
2. They can be also distinguished from *fictions* (for details, see Frigg & Hartmann, 2020, Section 2.2).
3. Some omissions lead to a misrepresentation, for example models can omit so many aspects of their target phenomena that they do not qualify as accurate representations.
4. Norton focuses on idealised models. Idealisations are (or represent) '[. . .] a real or fictitious system, distinct from the target system [. . .]' (2012, 209). This is compatible with Jones' account. A gas model that assumes that gas molecules do not repel each other can be construed as a model of fictitious gases.
5. Multiple-models idealisation is analyzed by, for example Morrison, 1999; Weisberg, 2007; Potochnik, 2017.
6. This example is taken from Rice (2016).

7. *Cognitive* scientific progress differs from other forms of progress (Niiniluoto, 2019, Section 2.). For instance, developing more precise microscopes or securing more external funding are non-cognitive improvements.
8. Laudan's term 'scientific problem' is akin to a Kuhnian *puzzle*.
9. Laudan's term 'research tradition' roughly corresponds to a Kuhnian *paradigm* or *disciplinary matrix*.
10. Solving a problem is not an arbitrary matter. For instance, Laudan (1977, 22–34) claims that a theory needs to entail an approximate account of a problem to solve it.
11. This also seems to apply to Pincock (2021)'s view: idealisations have non-representational functions in explanatory practice, and only explanatorily relevant truths are part of explanations.
12. As Bird (2016, 548) and Niiniluoto (2019, Section 3.2) emphasise (and Kuhn and Laudan admit), finding a framework for identifying problems is difficult. Any theory arguably entails answers to infinite questions.
13. If the problem is 'Construct a model that matches the data to *at least* the proximity degree *x*', then all models offer solutions to the very same problem.
14. Dellsén (2021) discusses whose understanding is decisive. I leave this issue aside.
15. Others defend that understanding is a form of knowledge (see Grimm, 2021 for details).
16. Strevens (2019) does not address how the derivation proposal applies to asymptotic idealisations.
17. In a footnote, Bird (2007, 87) suggests that knowledge might not be analysable.
18. Others disagree that justification is necessary for progress (e.g. Rowbottom, 2008; Cevolani & Tambolo, 2013; Niiniluoto, 2014; Dellsén, 2016).

References

Alexandrova, A. (2008). Making Models Count. *Philosophy of Science*, 75(3):383–404.

Bangu, S. (2015). Scientific Progress, Understanding and Unification. In Toader, I. D., Sandu, G., and Prvu, I., editors, *Romanian Studies in Philosophy of Science*, pages 239–253. Springer Verlag.

Barnes, E. (1991). Beyond Verisimilitude: A Linguistically Invariant Basis for Scientific Progress. *Synthese*, 88:309–339.

Batterman, R. (2002). *The Devil in the Details: Asymptotic Reasoning In Explanation, Reduction, and Emergence*. Oxford University Press.

Batterman, R. and Rice, C. (2014). Minimal Model Explanations. *Philosophy of Science*, 81(3):349–376.

Bird, A. (2007). What is Scientific Progress? *Noûs*, 41(1):64–89.

Bird, A. (2008). Scientific Progress as Accumulation of Knowledge: A Reply to Rowbottom. *Studies in History and Philosophy of Science*, 39:279–281.

Bird, A. (2016). Scientific Progress. In Humphreys, P., editor, *Oxford Handbook in Philosophy of Science*, pages 544–563. Oxford University Press.

Bokulich, A. (2011). How Scientific Models Can Explain. *Synthese*, 180:33–45.

Bokulich, A. (2016). Fiction as a Vehicle for Truth: Moving Beyond the Ontic Conception. *The Monist*, 99(3):260–279.

Butterfield, J. (2011). Less Is Different: Emergence and Reduction Reconciled. *Foundations of Physics*, 41(6):1065–1135.

Cartwright, N. (1983). *How the Law of Physics Lie*. Clarendon Press.

Caruso, M. V., Gramigna, V., Rossi, M., Serraino, G. F., Renzulli, A., and Fragomeni, G. (2015). A Computational Fluid Dynamics Comparison Between Different Outflow Graft Anastomosis Locations of Left Ventricular Assist Device (LVAD) in a Patient-Specific Aortic Model. *International Journal for Numerical Methods in Biomedical Engineering*, 31(2):e02700.

Cevolani, G. and Tambolo, L. (2013). Progress as Approximation to the Truth: A Defence of the Verisimilitudinarian Approach. *Erkenntnis*, 78:921–935.

Cohen, L. J. (1980). What Has Science to Do With Truth? *Synthese*, 45:489–510.

Dellsén, F. (2016). Scientific Progress: Knowledge Versus Understanding. *Studies in History and Philosophy of Science Part A*, 56:72–83.

Dellsén, F. (2018a). Scientific Progress: Four Accounts. *Philosophy Compass*, 13(11):e12525.

Dellsén, F. (2018b). Scientific progress, Understanding, and Knowledge: Reply to Park. *Journal for General Philosophy of Science*, 49:451–459.

Dellsén, F. (2021). Understanding Scientific Progress: The Noetic Account. *Synthese*, 199:11249–11278.

Elgin, C. (2017). *True Enough*. MIT Press.

Frigg, R. and Hartmann, S. (2020). Models in Science. In Zalta, E. N., editor, *The Stanford Encyclopedia of Philosophy*. Stanford University.

Gelfert, A. (2016). *How to Do Science with Models. A Philosophical Primer*. Springer.

Gettier, E. L. (1963). Is Justified True Belief Knowledge? *Analysis*, 23(6):121–123.

Greco, J. (2014). Episteme: Knowledge and Understanding. In Timpe, K. and Boyd, C., editors, *Virtues and Their Vices*, pages 287–302. Oxford University Press.

Grimm, S. (2021). Understanding. In Zalta, E. N., editor, *The Stanford Encyclopedia of Philosophy*. Stanford University.

Hannon, M. (2021). Recent Work in the Epistemology of Understanding. *American Philosophical Quarterly*, 58(3):269–290.

Ichikawa, J. J. and Steup, M. (2018). The Analysis of Knowledge. In Zalta, E. N., editor, *The Stanford Encyclopedia of Philosophy*. Stanford University.

Jones, M. (2005). Idealisation and Abstraction: A Framework. In Jones, M. and Cartwright, N., editors, *Idealisation XII: Correcting the Model. Idealisation and Abstraction in the Sciences*, pages 173–217. Rodopi.

Kennedy, A. (2012). A Non Representationalist View of Model Explanation. *Studies in History and Philosophy of Science, Part A*, 43(2):326–332.

Kuhn, T. S. (1970). *The Structure of Scientific Revolutions*, 2nd edition. University of Chicago Press.

Kuipers, T. A. F. (2009). Empirical Progress and Truth Approximation by the 'Hypothetico-Probabilistic Method'. *Erkenntnis*, 70(3):313–330.

Laudan, L. (1977). *Progress and its Problems: Toward a Theory of Scientific Growth*. University of California Press.

Laudan, L. (1981). A Problem-Solving Approach to Scientific Progress. In Hacking, I., editor, *Scientific Revolutions*, pages 144–155. Oxford University Press.

Lawler, I. (2021). Scientific Understanding and Felicitous Legitimate Falsehoods. *Synthese*, 198(7):6859–6887.

Levins, R. (1966). The Strategy of Model Building in Population Biology. *American Scientist*, 54(4):421–431.

McMullin, E. (1985). Galilean Idealisation. *Studies in History and Philosophy of Science*, 16:247–273.

Morrison, M. (1999). Models as Autonomous Agents. In Morgan, M. and Morrison, M., editors, *Models as Mediators: Perspectives on Natural and Social Science*, pages 38–65. Cambridge University Press.

Niiniluoto, I. (1984). *Is Science Progressive?* Reidel: Springer.

Niiniluoto, I. (1987). *Truthlikeness*. Reidel: Springer.

Niiniluoto, I. (2014). Scientific Progress as Increasing Verisimilitude. *Studies in History and Philosophy of Science Part A*, 46:73–77.

Niiniluoto, I. (2019). Scientific Progress. In Zalta, E. N., editor, *Stanford Encyclopedia of Philosophy*. Winter 2019 edition. https://plato.stanford.edu/archives/win2019/entries/scientific-progress/.

Norton, J. (2012). Approximations and Idealisations: Why the Difference Matters. *Philosophy of Science*, 79:207–232.

Park, S. (2017). Does Scientific Progress Consist in Increasing Knowledge or Understanding? *Journal for General Philosophy of Science*, 48:569–579.

Pincock, C. (2014). How to Avoid Inconsistent Idealisations. *Synthese*, 191:2957–2972.

Pincock, C. (2021). A Defense of Truth as a Necessary Condition on Scientific Explanation. *Erkenntnis*:1–20.

Popper, K. R. (1963). *Conjectures and Refutations: The Growth of Scientific Knowledge*. Hutchinson.

Potochnik, A. (2017). *Idealisation and the Aims of Science*. University of Chicago Press.

Rice, C. (2016). Factive Scientific Understanding Without Accurate Representation. *Biology & Philosophy*, 31(1):81–102.

Rice, C. (2018). Idealized Models, Holistic Distortions, and Universality. *Synthese*, 195(6):2795–2819.

Rice, C. (2019). Understanding Realism. *Synthese*, 198(5):4097–4121.

Rohwer, Y. and Rice, C. (2016). How are Models and Explanations Related? *Erkenntnis*, 81(5):1127–1148.

Rowbottom, D. P. (2008). N-rays and the Semantic View of Progress. *Studies in History and Philosophy of Science*, 39:277–278.

Saatsi, J. (2019). What is Theoretical Progress of Science? *Synthese*, 196(2):611–631.

Schmid-Hempel, P., Kacelnik, A., and Houston, A. (1985). Honeybees Maximize Efficiency By Not Filling Their Crop. *Behavioral Ecology and Sociobiology*, 17:61–66.

Shan, Y. (2019). A New Functional Approach to Scientific Progress. *Philosophy of Science*, 86(4):739–758.

Shech, E. (2018). Infinite Idealisations in Physics. *Philosophy Compass*, 18:e12514.

Strevens, M. (2008). *Depth. An Account of Scientific Explanation*. Harvard University Press.

Strevens, M. (2013). No Understanding without Explanation. *Studies in History and Philosophy of Science*, 44(3):510–515.

Strevens, M. (2017). How Idealisations Provide Understanding. In Grimm, S., Baumberger, C., and Ammon, S., editors, *Explaining Understanding: New Essays in Epistemology and Philosophy of Science*, pages 37–49. Routledge.

Strevens, M. (2019). The Structure of Asymptotic Idealisation. *Synthese*, 196(5):1713–1731.

Teller, P. (2001). Twilight of the Perfect Model. *Erkenntnis*, 55(3):393–415.

Wayne, A. (2011). Expanding the Scope of Explanatory Idealisation. *Philosophy of Science*, 78(5):830–841.

Weisberg, M. (2007). Three Kinds of Idealisation. *The Journal of Philosophy*, 104(12):639–659.

18 Scientific Speculation and Evidential Progress

Peter Achinstein

Is speculating in science valuable, as some have claimed? Or, as others have said, is it generally to be avoided because it is unempirical and often leads to the acceptance of unsubstantiated claims and to scientific decline rather than progress? To develop answers, I divide the chapter into two parts. The first (Sections 1–4) focuses on speculations, the second (Sections 5–7) on progress. In Section 1, I will summarise two opposing views about the legitimacy of speculating in science. In Section 2, I turn to the question of what is to count as a speculation, and offer a definition. In Sections 3 and 4, under the assumption that, even if speculating is risky, at least some scientific speculations have been valuable, I will ask how, in general, such value is to be understood. In Sections 5–7, I will focus on one particularly thorny question that has been raised about scientific progress, namely whether what I will call evidential progress in science (or the absence of it) should be invoked in assessing the credibility of a speculative theory.

Part I: Scientific Speculations

1. *Two Opposing Views About Scientific Speculations*

In his 'General Scholium' at the end of the *Principia*, Isaac Newton (1999, p. 943) says that although he has proved the law of gravity, he does not know the reason why gravity has the properties it has, including, that of acting over such immense distances. He adds, famously, that he will not 'feign' hypotheses about this matter, because hypotheses 'have no place in experimental philosophy'. For Newton, a 'hypothesis' is any proposition 'not deduced from the phenomena' by causal-inductive reasoning governed by his four 'Rules for the Study of Natural Philosophy' that begin the third book of the *Principia*. In accordance with Newton's official methodology, empirical scientists should eschew 'hypotheses', or, as I will call them in what follows, 'speculations'.

Newton is by no means alone here. Thomas Young (1802) published a paper resuscitating Huygens' wave theory of light by introducing the

DOI: 10.4324/9781003165859-22

assumption that a highly elastic ether pervades the universe and that a luminous body produces waves in this ether that can explain various known optical phenomena. In 1803, Henry Brougham, a defender of Newton's particle theory of light, as well as of his methodology, wrote a scathing review of Young's paper, saying:

> As this paper contains nothing which deserves the names either of experiment or discovery, . . . it is in fact destitute of every species of merit. . . . A discovery in mathematics, or a successful induction of facts, when once completed, cannot be too soon given to the world. But . . . an hypothesis is a work of fancy, useless in science, and fit only for the amusement of a vacant hour.
>
> (Brougham 1803, pp. 450, 455)

Views about speculation of the sort expressed by Newton and Brougham I will classify as views hostile to speculating. Perhaps with more charity than they deserve, I shall understand them to be saying this: speculations in science are to be avoided, at least in one's most serious scientific publications. Perhaps it is okay to speculate in private, or occasionally to do so in public. But if you do the latter, then make sure to label it as a speculation or the equivalent so that readers will not be misled.

There is, however, a friendlier view, which comes in two versions. In the standard version, associated with 'hypothetico-deductivism', there are two stages in scientific activity: the speculative stage and the testing stage, both scientific activities. According to William Whewell (1840, reprinted in Achinstein 2004, p. 155), 'advances in knowledge are not commonly made without the previous exercise of some boldness and license in guessing'. Scientists, he claims, are usually capable of putting forth different speculations to explain some set of phenomena, which is a good thing: 'A facility in devising [different] hypotheses, therefore, is so far from being a fault in the intellectual character of a discoverer, that it is, in truth, a faculty indispensable to his task' (Achinstein 2004, p. 154). The only constraint Whewell imposes on speculations is that they offer a possible explanation of the phenomena in question, whether or not that explanation is correct or even probable.[1]

There is, however, a view that is even friendlier than the standard h-d view. It is a radical position taken by Paul Feyerabend (1970), who expresses it in a principle he calls the *Principle of Proliferation*: 'Invent and elaborate theories which are inconsistent with the accepted point of view, even if the latter should happen to be highly confirmed and generally accepted . . ., such a principle would seem to be an essential part of any critical empiricism' (reprinted in Achinstein 2004, p. 377). For Feyerabend, these 'proliferations' are to be 'inconsistent either with well-established theories or with well-established facts'. Typically, then, they will be speculations. According to Feyerabend, introducing such contrary theories is the best way to test accepted theories by exploring alternative explanations.

In this chapter I will defend a position friendly to speculating, but one that rejects the idea present in both hypothetico-deductivism and Feyerabend that few if any constraints are to be imposed on speculations. Before turning to this, however, something needs to be said about what is to count as a speculation.

2. *What Is a Speculation?*

A general definition of 'speculation' offered by the *Oxford Living Dictionaries* (OLD) is this:

The forming of a theory or conjecture without firm evidence

But this leaves open two questions: what counts as 'evidence', and what counts as 'firm'? In what follows I will make use of a definition of 'evidence' (or 'explanatory evidence') that I developed years ago (Achinstein 2001). It says that e is evidence that h only if, given e, it is more probable than not that there is an explanatory connection between e and h. There is an explanatory connection between e and h if h correctly explains why e is true, or e correctly explains why h is true, or something correctly explains why both e and h are true. To make the definition sufficient as well as necessary I add the condition that e is true and that e does not entail h. (For details about, and defence of, this definition, and the concept of explanation used for the definition, see Achinstein 2001.) In what follows I will use the expression 'explanatory-evidence' (e-evidence, for short) to refer to evidence satisfying these conditions.

My claim is that this definition has advantages over more standard Bayesian ('increase in probability') and 'hypothetico-deductive' accounts. It avoids counterexamples that can be raised against these and other accounts. And, I argue, it better reflects what scientists seek when they seek evidence for their theories.[2] One consequence of this definition is that it is possible for a theory to be assigned high probability without its having e-evidence to support it. (A historical example is given in endnote 5, and its implications are discussed in Section 7.)

Finally, then, I will understand 'firm evidence' to be e-evidence that makes it highly probable that there is an explanatory connection between h and e, given e, not just that it be more probable than not. As I am construing it, the firmness of the evidence refers to the extent to which e supports h, not to whether e itself is sufficiently established. (The concept of evidence I am using requires e to be true, not that it be established to be true.) On the present definition, a theory can have e-evidence and still be a speculation. It can be more probable than not that, given e, there is an explanatory connection between h and e, without it being highly probable.

Building upon the OLD definition, we might begin by understanding speculating as introducing assumptions without knowing that there is firm e-evidence for them. If there is such evidence the speculator does not know there is. If there is no such evidence, the speculator may or may not know that. The term 'speculation' will be used to refer both to the activity of speculating and to the product of that activity – the assumptions themselves. Without knowing that there is firm e-evidence for those assumptions (if there is), speculators, particularly in science, introduce assumptions usually under these broad conditions: (a) (a truth-related condition) They believe that the assumptions are either true, or close to the truth, or a first approximation to the truth, or a possible candidate for truth that is worth considering.[3] (b) (a scientific-activity condition) They introduce such assumptions when explaining, predicting, unifying, calculating, and the like, even if the assumptions in question turn out to be incorrect. I will call (a) and (b) 'theorising' conditions. We might then give the following simple characterisation of a speculation, especially a scientific one:

> Where h is a claim that is introduced by a scientist S in a way that satisfies the theorizing conditions (a) and (b), h is a speculation for S if and only if S does not know that there is firm e-evidence for h.[4]

Isaac Newton begins Query 29 of his *Opticks* as follows:

> Are not the Rays of Light very small bodies emitted from shining Substances? For such Bodies will pass through uniform Mediums in right lines without bending into the Shadow, which is the Nature of Rays of Light.
>
> (Newton 1979, p. 370)

Newton is assuming that light consists of material particles projected from luminous bodies, that these particles obey his three laws of motion, and that they are subject to short-range forces of attraction and repulsion. In doing so he satisfies conditions (a) and (b). By introducing the theory using the rhetorical question here it is clear that he believes, at least, that his assumptions are possible candidates for truth that are worth considering. (Indeed, his belief seems even stronger.) And during his discussion in Query 29 he shows how these assumptions explain various known phenomena involving light, including rectilinear propagation, reflection, and refraction. He regarded it more probable that they are correctly explained by his particle theory than by the rival wave theory of Huygens. But he did not regard his explanations as highly probable. He lacked 'firm evidence' in the sense I am giving to that expression.[5]

How does Query 29 square with Newton's rejection of 'hypotheses' generally in science ('hypotheses have no place in experimental philosophy')?

It doesn't. But I believe that the following resolution would be satisfactory to Sir Isaac. If you are going to speculate, or introduce 'hypotheses', do so explicitly calling them such or introduce them as 'Queries' for further study. (Newton does both. In the *Principia* two 'hypotheses' are introduced and labelled as such. In the *Opticks* there are 31 'Queries'.) What is most important to Newton is that if you do speculate, you must not conclude from the fact that your speculation, if true, explains various phenomena that it is true or highly probable. This is the 'method of hypothesis', which Newton emphatically rejects.

3. How to Evaluate a Speculation: The Usual Answers

Whether speculations are to be just tolerated (as with Newton, if you properly label them), or to be required first steps in an investigation before experimental testing is initiated (as with h-d theorists), or to be necessary even after testing (as with Feyerabend), how are they to be evaluated? How is one to decide whether a speculation is any good? Philosophers as well as scientists have devoted a lot of attention to the question of how theories are to be evaluated in terms of evidence supplied by tests. When such tests are largely absent, however, there is very little discussion of how speculations are to be evaluated *as speculations*. Are scientists simply to wait for the tests to occur, and then judge the speculations to be good or bad, valuable or not, depending what the tests reveal? In the absence of testing, what constraints, if any, should be imposed on a scientific speculation such that if and perhaps only if it satisfies those constraints, it is a good or valuable speculation?

What philosophers (and some scientists) usually say in response to these questions is: impose minimal constraints. One of the favourite phrases of those who support speculating is that when one is speculating (in what used to be called the 'context of discovery') one should give 'free rein to the imagination'. The constraints, if any, should be few. (The important ones are left for the 'context of justification' when considering how strongly the evidence supports the theory.) Hypothetico-deductivists such as Whewell and Popper require that a speculation offer a potential explanation of the phenomena for which a theory is being sought. In doing so, the speculator organises otherwise disconnected phenomena in a more intelligible way than without the speculation, a process Whewell calls 'colligation'. There may be several different speculative colligations. For Whewell the best is 'that one which most agrees with what we know of the observed facts' (Achinstein 2004, p. 154). For Popper, it is the boldest or most general. Perhaps Newton in his grudging use of speculations would agree with that.[6] Feyerabend doesn't even require this much. A speculation may be a good one even if (and perhaps especially if) it contradicts the observed facts with which the speculator began, and so doesn't explain those facts, but perhaps explains why observer took them to be facts.[7]

Finally, perhaps all such writers would require a speculation to be empirically 'testable', even if at the moment it cannot be tested, and if even these writers disagree with each other about what counts as testable.

Such constraints on speculations are not only few, but more importantly, very widely applicable. Constraints such as 'explanatory power', 'agreeing with observed facts', 'bold', and 'testable' are to hold for scientific speculations generally, whatever the theory or the field of science. This, I will argue, is bound to miss some of the most illuminating ways to evaluate speculative theories, which depend on the particular theory being evaluated and can vary from one theory to the next. They also depend on the perspective of the evaluation, which can vary from one context of evaluation to another. I will illustrate what I mean by turning to one of the greatest speculators in the history of physics, James Clerk Maxwell, and to how he and others evaluated one of his most important speculations.

4.   How to Evaluate a Speculation: Learn From Maxwell

Maxwell not only developed major speculations in molecular and electrical theory, but also had important things to say about speculating in science, particularly in his work on molecules. There he was concerned with the question of how one should proceed when one has at least the beginnings of a molecular theory to explain gaseous phenomena but no firm experimental e-evidence to support its basic ideas. How, if at all, should such theory be developed and evaluated in the absence of such evidence?

In 1860, Maxwell (1965, vol. 1, pp. 377–409) published 'Illustrations of the Dynamical Theory of Gases', his first paper on kinetic theory. In it he assumes that gases are composed of spherical molecules that move with uniform velocity in straight lines, except when they strike each other or the sides of a container; that they obey the other laws of dynamics; that they exert forces only at impact, not at a distance; and that they make perfectly elastic collisions. He then works out these fundamental ideas to explain known gas laws (e.g. Boyle's law, ideal gas law, Avogadro's law), and most importantly to derive new ones (such as the distribution of molecular velocities law, named after him, but not capable of testing until the 1920s). He does so in a series of what he calls 'propositions', which are tasks he sets himself in order to answer questions he raises about gases. In 1859, just before publishing the paper, Maxwell wrote to Stokes, saying:

> I do not know how far such speculations may be found to agree with facts, . . . and at any rate as I found myself able and willing to deduce the laws of motion of systems of particles acting on each other only at impact, I have done so as an exercise in mechanics. Now do you think there is any so complete a refutation of this theory of gases as would make it absurd to investigate it further so as to found arguments upon

measurements of strictly "molecular" quantities before we know whether there be any molecules.

(Garber, Brush, and Everitt 1986, p. 279)

How should Maxwell's grand speculation (it is 32 pages) be evaluated? From what perspective? We might consider it from the perspective of a contemporary historian of science, asking what impact it had on the physics community at the time of Maxwell, or during the period from 1860 until the beginning of the twentieth century, or at the present time (his distribution law is still taught in classes on statistical mechanics), or others. Alternatively, we might consider Maxwell's theory from his own perspective in 1860, asking questions such as these: what problems was Maxwell trying to solve with his kinetic theory, why, and how far did he get? What were the stumbling blocks? What general and what specific constraints did he impose on the answers, why, and was he justified? And we might add a broader question which evaluates the perspective itself, given what was known at the time: was what he was trying accomplish – a mechanical explanation of gaseous phenomena in terms of the motions of particles subject to Newtonian laws of dynamics – a reasonable one, given what he and others knew in 1860? In what follows I will consider a Maxwellian perspective, one that he himself took towards the end of the paper in evaluating what he had accomplished. It is also a perspective a physicist might take who was a contemporary of Maxwell, particularly one reviewing the paper for a professional journal.

Let's start with the question: are there any general constraints Maxwell himself imposed on the theory, and if so why? Maxwell had strong views on how to theorise in physics when attempting to explain phenomena established by experiment. In an earlier 1855 paper,[8] he presents three requirements for doing so, especially if one is publishing a technical paper in a professional journal. First, a physical, rather than a purely mathematical, way of understanding the phenomena must be provided. (Don't just give a set of equations from which other equations representing the phenomena can be derived mathematically. Say what these equations mean in physical terms.) Second, do express the physical ideas using mathematics. (Don't give simply a qualitative physical understanding of the phenomena, which you might do for an encyclopaedia article; make it precise using mathematics.) Third, the theorising should not be sketchy but worked out in considerable detail. (You are not writing an encyclopaedia article, although Maxwell did quite a few of those.) In the 1855 paper Maxwell is concerned not with molecular theory but with electricity and magnetism. And the theorising in this paper consists in working out an elaborate physical analogy between the electromagnetic field and an imaginary incompressible fluid flowing through tubes of varying section. Maxwell believed that one way to understand phenomena was using what he called a 'physical analogy' between two distinct systems. In his 1860 kinetic theory paper he also says he is providing a physical analogy. But it is soon clear that this is not an analogy. It is a possible explanation of observed gaseous properties by making claims about what gases are actually composed of. In the 1860 paper one system is being described, not two. The important point, however, is that the three general theorising constraints he imposes are the same in both cases: provide a physical description, express it using mathematics, and spell it out in detail.

Maxwell's aim is to determine whether a mechanical theory of gases is possible, not necessarily to provide one that he believed is literally true. He wants to see whether he can explain known properties of gases and even generate some new interesting results by assuming that gases are composed of moving particles (molecules) which obey Newtonian laws of dynamics and satisfy the mathematically simplifying assumptions given in the second paragraph of this section. Why does he search for a mechanical explanation? In a later paper he spells this out:

> When a physical phenomenon can be completely described as a change in the configuration and motion of a material system, the dynamical explanation of that phenomenon is said to be complete. We cannot conceive any further explanation to be either necessary, desirable, or possible.
>
> (Maxwell 1965, vol. 2, p. 418)

In addition to the a priori metaphysical/semantical idea expressed in the previous quote, Maxwell cites the success of mechanical theories in astronomy as part of his support for the search for a mechanical theory of gases. Before settling on a mechanical theory that he thinks is correct in its details, let's see how far one can get with a simplified version. This is his aim. How far does he get, and what are the stumbling blocks?

Thirty out of 32 pages of his paper consist of problems to be solved using the aforementioned assumptions, mathematics, solutions to previous problems, and new assumptions introduced in order to solve those problems. These are the 'exercises in mechanics' he mentions in his letter to Stokes. Performing one of these 'exercises' (his Proposition 4) led to the most important result in the paper, the Maxwell velocity distribution law. Towards the end of the paper Maxwell summarises what he has accomplished and what the difficulties are. In the former category he mentions that the molecular theory yields the known relationship between pressure, temperature, and density of a gas (a form of the ideal gas law), as well as the law that at equal temperature and pressure, equal volumes of gases contain the same number of molecules (Avogadro's law). In addition, he has explained the internal friction of gases as well as gaseous diffusion in molecular terms. But he also gets some results which, as he puts it, seem 'decisive against the unqualified acceptation of the hypothesis that gases are such systems of hard elastic particles' (p. 409). These include the ratio of specific heat of a gas at constant pressure to that at constant volume. The theoretically derived value is too far off the experimental one. Also, he theoretically derives what he regards as the 'remarkable result' that the coefficient of internal friction of a gas is independent of the density of the gas. But he adds: 'The only experiment I have met with on the subject does not seem to confirm it' (p. 391).

Maxwell's theory is, by his own admission, a speculation. It lacks 'firm evidence', even though it has some successes, as well as some failures. But evaluating this theory as a speculation is not, or at least not always, the simple matter that some philosophers have made it out to be. It is not confined to establishing whether it is consistent with observed phenomena, whether it offers an explanation of those phenomena, whether it is bold, and whether it is testable. How it is evaluated depends crucially on the specific questions it raises, on the answers it is proposing, and on the methodological and empirical constraints being imposed on those answers. To offer this kind of evaluation, one needs to get into the nitty-gritty of the theory. That is why it is done best by professional scientists, whether by the speculator himself or a reviewer – not by a philosopher armed only with generalities. It is the type of evaluation most useful to the scientist and the scientific community at the time.

Part II: Evidential Progress and Credibility

5. The Pessimistic Induction

I turn now to a difficult question that has been raised about determining the credibility of a theory in the light of evidential progress (or the lack of it) in science. In Achinstein (2001), I define e-evidence in terms of probability, understood in an objective epistemic sense as representing the degree to which it is reasonable to believe a proposition. (For this account of probability, see Chapter 5 of that work.) In what follows, 'credibility' will be employed in this sense. Not all the ways of positively evaluating a theory contribute to its credibility. In the context in which Maxwell was evaluating his kinetic theory in 1860, this was not important in the evaluation. He was not attempting to show that his particular mechanical theory (involving spherical molecules and only contact forces) is true or credible, only that it is possible, that it is consistent with mechanical principles.

Suppose, however, that, unlike with Maxwell, we are attempting to evaluate a theory in terms of its credibility. The theory may have little, if any, e-evidence, and still have some credibility. Or it could have some e-evidence, just not 'firm evidence'. (An example to be given later is Rutherford's 1911 solar theory of the atom.) In either case the theory is speculative. In what follows, I will not restrict my attention only to speculative theories, since the philosophical issue to be discussed is a general one.

However, since speculative theories are much more vulnerable than ones for which firm evidence has been provided, the following question, which I will now focus on, is especially important for them: in determining the credibility of a theory do we need to take into account not only any available e-evidence, but also historical facts about evidential progress in science? This question needs clarification and motivation.

Let us say that a theory T exhibits evidential progress from time t1 to t2 if, as e-evidence is amassed during this period, the total available e-evidence for T makes T more and more credible. And let us say that a theory T exhibits evidential decline from t1 to t2 if, as e-evidence is amassed, the total available e-evidence for the negation of T makes the negation of T more and more credible, and hence T less and less credible. (On this definition, over time a theory may exhibit neither evidential progress nor decline, or of course both.) Let us also say that science as a whole exhibits evidential progress (or decline) during a period of time (say from the seventeenth century until now) if, in general, many or most of its theories exhibit evidential progress (or decline) during that period. In introducing these concepts, I am not supposing that evidential progress is the only, or even the most important, type of progress in science, only that it is one type, and that it is important for dealing with the question I propose to discuss.

Suppose that historically speaking science, as a whole, has exhibited evidential progress. Should this historical fact be some reason to believe that a current theory T that has at least some credibility will retain that credibility, and even increase it, as new e-evidence is amassed? By contrast, suppose that, historically speaking, science, as a whole, has exhibited evidential decline. Should this historical fact, if it is one, be some reason to believe that a current theory T, even if highly credible given the current e-evidence, will meet the same fate?

Let me use the expression 'historical-progress (HP-) evidence' to refer to historical information about the evidential progress of science as a whole (or lack of it), during whatever period is chosen. HP-evidence is not e-evidence, as I have defined these terms. (Given HP-evidence that science as a whole has – or has not – exhibited evidential progress, it is not probable that there is an explanatory connection between this historical fact about scientific progress generally and the particular fact, if it is one, that all bodies obey Newton's law of gravity.) The question I am raising is this: should both HP-evidence and e-evidence be considered in deciding what credibility to assign to a particular scientific theory?

There is a requirement that many who speak about evidence believe is fundamental and inviolable for scientists. It is called the *Requirement of Total Evidence* (RTE). It says that in determining how credible a proposition is, you must consider the total evidence available, not just part of it. Otherwise, you are biased, lazy, or irrational. This seems to require that in determining the credibility of a particular scientific theory you need to consider not just e-evidence for that theory, but the HP- evidence as well, assuming it affects the credibility of that theory. So, three questions need answering: (i) Is science as a whole evidentially progressing or declining? (ii) If it is, does this fact affect the credibility of a particular theory? (iii) If it does, should we follow the RTE and consider both e-evidence and HP-evidence in determining that credibility? Or should we violate RTE and disregard HP-evidence?

Conflicting answers to the first question are given by optimists and pessimists who draw different lessons from the history of science. Larry Laudan (1981) offers a pessimistic view, the physicist Steven Weinberg (1994, pp. 231–232), a contrasting optimistic historical view. (In what follows, I will concentrate on the former, but what I say, with appropriate changes, is applicable to the optimistic one also.) According to Laudan, history shows that generally speaking scientific theories once evidentially progressing, indeed once highly credible on the basis of the available e-evidence, with new e-evidence, or new ways of thinking about the evidence, or discoveries that the original evidence is faulty, exhibit evidential decline and lose their credibility, even to the point of being considered refuted. This happens even when the theory gets modified in the light of new information. This gives rise to the 'pessimistic induction' that in the future, new theories, which may be initially credible, will eventually lose that credibility as more e-evidence is obtained.

Does this historical pessimism have consequences for how we should determine the credibility of particular theories we are developing and testing? Let me mention some possible ways of responding.

1. We should argue that it has not been shown that, on the whole, history favours pessimism (or optimism). Some theories have become and remained evidentially successful until the present (e.g. the heliocentric theory), while others have not (e.g. the geocentric). Without a more careful historical analysis of different types of theories (not just the major ones), and a more careful statistical analysis of the historical data, we cannot make a legitimate induction from general evidential decline (or progress) to the future evidential decline (or progress) of any specific theory. Until this is done we should disregard HP-evidence completely in determining the credibility of a particular theory and use just e-evidence for doing so.

2. If we are pessimists about history, and believe that this pessimism should be reflected in our assessment of the credibility of a theory, we should give a 'mixed review' to that theory. Assuming that the e-evidence is or becomes very positive for the theory and that the HP-evidence concerning theories in general (or for theories that cover some of the same ground as the one in question) is quite negative, we should say that the total evidence, containing both

e- and HP-evidence, justifies us in assigning a so-so credibility to the particular theory in question. This is in keeping with the RTE.

3. Assuming that the theory under consideration gets a 'mixed review', given negative HP-evidence and positive e-evidence, we should be unwilling to assign an overall credibility to the theory, but remain agnostic about this. Just give the positive and negative evidence, without drawing a general conclusion about credibility.

4. In determining the credibility of a particular theory, we should disregard the HP-evidence completely, and use just e-evidence for doing so. But we should do so in a way different from option 1, a way that recognises negative (or positive) HP-evidence without including it in the total evidence. If this violates the sacred RTE, so be it. Such a violation can be justified.

In what follows, I want to explore option 4, by arguing that it is, or can be, a legitimate option. (In my discussion I will assume for the sake of argument that the pessimist is right about the facts of history. I will not attempt to argue for or to dispute this pessimistic assumption.)

My first question is this: is it ever legitimate to violate RTE? I think it is. In Section 6, I will consider a scientific case in which e-evidence (not HP-evidence) was disregarded in giving a credibility assessment, RTE was violated, and this violation seems reasonable. In Section 7, I will consider whether what is said here is applicable to disregarding HP-evidence in a manner proposed by option 4.

6.   Disregarding e-Evidence: Learn From Rutherford

In 1911, Ernest Rutherford (1911) introduced his revolutionary solar model of the atom. He did so, based on experimental results of Geiger and Marsden involving the scattering of alpha particles by gold atoms. Most of the alpha particles went through the gold atoms without much deflection, but a few, remarkably, were scattered through an angle of more than 90 degrees. Based on these experiments Rutherford proposed that an atom contains a charge concentrated in a nucleus whose volume is small compared with that of the atom. The charged nucleus is surrounded by oppositely charged electrons revolving around the nucleus, which are too small to deflect the alpha particles. But when the alpha particles get close to the nucleus, they are deflected at large angles; hence, the remarkable scattering observed.

Now, there is a serious problem with this model. According to classical electromagnetic theory, and the evidence on which it is based, such an atom cannot be stable, since moving charged particles in such a system would collapse into the nucleus. But gold and most other atoms are stable. So there is a stability problem with the model. (For details, see Boorse and Motz 1966, p. 706; Heilbron 2003, p. 69.) Rutherford chose to ignore this problem until further experiments on the scattering itself could be performed. He wrote:

> The question of the stability of the atom proposed need not be considered at this stage, for this will obviously depend upon the minute structure of the atom, and on the motion of the constituent charged parts.
> (Rutherford 1911, p. 671)

Commenting on Rutherford's 1911 paper, Boorse and Motz (1966, vol. 1, p. 706) write

> Rutherford was aware of this [contradiction between his model and classical electrodynamics and all the evidence for the latter] but chose to ignore the difficulty for the time being.

Rutherford in 1911 did not believe that his model was established by the scattering experiments. But he did believe that they provided a reason to assign some credibility to the model, even if additional e-evidence to fill out more details and additional support was needed to make the evidence 'firm'. (He treated his theory as a speculation, in the sense I am giving to that term.) But in assigning whatever credibility to the model he did, he disregarded the negative evidence from classical electrodynamics and the theory based on that evidence, which is a violation of RTE. He disregarded that evidence not because he considered it bogus or because he had a solution to the problem, but because he lacked a solution. Can this be justified?

I believe it can. As Boorse and Motz say, Rutherford 'chose to ignore the difficulty for the time being', and did so to work out other details about the structure of the atom, such as the whether the central nucleus is positively or negatively charged, and, in his words, to discover the 'extent of the distribution of the central charge'. One question at a time, please. 'Okay', you may ask,

> [I]n view of the stability problem, why didn't Rutherford give the model a mixed review, saying that some evidence (classical electrical evidence) is negative, and some (scattering evidence) is positive? Or, why didn't he remain agnostic, saying that he will give no credibility assessment until the stability problem is solved or resolved? Doing either of the latter would not be violating RTE. What justification can be given for disregarding the stability problem for the present in his credibility assessment rather than giving a mixed review or no review at all?

My answer is pragmatic. Disregarding the problem for the moment while saying you are, and basing the positive assessment on the striking Geiger-Marsden experimental results, is likely to encourage further research on the stability problem (as well as on the more tractable 'charge' problem Rutherford mentions) in a way that the more discouraging mixed review or the neutral no-review will. This it certainly did. Two years later, in 1913, the young Niels Bohr, who had visited Rutherford's laboratory in Manchester, published three papers that adopted Rutherford's model and solved Rutherford's stability problem by introducing the first ideas of the quantum theory of the atom. To be sure, this 'solution' was temporary, and led to new problems for Bohr (such as discontinuous motion). My point is only that there are occasions on which contrary, or seemingly contrary, evidence is at least temporarily disregarded, and the RTE violated, for justifiable pragmatic reasons. In making this claim I do not mean to say that in addition to evidential reasons for believing a theory there are non-evidential pragmatic ones and that one must balance the former against the latter in deciding what to believe or how much to believe it. It is to say only that in some cases one is justified in using non-evidential pragmatic considerations in selecting the evidence on which to base one's assignment of credibility.

7. *Disregarding HP-Evidence*

Is what has been said about disregarding e-evidence in the Rutherford case applicable to disregarding HP-evidence? There are some big differences here. In the Rutherford case what is being disregarded is one puzzling piece of the total e-evidence (known phenomena involving motions of charged particles), and it is being disregarded just temporarily until a solution can be found and other problems worked out. In the case of HP-evidence what is being disregarded is the entire history of science, or at least parts of it, and there is nothing temporary about it, as long as evidentially successful theories eventually become unsuccessful. So we need to proceed with care.

Concentrating just on negative HP-evidence, what do scientists themselves do with such evidence, assuming (a) that history is largely negative, (b) that scientists are aware of this and even think about it, and (c) that the e-evidence for their theory is very positive? They don't give the theory a mixed overall evidential review, saying that some evidence is very positive and some negative. Nor do they remain epistemically agnostic about how credible their theory is on the basis of all the evidence. Nor do they say, in a manner analogous to one I ascribed to Rutherford, that *until the 'history of science' problem is solved,* they are justified in assigning a credibility to the theory based only on the e-evidence. For them, there is no 'history of science' problem. What they do, in their professional papers and reports, is disregard the problem and the HP-evidence in their credibility assessments. They believe they are justified in assigning a credibility to the theory based on the e-evidence alone, without considering the HP-evidence (or at least they act that way). How can they do so, assuming that the pessimistic view of history is correct?

To speak paradoxically, in determining how credible a theory is one can disregard the HP-evidence without ignoring that evidence. There are different ways to take the HP-evidence into account, not all of which require you to combine it with the e-evidence in determining the credibility of a theory. How does this work, and why do it? On the present proposal – to expand on option 4 in Section 5 – one uses only e-evidence, not HP-evidence, in determining how credible a particular theory is at a given time, while also recognising, explicitly or implicitly, that, as history demonstrates (according to the pessimist), at some later time new e-evidence may well show that the theory is not credible, or new information may show that the facts that we took to be e-evidence are shaky or false. Until this happens (if it does) in the present case, we will assess a theory's credibility at a given time on the e-evidence available at that time.

Is this a violation of the RTE? Yes, it is, assuming that the HP-evidence is what the pessimist says it is. (For the sake of argument, I am assuming it is. Those adopting option 1 will argue otherwise.) One might respond that the RTE should be understood as applying only to e-evidence, not to HP-evidence. But those who defend the RTE don't seem to have that exclusion in mind. They want a very general principle that applies to all evidence, not just to what I have been calling e-evidence. So, how, if at all, could disregarding HP-evidence and violating RTE be defended? Why favour option 4 over the others? Let me offer four considerations in favour of this option.

First, as noted, it is what scientists actually do in their articles, reports, and books. In such places mixed reviews are confined to ones in which there is mixed e-evidence; they do not include ones in which the e-evidence and the HP-evidence point in different directions. Also, it is what they frequently preach when they are in a philosophical preaching mode. This attitude is reflected in Rule 4 of Newton's 'Rules for the Study of Natural Philosophy', which he employs in defence of his law of gravity:

> Rule 4: In experimental philosophy, propositions gathered from phenomena by induction should be considered either exactly or very nearly true notwithstanding any contrary hypotheses, until yet other phenomena make such propositions more exact or liable to exceptions.

In accordance with his prior inductive Rule 3, Newton's inductions are inferences from 'the qualities of bodies . . . known only through experiments' to the claim that all bodies have such qualities. They are not inferences from the success, or lack of it, of known theories to success, or lack of it, of theories generally. Newton's 'phenomena' do not include facts about the evidential success of theories. In Rule 4, Newton recognises that theories that have high inductive support can be shown to be faulty. He was well aware of this risk, and of previous theories which had to be modified (or discarded) when 'other phenomena' were discovered. But until they are, he is telling us, we should base our epistemic judgements only on the 'phenomena' we have, not on those we don't have. And these 'phenomena' do not include historical facts of the sort the pessimist invokes.

What I have said here is applicable to another type of non-e evidence as well, namely 'authoritative evidence', by which I will mean simply what other authorities believe (not why they believe it). Shouldn't the fact that leading authorities disbelieve (or believe) your theory be used, together with e-evidence, in determining the overall credibility of your theory? As with historical evidence, scientists may be aware of such evidence, and may even incorporate into their work a reference to believers and disbelievers. But, in accordance with option 4, in drawing an inference about the truth of their theory one is to use only e-evidence. This is implicit in Newton's methodological rules.

A second consideration concerns the aim of science with respect to credibility. Yes, scientists prize credibility in their theories. But it needs to be credibility established (or refuted) in a certain way and only in that way. In science the extent to which a proposition should be considered worthy of belief is to depend on, and only on, the extent of support given to it by e-evidence. For Newton in what he calls 'experimental philosophy', this requires a 'deduction from phenomena', which I have claimed can be understood in terms of e-evidence, not HP-evidence or authoritative evidence. And for Newton the only way to show that a proposition is not credible and needs revision or rejection is by producing new phenomena (ones which are such that the probability of an explanatory connection between h and e is not high). HP-evidence and authoritative evidence are not the sort of evidence needed or wanted to satisfy the aim of science in establishing or refuting credibility.

Third, as noted, option 4, by contrast with option 1, allows us to take into account HP-evidence, as well as authoritative evidence. If we are historical pessimists we simply say that, as history shows, new e-evidence often appears that undermines a theory and weakens or destroys the efficacy of the previous e-evidence for it; so even if all the available e-evidence makes our theory very credible at present, that theory may turn out to be weakened or refuted by new e-evidence. And if leading authorities don't agree with us, we can admit this. We can do these things, however, without combining negative HP- or authoritative evidence with positive e-evidence in forming an overall judgement about a theory's credibility. We can follow Newton's idea that in 'experimental philosophy' credibility judgements are to be based entirely on physical 'phenomena' we are aware of, not on the past success rate of theories, or on what other authorities believe. If these are negative, that fact may motivate us to reconsider the e-evidence to see whether, in the light of past scientific failures, or of specific objections by other authorities, that e-evidence does really give the amount of credibility we have assigned to the theory. If after addressing such objections, and thinking more about it, we arrive at the same credibility assignment, then we can say that we have taken into account the negative evidence, just not in the manner required by the RTE.

Fourth, as in the Rutherford case, I want to add a pragmatic reason for favouring option 4. In assessing the credibility of a theory, disregarding the negative HP-evidence (if, as the pessimist says, it exists), or disregarding negative authoritative evidence (if it exists), encourages scientists who believe in the importance of their theories to further develop them by asking and answering new questions, finding more e-evidence, and answering objections whether those theories are speculations or theories supported by firm scientific evidence. Imagine the pessimistic historian, or the sceptical authority, saying to Rutherford:

> Ernest, the situation is more dire than you think. Not only is the e-evidence you have for your theory (in 1911) not very firm (you are speculating); not only do you have an atomic stability problem; and not only have you not figured out whether the central nucleus is positively or negatively charged; but you have two problems very different from these, but just as serious: History shows that most theories, even ones better supported by scientific evidence than yours, eventually turn out to be in need of serious modification or are shown to be just plain false, as more scientific evidence is discovered. And the scientific community isn't at the moment buying your theory (because of the problems above). So, with all this negativity, why pursue your theory?

With option 4, Rutherford can reply:

> Not so fast. I recognize that, as history shows, in the light of new e-evidence, lots of theories turn out to be in need of modification or rejection. And I recognise that my atomic theory has a stability problem, and needs further development to answer certain questions as well as further e-evidence in its support. And I recognize that the scientific community hasn't yet bought my ideas. But at the moment I have some extraordinary e-evidence – the results of the Geiger-Marsden experiments which lends at least some support to my solar model of the atom. I am going to base my epistemic judgment of credibility – although tentative and not completely positive – on the e-evidence I have, not on historical facts about scientific success generally, or on the scepticism of the scientific community. This encourages me as well as others, much more than does your historical or authoritative pessimism, to develop my theory to address questions of the sort you note, and to search for new e-evidence that

will support the central ideas in my theory. Furthermore, you, the historical pessimist, have not shown me that I must combine historical, authoritative, and scientific evidence to get a legitimate credibility judgment. All you have really done is said that this is demanded by the Requirement of Total Evidence. But in section 6, I gave you a pragmatic reason for legitimately violating that requirement in a case in which there is conflicting e-evidence. Now I am giving you a pragmatic reason for legitimately violating that requirement in a case in which the HP- and authoritative evidence conflicts with the e-evidence.

Finally, Rutherford, an optimist who was remarkably sure of himself, might conclude:

You philosophers ask questions that are of little interest to scientists. Why should I care whether science as a whole is progressing? I think it is, on an intuitive level. We have lots of e-evidence that Newton's physics is closer to the truth than Aristotle's, and Einstein's even closer than Newton's. That's progress. But I do science, not philosophy or history, which means that when I assess the credibility of a theory I consider only e-evidence, not HP-evidence, or authoritative evidence. Here I follow the methodological sermon of Sir Isaac, which, in effect, tells me to disregard non-e-evidence in making inferences to the truth or approximate truth of scientific claims. And if the author of the present paper has provided a plausible option for doing so that allows me to recognise that, historically speaking, scientists often get it wrong, and that authorities often disagree with pioneers, I believe that this is the only path for me to follow.[9]

Notes

1. In the twentieth century, the view was endorsed by Karl Popper (1959), among others. Popper's only constraint on hypotheses in the speculative stage of science is that they be bold.
2. For counterexamples, and a defense of the second claim, see Achinstein (2001).
3. Those wedded to some form of anti-realism can substitute 'empirically adequate' for 'true'.
4. For a discussion and defense of this definition, see Achinstein (2018).
5. Historically speaking, however, despite the fact that Newton clearly regarded this theory as lacking firm evidence, many of Newton's followers took the fact that Newton introduced the particle theory and preferred it over the wave theory to imply that he believed it (which let us say he did). We may suppose that given that he did believe it, and that he was (in Locke's terms) the 'incomparable Mr. Newton', the probability they assigned to the particle theory was very high, even though, these followers might agree, the particle theory was a hypothesis in Newton's sense.
6. Newton certainly regards more general inductions as better than more limited ones. See *Opticks*, pp. 404–405.

7. Feyerabend praises Galileo for introducing a speculative theory about the true motion of the stone falling from the tower that contradicts what is claimed to be observed.
8. 'On Faraday's Lines of Force', in Maxwell (1965, vol. 1), pp. 155–229.
9. I thank Yafeng Shan for very helpful suggestions.

References

Achinstein, Peter. 2001. *The Book of Evidence*. New York: Oxford University Press.

Achinstein, Peter, ed. 2004. *Science Rules*. Baltimore: Johns Hopkins University Press.

Achinstein, Peter. 2018. *Speculation*. New York: Oxford University Press.

Boorse, H. and Lloyd Motz, eds. and commentators. 1966. *The World of the Atom*, vol. 1. New York: Basic Books.

Brougham, H. L. 1803. "Review of Thomas Young Bakerian Lecture," *Edinburgh Review*, 1.

Feyerabend, P. 1970. "Against Method: Outline of an Anarchistic Theory of Knowledge," *Minnesota Studies in the Philosophy of Science*, 4.

Garber, Elizabeth, Stephen Brush, and C.W.F. Everitt, eds. 1986. *Maxwell on Molecules and Gases*. Cambridge, MA: MIT Press.

Heilbron, J.L. 2003. *Ernest Rutherford, and the Explosion of Atoms*. New York: Oxford University Press.

Laudan, Larry. 1981. "The Confutation of Convergent Realism," *Philosophy of Science*, 48: 19–49.

Maxwell, J.C. 1965. *Scientific Papers*, W.D. Niven, ed. New York: Dover.

Newton, I. 1979. *Opticks*. New York: Dover Publications.

Newton, I. 1999. *The Principia*. Trans. I. Bernard Cohen and Anne Whitman. Berkeley: University of California Press.

Popper, K. 1959. *The Logic of Scientific Discovery*. New York: Basic Books.

Rutherford, E. 1911. "The Scattering of Alpha and Beta Particles by Matter and the Structure of the Atom," *Philosophical Magazine*, 21: 669–688. Reprinted in Boorse and Motz (1966).

Weinberg, S. 1994. *Dreams of a Final Theory*. New York: Vintage.

Whewell, W. 1840(1996). *The Philosophy of the Inductive Sciences*. London: Routledge.

Young, T. 1802. "Outlines of Experiments and Inquiries Respecting Sound and Light," *Philosophical Transactions of the Royal Society*: 106–150.

19 Scientific Progress and Interdisciplinarity

Hanne Andersen

The time is upon us to recognize that the new frontier is the interface, wherever it remains unexplored.

(Kafatos and Eisner 2004, 10)

1. Current Discourse on Interdisciplinary Progress

A persisting narrative in current science policy is that, in contemporary science, major progress is produced primarily through research that is interdisciplinary. For example, the National Academy of Sciences states in the opening of its 2005 report *Facilitating Interdisciplinary Research* that

> Interdisciplinary research (...) can be one of the most productive and inspiring of human pursuits – one that provides a format for conversations and connections that lead to new knowledge. As a mode of discovery and education, it has delivered much already and promises more – a sustainable environment, healthier and more prosperous lives, new discoveries and technologies to inspire your minds, and a deeper understanding of our place in time and space.
>
> (Committee on Facilitating Interdisciplinary Research and Committee on Science, Engineering and Public Policy 2005, 1)

Similarly, the National Science Foundation stresses in its 2008 report *Impact of transformative interdisciplinary research and graduate education on academic institutions* that

> From global sustainability to renewable energy to the origins of life in the cosmos to forecasting and potentially mitigating economic upheavals, the largest scientific challenges – and those that may hold the greatest opportunity for transformative technological solutions into the 21st century – are interdisciplinary in nature.
>
> (Hartesveldt and Giordan 2008, 7)[1]

DOI: 10.4324/9781003165859-23

Because interdisciplinarity is seen as harbouring special opportunities for producing scientific progress, funding agencies around the globe have created funding programmes directed specifically at promoting interdisciplinarity. For example, in Europe, the ERC Synergy scheme is directed at supporting 'substantial advances at the frontiers of knowledge, stemming, for example, from the cross-fertilization of scientific fields, from new productive lines of enquiry, or new methods and techniques, including unconventional approaches and investigations at the interface between established disciplines' (https://ec.europa.eu/info/funding-tenders/opportunities/portal/screen/opportunities/topic-details/erc-2022-syg). Similarly, NSF's Growing Convergence Research programme in the United States is aimed at bringing 'together intellectually diverse researchers and stakeholders to frame the research questions, develop effective ways of communicating across disciplines and sectors, adopt common frameworks for their solution, and, when appropriate, develop a new scientific vocabulary' (https://beta.nsf.gov/funding/opportunities/growing-convergence-research-gcr).

In policy documents on interdisciplinarity, it is often stressed that in order to fulfil its promise as especially rewarding, progressive or transformative, interdisciplinary research needs to integrate the disciplines involved. Hence, truly interdisciplinary research is expected to somehow merge disciplinary perspectives in a way that changes these disciplines, in contrast to multidisciplinary research that merely juxtaposes different disciplines (or specialties) without merging them, questioning their structure of knowledge, or modifying their identity.[2] This distinction between interdisciplinarity and multi- or pluridisciplinarity on the basis of integration have become standard in many handbook and encyclopaedia articles (e.g. Klein 2010) as well as in research and review papers (e.g. Hoffmann et al. 2013; Holbrook 2013; Brigandt 2013a) on interdisciplinarity, often together with numerous other distinctions between various ways in which disciplines or specialties can be relate (see, e.g. Klein 2010). This has led some scholars to describe integration as 'the litmus test' of interdisciplinarity (Lattuca 2001, 77). However, what exactly this integration means is often left open.

In contrast to this traditional focus on integration as the defining feature of interdisciplinarity, I shall argue that interdisciplinary research can vary in many different ways, and for the rest of this chapter I will discuss how, in the midst of all this diversity, anything general be said about how interdisciplinary research can contribute to scientific progress.

2. The Outcomes of Interdisciplinarity

The claim that interdisciplinarity increases science's ability to identify, propose, refine, specify, or solve problems is in policy documents often justified by pointing to some of the many great scientific triumphs of the past that have resulted from interdisciplinary research. For example, in the US National Academy of Sciences 2004 report quoted earlier, this kind of anecdotic evidence includes 'discovery of the structure of DNA, magnetic resonance imaging, the Manhattan Project, laser eye surgery, radar, human genome sequencing, the "green revolution" and manned space flight' (p. 17).

Scientometric studies have attempted to provide a more systematic vindication of the claims about the increasing popularity and impact of interdisciplinarity. Many of these studies draw on databases such as Web of Science that cover a large share of the world's research publications.[3] Some studies have focused on in how far research declares itself interdisciplinary. For example, a study by Braun and Schubert (2003) shows an overall exponential growth of the use of the term interdisciplinarity or multidisciplinarity in the title of scientific papers between 1980 and 1999, but with notable differences between disciplines in the usage of the terms. Other studies have measured interdisciplinarity not by direct usage of the term, but instead by considering the extent to which the references in a publication include multiple disciplines or specialties, or the extent to which a paper is later referred to by publications from multiple disciplines or specialties. Using this methodology on publications from 1975 to 2005, Porter and Rafols (2009) show that the degree of interdisciplinarity has increased, but that the increase is linked primarily to increasing collaboration between neighbouring specialties or fields. Others again have measured interdisciplinarity by examining whether scientific publications have co-authors affiliated with different disciplinary departments (Porter et al. 2007; Schummer 2004; Choi and Pak 2007). Comparing the different results that have been obtained from measuring interdisciplinarity by diversity in references or by diversity in author affiliation, respectively, Zhang et al. (2018) argue that studies of interdisciplinarity need to distinguish between highly (interdisciplinary research) that has a high disciplinary diversity in both references and author affiliations, specialised disciplinary research with interdisciplinary affiliations that has a high disciplinary diversity in author affiliations but a low disciplinary diversity in references, and interdisciplinary research with specialised affiliations that has a high disciplinary diversity in references but a low disciplinary diversity in author affiliations.

To measure the scientific importance of interdisciplinarity many scientometric studies draw on citation counts as an indicator for impact. On the basis of this methodology, Larivière and Gingras (2010) have argued that for research in general, there is no clear correlation between interdisciplinarity and impact, but if analysed at the disciplinary level, for some disciplines a higher level of interdisciplinarity is related to higher impact, while for other disciplines the opposite hold. A later study by Chen et al. (2015) has argued that, for most specialties, the top 1% most cited scientific papers exhibit a higher degree of interdisciplinarity than the papers that are less cited.

While the scientometric studies to some extend vindicate the claims about an increasing popularity and impact of research that draws on, or is later used by, multiple disciplines or fields, they still reveal little about how the various disciplines or specialties relate in the research process. In contrast, the literature in philosophy of science is abundant with approaches to how disciplines or specialties may interact in identifying, proposing, refining, specifying, and solving scientific problems. Some philosophers have focused on the exchange, transfer, or interlocking of specific resources, such as, for example, how methods can be transferred from one discipline to another (e.g. Herfeld and Doehne 2015, 2019), how concepts from different disciplines can be connected (e.g. Nersessian 2009; Andersen 2011), how data can be integrated (e.g. Leonelli 2013), or how problems can be transferred across disciplines or specialties (e.g. Thoren and Persson 2015; Thorén et al. 2021). Others have examined how theories (e.g. Darden and Maull 1977; Mitchell 2002, 2003), explanations (e.g. Green et al. 2015; Green and Andersen 2019; Love and Lugar 2013), or models may be coupled, integrated, or exchanged (e.g. MacLeod and Nagatsu 2016, 2018; Brigandt 2013b; Plutynski 2013; Grüne-Yanoff and Mäki 2014; Grüne-Yanoff 2015, 2016; MacLeod and Nersessian 2016). Above all, this rich literature shows that interdisciplinarity is a many-faceted activity that is difficult to describe by a single, generalised account.

Instead, it must be recognised that interdisciplinarity varies from case to case, and that it can vary in many different ways. It may vary from narrow interdisciplinarity between closely related fields, to wide interdisciplinarity across the domains of the natural and health sciences, social sciences and humanities. It may vary from the focused import into one discipline of a particular method from another discipline in order to solve a specific problem, to the import of a broad methodological repertoire that also opens for the definition of a whole range of new problems. It may vary from a narrow interlocking of a few selected concepts across disciplinary boundaries, to a major integration of conceptual resources across disciplines. It may vary from feeding a particular problem from one discipline to another, to the recognition of problems that can only be defined and addressed by drawing on multiple disciplines. It may vary from one discipline calling upon the mere services of another discipline, to a full-fledged and equal collaboration between the two.

Along all these dimensions, variation is gradual. Further, consensus within a community of researchers on the problems to solve and on the concepts, methods, and theories by which to solve them may be found at the level of disciplines, subdisciplines, or fields. Hence, an analysis of progress in interdisciplinary research may not only examine research that cuts across disciplines, but also examine research that cuts across subdisciplines or fields.[4] Finally, while interdisciplinary research is not necessarily collaborative, an increase in the ability to identify, propose, refine, specify, and solve problems can only be fully understood in relation to the community of experts that identifies and assesses a problem as a relevant research problem, or assesses, accepts, or rejects developed solutions to the problems that have been seen as relevant for pursuit.

3.   Two Examples of Interdisciplinary Progress

To provide an initial idea of some of the many ways in which interdisciplinary research can produce scientific progress, let me first present two cases in which interdisciplinary or interspeciality research has led to major progress in the form of important new discoveries, namely, the discovery of nuclear fission, and the discovery of fullerenes.

The Discovery of Nuclear Fission

Nuclear fission is a process in which an atomic nucleus splits into smaller fragments. The process was discovered in the 1930s when the specialties of nuclear physics and radiochemistry were developing rapidly.[5] One of the first steps towards the discovery was taken by the chemist Irene Curie and her husband, the physicist Fréderic Joliot, when they bombarded light elements with alpha particles and discovered that the bombarded elements transmuted into radioactive isotopes of near-by elements. Because of the positive charge of the alpha particle, only light elements could be success-fully bombarded. However, as the electrically neutral neutron had been discovered only a few years previously, the physicist Enrico Fermi soon developed a similar experiment in which nuclei were instead bombarded with neutrons. From these experiments, Fermi and his team of physicists and chemists reported that for a large number of elements, neutron bombardment led to an unstable element that decayed by beta-emission and which therefore resulted in an element with a higher atomic number. By a systematic investigation of as many elements from the periodic system as they could get hold of, the experiments step by step added new pieces to the physical understanding of nuclear transmutation, and it revealed the existence of radioactive isotopes of almost all chemical elements. While the use of radioactive tracers to investigate a wide range of questions in physics, chemistry, biology, and medicine had been initiated by Hevesy and Paneth two decades earlier, the possibility of producing radioactive isotopes of almost any chemical element increased the range of applications of this technique enormously. Further, Fermi's team also reported that by bombarding uranium, the last known element in the periodic table at the time, they had induced a decay process that led to a transuranic element – in other words, a new element that did not exist in nature. Admittedly, their chemical identification of the new transuranic element was questioned by a single chemist, Ida Noddack, who had worked exten-sively on elements from the group in the periodic table to which Fermi's team expected that the new element belonged. However, she was not very specific in which characteristics she would have expected instead, and she did not seem to consider that if her criticism was right it would have very wide-ranging implications for nuclear physics. Hence, her criticism was silently ignored, and the discovery of a transuranic element was generally accepted and celebrated as a major scientific advance.

At this time, nuclear disintegration has been treated theoretically primarily as a tunnelling phenomenon that allowed only particles up to the size of the alpha particle to tunnel through the potential barrier. Hence, the expectation provided by physical theory was that the elements produced by the neutron bombardment would not deviate much in atomic number from the element bombarded. The various groups of collaborating physicists and chemists that went into the race of discovering new transuranic elements therefore made their chemical analysis based on the assumption that they could focus on just a few, heavy elements. As they discovered more and more transuranic elements, they also encountered more and more strange results. For example, some transmutations led to very long beta decay series that were difficult to understand, and too many decay series seemed to originate from the same isotopes.

Finally, the two chemists Otto Hahn and Fritz Strassmann conducted an experiment from which they felt compelled to conclude that they had produced an element that, contrary to their expectation, was not the heavy element radium, but instead the much lighter element barium. They therefore communicated to their collaborator, the physicist Lise Meitner, that as chemists they had to conclude that the nucleus had split, although they also knew that for a physicist this would sound impossible. Together with another physicists, Otto Frisch, Meitner quickly realised that if adopting Bohr's relatively new droplet model of the nucleus, it could be explained how the bombardment with neutrons could lead to violent oscillations that would cause the drop to divide into two smaller drops.

This re-interpretation of the transmutation process had far-reaching consequences for all the results that the groups working on transuranic elements had produced during the preceding years. All the examined processes had to be re-examined, including the first discovery of a transuranic element for which Fermi had received the Nobel Prize, and many of results on transuranic elements had to be retracted. The possibility of producing transuranic elements was still an important topic, but it now required measuring whether there was any recoil of the transmuted nuclei. Further, new avenues of research also opened up. The splitting of a heavy nucleus into light nuclei produced free neutrons, and that opened for the possibility of chain reactions. Due to the differences in binding energy between the original heavy nucleus and the much lighter fission products, energy is released in the process. Hence, a research area emerged that became one of the prime examples of modern big science: the Manhattan projects' creation of the first atomic bomb as well as the development of peaceful exploitation of the process to generate energy in atomic power plants.

The Discovery of Fullerenes

Fullerenes are hollow carbon molecules in the shape of either a closed cage (also known as buckyballs) or a cylinder (also known as carbon nanotubes). They were discovered as a result of a collaboration between three chemists specialised in physical chemistry, spectroscopy, organic chemistry and astrochemistry, Robert Curl, Richard Smalley, and Harry Kroto, together with the three graduate students Sean O'Brien, James Heath, and Yan Liu.[6]

Prior to their collaboration, Smalley and Curl had been investigating the formation of small aggregates of atoms, especially inorganic semiconductor clusters, using a particular apparatus called a cluster-beam generator that first creates a very hot vapour of an element and then cools it so that the atoms align in clusters that are analysed in a mass spectrometer. Kroto had not worked on inorganic semiconductors, but instead on the formation of some long linear carbon chain molecules that he had previously discovered in interstellar space, and which he hypothesised were created around red giants. Kroto met Curl at a conference, and upon learning of Kroto's hypothesis on the formation of long linear carbon chain molecules, Curl suggested that the cluster-beam generator offered a possibility for testing this. Further, as a microwave spectroscopist, Curl also saw a possibility for recording the spectra of these molecules as this would show whether they could be responsible for some unidentified emission band that had been observed in space. At first, Smalley was hesitant because the team was already busy working on semiconductors, but after some time he agreed to the experiment. During ten days of experimentation in 1985, the team found the long linear carbon chains that they had set out to produce, but they also discovered some peculiar peaks in the mass spectra corresponding to 60 and 70 carbon atoms. In trying to figure out what could be the structure of these 'wadges', as they initially referred to them, they noticed that they did not react with other molecules. They considered if it could be carbon stacked in hexagonal sheets, similar to graphite, but that would require dangling bonds to be tied up in some way. Another option would be a spherical form where the graphite sheet curled around into a closed cage. Discussing this latter option, some of the team members mentioned the architect Buckminster Fuller and the geodesic domes that he had designed as structure of hexagons and pentagons. This prompted Smally to build a paper model of a geodesic dome with 60 vertices – an almost perfect sphere that matched the 60 carbon atoms they were looking for – or, as a mathematician pointed out to them, also the familiar shape of a soccer ball. This spherical carbon molecule, which they termed Buckminster Fullerene, or fullerene for short, added a new allotrope of carbon in addition to graphite, diamonds, and amorphous carbon.

How the spheres were formed was a question that overlapped with existing research within carbon chemistry, especially research on soot and combustion. However, research on soot and combustion usually focuses on large quantities of material and on particles. Further, these particles are usually not pure, but contain also small amounts of other elements than carbon. From this perspective, the small quantities of fullerenes, produced under very special conditions, was perceived as esoteric. Hence, the initial reaction to this new research on the creation of very small quantities of pure carbon particles was at first indifference. It was not until later, when fullerenes had been observed in flames, that it was proposed that fullerenes could be seen as an important piece in the puzzle of understanding the process of soot formation.

Discovering a new allotrope opened a completely new field of fullerene chemistry. Among the new questions that arose were – apart from the question of their formation – if atoms could be trapped inside, how large the structures could become, and if they could be so elongated that they formed tubes. As such cylindrical tubes were produced and turned out to have quite extraordinary macroscopic properties, including high tensile strength, high ductility, and high electrical conductivity, research on their potential technical applications soon became a major field within the emerging discipline of nanoscience.

4. Incremental, Transformative, and Quasi-transformative Progress

Together, the two cases briefly presented in the previous sections show some of the various ways in which interdisciplinary (or interspeciality) research can contribute to increasing the capacity of science for identifying, proposing, refining, specifying, or solving problems.

First, both cases display phases in which research progresses incrementally, such as, for example, when the systematic bombardment of nuclei was used to reveal new details about the physical transmutation process as well as to create new radioisotopes, or when the expected, long carbon chains were formed in the cluster-beam generator.

Second, both cases also show how unanticipated result can sometimes open for a multitude of new research problems, or for substantial refinement of existing ones. In some cases, this happens when recalcitrant results reveal that something is wrong, and that the way in which research is performed needs to be transformed, such as, for example, when it was discovered that the nucleus could split and that all previous results on transuranic elements needed to be re-assessed. In other cases, there is no need to fundamentally reassess previous results, but some new insight turns out to harbour a multitude of new questions that need to be examined, such as, for example, when it was discovered that carbon can form large, spherical molecules.

In the following, I shall discuss these three types of incremental, transformative, and quasi-transformative progress in more detail. In doing so, I shall adopt the functional account of progress recently advanced by Shan (2019; see also Chapter 3 this volume). On his account, progress is defined in terms of usefulness of both problem defining and problem-solving. Importantly, problem defining should here be understood not only as the initial proposal of a problem, but also as including the subsequent and often iterative process of refinements and specifications of the problem that forms part of developing a solution to it. Vice versa, problem solution should also be understood as an extended and often iterative process of reasoning and conceptualisation, hypothecation, and experimentation. On such a view, science progresses where these processes of problem defining and problem solution consist in 'a reliable framework for further investigation to solve more unresolved problems and to generate more testable research problems across additional different areas (or disciplines)' (ibid., p. 746).

4.1.   Incremental Progress in Interdisciplinary Research

As illustrated by the two cases presented earlier, sometimes, interdisciplinary research activities are directed primarily at solving questions that are well defined, but whose solution just happen to require contributions from more than one discipline or speciality. Interdisciplinary research of this kind is similar to how normal science is conducted within a discipline. It is not aimed at calling forth new sorts of phenomena or inventing new theories, but instead at refining and solving problems by the use of already accepted concepts, methods, and theories. These may be further refined, and additional concepts, methods, and theoretical details may be added, but not fundamentally changed. The firm consensus in the scientific community on the foundation of this activity also implies that it remains clear to all participants involved when science has progressed by identifying and solving the problems that the accepted theories define, and where science can still progress by identifying and solving additional problems of the same kind.

In some cases, the primary problems to be solved are defined primarily within one of the involved disciplines, although the solution requires input from another discipline or speciality that may contribute to the process of problem-solving in a service-like function. For example, as described earlier, exploring which decay processes could be induced by bombarding nuclei with small particles was initially interesting primarily for nuclear physics, but identifying the individual decay processes required detailed chemical analyses of the decay products. In other cases, the identification, refinement, and solution of problems may be distributed among multiple specialties on more equal footing. Sometimes, research may drift between the two situations. For example, when focusing on induced radiation in heavy nuclei, the decay products included new elements in the periodic table and hence new research questions arose about the chemical characteristics of these elements.

Importantly, incremental progress does not necessarily require any fundamental integration of the involved disciplines or specialties. Nevertheless, incremental progress can still constitute very important developments that should not be underestimated. When assessing the merits of interdisciplinary research, it is often overlooked that incremental progress can be unevenly distributed among the involved disciplines, but still be a very valuable outcome of an interdisciplinary process. For example, studies that assess interdisciplinary progress by examining whether published articles are cited by multiple disciplines will overlook research that contributes primarily to one of the involved disciplines. Similarly, studies that examine whether published articles cite multiple disciplines will overlook research in which one disciplines draws on services from another discipline in the form of, for example, well-developed methods that typically require only limited references to literature from this discipline.

4.2. *Transformative Progress in Interdisciplinary Research*

As described originally by Kuhn as an 'essential tension' within normal science (Kuhn 1959), the focused effort of normal science to investigate in detail each problem that a theory poses also entails a strong focus on the problems that turn out to be recalcitrant. When recalcitrant problems reveal shortcomings of the accepted concepts, methods, and theories, changes in this foundation may be needed in order to resolve the situation. However, making substantial changes to the accepted concepts, methods, and theories may often have the consequences of not only solving the recalcitrant problems that prompted these changes, but also of re-opening other problems that had already been solved so that previously achieved progress will go lost. Hence, despite the fact that the suggested changes may solve the recalcitrant problems at hand, due to the loss of other solutions, the suggested change may nevertheless appear to be a move backwards rather than progress. In philosophical analyses of scientific progress, this has often been seen as presenting a problem. However, on Shan's (2019) new functional approach to scientific progress, a new theory is only adopted if it can solve the recalcitrant problems on which the old theory had to give up, and if it overall offers the opportunities of identifying, refining, and solving more, or more interesting, problems than its predecessor. The perception of overall progress in the sense that a new consensus offers more opportunities for identifying and resolving more research puzzles is therefore a transformative rather than an incremental kind of progress.

What is special about transformative progress in interdisciplinary research, compared to mono-disciplinary research is that, in interdisciplinary research, what is perceived as a recalcitrant problem from one disciplinary perspective may not necessarily be recognised as such from another disciplinary perspective. Sometimes, a problem may be discovered by one discipline, but the resolution of this situation can only be achieved by making profound changes in another theory. Thus, an important ingredient in transformative progress in interdisciplinary research is that recalcitrant problems can be transferred among the involved disciplines. Hence, the ability to transfer recalcitrant problems from the discipline in which it is discovered to another discipline in which it must be solved depends on the involved scientists' willingness to trust each other's judgements, and thereby to defer to each other's expertise. For example, in the case of nuclear fission described earlier, it was chemical analysis of the produced decay products that revealed that something was seriously wrong with the physical theories of nuclear decay. The first indications that something might be wrong were noticed by a chemist who came from outside the groups working on transuranic elements, and who seemed ignorant or uninterested in which implications her warnings would have for nuclear physics. This first warning was therefore ignored. In contrast, it was taken seriously at once when two chemists who had collaborated with physicists on transuranic elements for several years declared that they had a result, which they knew would create problems for nuclear physics. Cases such as this point to the importance of communication and trust across disciplinary boundaries and calls for research in social epistemology on how researchers involved in interdisciplinarity communicate results across disciplinary boundaries; and how they assess the expertise and calibrate their trust in researchers from other specialties or disciplines.

4.3. *Quasi-transformative Progress in Interdisciplinary Research*

In addition to incremental and transformative progress, which are well-known types of progress within both mono-disciplinary and interdisciplinary research, interdisciplinary research also harbour the possibility for another type of progress which is similar to transformative progress in constituting a sudden, substantial increase in the ability to identify, propose, refine, specify, and solve problems, but at the same time different from transformative progress in not implying a loss of previously obtained results. In the following, I shall describe this kind of quasi-transformative progress in more detail.

When concepts, methods, and theories from one discipline are brought together with the concepts, methods, and theories from another discipline in the context of problem-solving, these resources may not only solve the problem at hand, but they may also open up for new perspectives and define new and hitherto unanticipated research problems. Sometimes, such new perspectives may even develop into a new speciality or discipline of its own, rather than being a combination of the disciplines from which the research originally emerged. Sudden and substantial progress in the form of not only solutions to already recognised problems, but also the proposal, refinement, and specification of entirely new problems resembles the kind of progress achieved through revolutions. In this sense, it is transformative. However, at the same time it does not include the same kind of loss of previously accepted results as revolutions do, and in this sense, it is more adequately described as quasi-transformative.

The broader the methodological repertoire, the richer the conceptual resources, or the more detailed and refined the theories that are brought together in an interdisciplinary research endeavour, the more opportunities may seem to arise for the identification of new problems or the development of new types of solutions. Hence, if scientific progress is understood as an increase in the ability to identify, define, refine, and solve scientific problems, this means that interdisciplinarity that brings together diverse resources seems to offer especially beneficial conditions for such sudden and substantial progress to arise. At the same time, the more diverse the resources brought together, the more difficult and time-consuming will it be for the researchers involved to communicate across disciplinary boundaries, assess each other's expertise, and calibrate their mutual trust. Hence, a tension exists between the opportunities for progress that interdisciplinarity offers, and the detailed assessment of results that is provided within the individual disciplines (see Andersen 2013, 2016 for further details).

In science policy, the possibility for interdisciplinary research to open for sudden identification, refinement, specification, and solution of new problems on a large scale is often used to justify infrastructures or events that bring researchers together across disciplinary boundaries. The expectation seems to be that if only scientists from different disciplines meet, the right combinations of expertise will arise by chance. In this sense, it seems similar to traditional views of discovery as the result of serendipity, only here what is serendipitous is not a cognitive process of getting the right idea, but the social process of bringing the right people together.[7]

5. Acknowledging, Ignoring, or Rejecting Interdisciplinary Progress

An important part of the process through which scientific progress is established is scientists' mutual scrutiny of each other's results. However, participating in this mutual, critical scrutiny of new proposals or new results requires competences of a very specialised kind. In much research conducted within a given speciality or discipline, researchers working on similar problems have very similar expertise and they are therefore in general all well attuned to scrutinise each other's achievements critically. That is exactly the reason why the mutual, critical scrutiny of new claims advanced within in the scientific community is so immensely efficient in identifying any loci of trouble, and in focusing research on recalcitrant problems when they arise. Further, in those cases were scientists need to accept the testimony from a peer, their shared expertise provides a privileged position from which to calibrate their trust in the peer to whose testimony they defer.

The situation is different in interdisciplinary research. Researchers involved in an interdisciplinary research endeavour have different areas of expertise, and they therefore often tend to pursue different goals using different means. That also means that they are attuned very differently for the task of critically scrutinising new results advanced within their interdisciplinary research endeavour, as well as for calibrating their trust in peers. This creates a special tension. On the one hand, bringing together concepts, methods, and theories from multiple disciplines in an interdisciplinary research endeavour may open for the identification of entirely new research problems or problem solutions, perhaps of a kind that could not be anticipated within any of the involved disciplines alone. As argued previously, the more disparate the resources that are brought together, the more opportunities may arise for identifying previously unanticipated problems or for developing new and unanticipated problem solutions. On the other hand, the more disparate the resources that are brought together are, the less will scientists involved in an interdisciplinary research endeavour be attuned to critically assess the contributions of each other. In effect, the more interdisciplinary research opens new directions, the more challenging will it be for the researchers involved it be to assess the relevance of the problems identified, as well as the quality of the problem solutions obtained.[8]

Not only can it be more challenging to assess the relevance of problems and the quality of proposed solution in interdisciplinary research. Because interdisciplinary research does not necessarily build on a firm consensus among the involved researchers about the foundation on which their research activities build, scientists engaging in interdisciplinary research can sometimes disagree substantially on what they identify as relevant research problems, on how to solve these, and on whether proposed solutions are acceptable. That means that what some recognise as progress in interdisciplinary research may not necessarily be recognised as such by others. Such situations can draw wedges between scientists who work on the same problems from different disciplinary perspectives, and who may therefore have very different views on which achievements represent scientific progress. Sometimes, it may even result in hostility, such as, for example, when early molecular biology was perceived from the perspective of biochemistry as 'practicing biochemistry without a license' (De Chadarevian and Gaudillière 1996). Some philosophers of science have described such attempts from one discipline at addressing problems that have previously been addressed primarily by another discipline as 'imperialism' (Mäki 2009, 2013; Mäki et al. 2017) or as 'intrusion in a niche' (Andersen 2006). In such cases, scientists from different disciplines or specialties have conflicting views about their disciplines' scope, such as when one discipline starts explaining phenomena that another discipline sees as its domain; about their disciplines' style, such as when methods and standards of one discipline are imposed upon another discipline that has not requested this; or about their disciplines' standing, such as when the power and prestige of one discipline are increased at the expense of another. A common trait of these situations is that there is a fundamental disagreement between scientists from different disciplines on whether an interdisciplinary relation even exists, and the relation may therefore be best described as pseudo-interdisciplinary. Similarly, scientists working on the same problem from different perspectives do not necessarily either collaborate or clash. They may also simply choose to ignore each other's work, such as, for example, the indifference that for a long time reigned between economists and econophysicists (Cho 2009). Finally, scientists from within a discipline or speciality may disagree with each other on how to perceive the situation. For example, among archaeologists, some expect that ancient DNA can 'solve everything', while others find that it is 'the devil's work' (cf. Callaway 2018).

6. Conclusion

In this chapter, I have argued that progress in interdisciplinary research can take several different forms, and that it can vary in its distribution of results among the involved disciplines. Hence, it does not hold for interdisciplinary research as such that it requires specific forms of integration between the involved disciplines to produce progress. I have also argued that this means that studies of interdisciplinary progress based on citations across disciplines may underestimate the importance of interdisciplinary research processes in which one discipline provide basic services to another, or in which the progress produced is important primarily in one of the involved disciplines. Finally, I have argued that some forms of development in interdisciplinary research require specific social relations between the researchers involved. Therefore, changes to research infrastructures aimed primarily at bringing researchers closer to each other may not in itself suffice to promote more interdisciplinary progress.

Notes

1. These quotations serve only as illustrations of a much broader narrative that can be found in science policy globally. For historical overviews the development as well as the discourse on interdisciplinarity see e.g. (Klein 1990, Chapter 1; Ash 2019; Graff 2015; Weingart and Stehr 2000; Weingart and Padberg 2014).
2. This distinction between interdisciplinarity and multidisciplinarity has its root in a wider distinction between multidisciplinarity, pluridisciplinarity, interdisciplinarity, and transdisciplinarity introduced by Michaud in a document prepared for OECD's Center for Educational Research and Innovation (CERI/HE/CP/69.04). This document provided the background for much of the thinking in the seminal and much cited 1972 volume on interdisciplinarity issued by the OECD, where pluridisciplinarity is defined as the juxtaposition of disciplines, while interdisciplinarity is defined as the integration of concepts and

methods in these disciplines (CERI 1972, 12). However, in the same volume, the distinction between multidisciplinarity and interdisciplinarity by means of integration was further elaborated by Jantsch from a system theoretic perspective that introduced a hierarchical organisational principle guiding the interdisciplinary coordination (Jantsch 1972). Often, Jantsch' interpretation is used as the generic reference for the distinction, without noting its system theoretic basis. For a philosophical criticism of the notion of integration, see for example Grüne-Yanoff (2016).

3. Currently, the Web of Science traces almost two billion cited references from more than 170 million publications (see https://clarivate.com/webofscience-group/solutions/web-of-science/, accessed 15/2–2022). However, it should also be noted that Web of Science has a stronger coverage of English language publications than of other languages, and that it has a stronger coverage in the natural and health sciences than in the social sciences and humanities.

4. See also, for example Collins (2018) and Abbott (2001) for the idea that the structure of disciplines, sub-disciplines, and fields should be seen as fractal.

5. See Andersen (1996, 2009, 2010) for detailed expositions of this case, on which this very brief summary builds.

6. This brief summary of the development builds on the detailed exposition presented in Aldersey-Williams (1995).

7. Over the last decades, work on model-based reasoning, cognitive history of science, creativity, etc. has showed that the scientific developments that the concept of serendipity has been used to describe, very often are the result of structured cognitive processes that in various ways have paved the way for these insights. In a similar way, current work in social epistemology may reveal in how far this special kind of social serendipity will also turn out to be the result of structured social processes that in various ways prepare researchers for fruitful exchange. However, this will have to be the topic for a separate paper.

8. See Andersen (2013, 2016) for further details.

References

Abbott, A. 2001. *Chaos of Disciplines*. Chicago: University of Chicago Press.

Aldersey-Williams, Hugh. 1995. *The Most Beautiful Molecule: The Discovery of the Buckyball*. New York: John Wiley.

Andersen, H. 1996. "Categorization, anomalies, and the discovery of nuclear fission." *Studies in the History and Philosophy of Modern Physics* 27:463–492.

Andersen, H. 2006. "How to recognize intruders in your niche." In *The Way Through Science and Philosophy: Essays in Honour of Stig Andur Pedersen*, edited by H.B. Andersen, F.V. Christiansen, K.V. Jørgensen, and V. Hendricks, 119–136. London: College Publications.

Andersen, H. 2009. "Unexpected discoveries, graded structures, and the difference between acceptance and neglect." In *Models of Discovery and Creativity*, edited by J. Meheus and T. Nickles, 1–27. Dordrecht: Springer.

Andersen, H. 2010. "Joint acceptance and scientific change: A case study." *Episteme* 7 (3):248–265.

Andersen, H. 2011. "Conceptual development in interdisciplinary research." In *Scientific Concepts and Investigative Practices*, edited by U. Feest and F. Steinle, 271–292. Berlin: Kluwer.

Andersen, H. 2013. "The second essential tension." *Topoi* 32 (1):3–8.

Andersen, H. 2016. "Collaboration, interdisciplinarity, and the epistemology of contemporary science." *Studies in History and Philosophy of Science A* 56:1–10.

Ash, Mitchell G. 2019. "Interdisciplinarity in historical perspective." *Perspectives on Science* 27 (4):619–642.

Braun, T., and A. Schubert. 2003. "A quantitative view on the coming of age of interdisciplinarity in the sciences 1980–1999." *Scientometrics* 58 (1):183–189.

Brigandt, Ingo. 2013a. "Integration in biology: Philosophical perspectives on the dynamics of interdisciplinarity." *Studies in History Studies in History and Philosophy of Biological and Biomedical Sciences* 44 (4):461–465.

Brigandt, Ingo 2013b. "Systems biology and the integration of mechanistic explanation and mathematical explanation." *Studies in History and Philosophy of Biological and Biomedical Sciences* 44 (4):477–492.

Callaway, Ewen. 2018. "Divided by DNA: The uneasy relationship between archaeology and ancient genomics." *Nature* 555 (7698):573–576.

CERI. 1972. *Interdisciplinarity. Problems of Teaching and Research in Universities.* Paris: Organisation for Economic Co-operation and Development.

Chen, Shiji, Clément Arsenault, and Vincent Larivière. 2015. "Are top-cited papers more interdisciplinary?" *Journal of Informetrics* 9 (4):1034–1046.

Cho, Adrian. 2009. "Econophysics: Still controversial after all these years." *Science* 325:408.

Choi, Bernard C.K., and Anita W.P. Pak. 2007. "Multidisciplinarity, interdisciplinarity, and transdisciplinarity in health research, services, education and policy: 2. Promoters, barriers, and strategies of enhancement." *Clinical Investigative Medicine*:E224–E232.

Collins, Harry. 2018. "Studies of expertise and experience." *Topoi* 37 (1):67–77.

Committee on Facilitating Interdisciplinary Research and Committee on Science, Engineering and Public Policy. 2005. *Facilitating Interdisciplinary Research.* Washington, DC: The National Academic Press.

Darden, L., and N. Maull. 1977. "Interfield theories." *Philosophy of Science* 44 (1):43–64.

De Chadarevian, Soraya, and Jean-Paul Gaudillière. 1996. "The tools of the discipline: Biochemists and molecular biologists." *Journal of the History of Biology*:327–330.

Graff, Harvey J. 2015. *Undisciplining Knowledge: Interdisciplinarity in the Twentieth Century.* Baltimore: Johns Hopkins University Press.

Green, Sara, and Hanne Andersen. 2019. "Systems science and the art of interdisciplinary integration." *Systems Research and Behavioral Science* 36 (5):727–743.

Green, Sara, Melinda Fagan, and Johannes Jaeger. 2015. "Explanatory integration challenges in evolutionary systems biology." *Biological Theory* 10 (1):18–35.

Grüne-Yanoff, Till. 2015. "Models of temporal discounting 1937–2000: An interdisciplinary exchange between economics and psychology." *Science in Context* 28 (4):675–713.

Grüne-Yanoff, Till. 2016. "Interdisciplinary success without integration." *European Journal for Philosophy of Science* 6 (3):343–360.

Grüne-Yanoff, Till, and Uskali Mäki. 2014. "Introduction: Interdisciplinary model exchanges." *Studies in History and Philosophy of Science Part A* 48.

Hartesveldt, Carol Van, and Judith Giordan. 2008. "Impact of Transformative Interdisciplinary Research and Graduate Education on Academic Institutions. Workshop Report." Arlington, VA: National Science Foundation.

Herfeld, Catherine, and Malte Doehne. 2015. "The diffusion of scientific innovations: Arguments for an integrated approach." *Emerging Trends in the Social Behavioral Sciences: An Interdisciplinary, Searchable, Linkable Resource*:1–14. https://doi.org/10.1002/9781118900772.etrds0462

Herfeld, Catherine, and Malte Doehne. 2019. "The diffusion of scientific innovations: A role typology." *Studies in History and Philosophy of Science Part A* 77:64–80.

Hoffmann, Michael H.G., Jan C. Schmidt, and Nancy J. Nersessian. 2013. "Philosophy of and as interdisciplinarity." *Synthese* 190 (11):1857–1975.

Holbrook, J Britt. 2013. "What is interdisciplinary communication? Reflections on the very idea of disciplinary integration." *Synthese* 190 (11):1865–1879.

Jantsch, Erich. 1972. "Inter-and transdisciplinary university: A systems approach to education and innovation." *Higher education* 1 (1):7–37.

Kafatos, Fotis C., and Thomas Eisner. 2004. "Unification in the century of biology." *Science* 303 (5662):1257–1258.

Klein, J.T. 1990. *Interdisciplinarity. History, Theory, & Practice*. Detroit: Wayne State University Press. Reprint, Not in File.

Klein, J.T. 2010. "A taxonomy of interdisciplinarity." In *The Oxford Handbook of Interdisciplinarity*, edited by R. Frodeman, J.T. Klein, and C. Mitcham, 15–30. Oxford: Oxford University Press.

Kuhn, T.S. 1959. "The Essential Tension: Tradition and Innovation in Scientific Research." In *Scientific Creativity: Its Recognition and Development*, edited by C.W. Taylor and F. Barron, 341–354. New York: John Wiley & Sons.

Larivière, Vincent, and Yves Gingras. 2010. "On the relationship between interdisciplinarity and scientific impact." *Journal of the American Society for Information Science and Technology* 61 (1):126–131.

Lattuca, Lisa R. 2001. *Creating Interdisciplinarity: Interdisciplinary Research and Teaching among College and University Faculty*. Nashville: Vanderbilt University Press.

Leonelli, Sabina 2013. "Integrating data to acquire new knowledge: Three modes of integration in plant science." *Studies in History and Philosophy of Biological and Biomedical Sciences* 44 (4):503–514.

Love, Alan C., and Gary L. Lugar. 2013. "Dimensions of integration in interdisciplinary explanations of the origin of evolutionary novelty." *Studies in History and Philosophy of Biological and Biomedical Sciences* 44 (4):537–550.

MacLeod, Miles, and Michiru Nagatsu. 2016. "Model coupling in resource economics: Conditions for effective interdisciplinary collaboration." *Philosophy of Science* 83 (3):412–433.

MacLeod, Miles, and Michiru Nagatsu. 2018. "What does interdisciplinarity look like in practice: Mapping interdisciplinarity and its limits in the environmental sciences." *Studies in History and Philosophy of Science Part A* 67:74–84.

MacLeod, Miles, and Nancy Nersessian. 2016. "Interdisciplinary problem-solving: Emerging modes in integrative systems biology." *European Journal for Philosophy of Science* 6 (3):401–418.

Mäki, U. 2009. "Economics imperialism: Concepts and constraints." *Philosophy of the Social Sciences* 39:351–380.

Mäki, U. 2013. "Scientific imperialism: Difficulties in definition, identification, and assessment." *International Studies in the Philosophy of Science* 27 (3):325–339.

Mäki, Uskali, Adrian Walsh, and Manuela Fernández Pinto. 2017. *Scientific Imperialism: Exploring the Boundaries of Interdisciplinarity*. London and New York: Routledge.

Mitchell, S.D. 2002. "Integrative pluralism." *Biology and Philosophy* 17:55–70.

Mitchell, S.D. 2003. *Biological Complexity and Integrative Pluralism*. Cambridge: Cambridge University Press. Reprint, Not in File.

Nersessian, N. 2009. "How do engineering scientists think? Model-based simulation in biomedical engineering research laboratories." *Topics in Cognitive Science* 1:730–757.

Plutynski, Anya. 2013. "Cancer and the goals of integration." *Studies in History and Philosophy of Biological and Biomedical Sciences* 44 (4):466–476.

Porter, A.L., Alex Cohen, J. David Roessner, and Marty Perreault. 2007. "Measuring researcher interdisciplinarity." *Scientometrics* 72 (1):117–147.

Porter, A.L., and I. Rafols. 2009. "Is science becoming more interdisciplinary? Measuring and mapping six research fields over time." *Scientometrics* 81 (3):719–745.

Schummer, Joachim. 2004. "Multidisciplinarity, interdisciplinarity, and patterns of research collaboration in nanoscience and nanotechnology." *Scientometrics* 59 (3):425–465.

Shan, Yafeng. 2019. "A new functional approach to scientific progress." *Philosophy of Science* 86:739–758.

Thorén, H., and J. Persson. 2015. "The philosophy of interdisciplinarity: Sustainability science and problem-feeding." *Journal for General Philosophy of Science* 44 (2):337–355.

Thorén, H., Niko Soininen, and Niina Kotamäki. 2021. "Scientific models in legal judgements: The relationship between law and environmental science as problem-feeding." *Journal of Environmental Science* 124:478–484.

Weingart, P., and B. Padberg. 2014. *University Experiments in Interdisciplinarity*. Bielefeld: Transcript Verlag. Reprint, Not in File.

Weingart, P., and N. Stehr. 2000. *Practicing Interdisciplinarity*. Toronto: University of Toronto Press. Reprint, Not in File.

Zhang, Lin, Beibei Sun, Zaida Chinchilla-Rodríguez, Lixin Chen, and Ying Huang. 2018. "Interdisciplinarity and collaboration: on the relationship between disciplinary diversity in departmental affiliations and reference lists." *Scientometrics* 117 (1):271–291.

20 A Human Rights Approach to Scientific Progress

The Deontic Framework

Michela Massimi

1. Introduction: A Different Way of Thinking About Scientific Progress

An expert's meeting in Venice organised by UNESCO in collaboration with the European Inter-University Centre for Human Rights and Democratisation (EIUC) took place in July 2009. The goal of the meeting was to examine the core content and legal national and international obligations concerning the 'right to enjoy the benefits of scientific progress and its applications', or, in brief, REBSP. Such right has been enshrined in a long list of legal documents: from Article 13 of the American Declaration of the Rights and Duties of Man in 1948 to Article 27(1) of the UN Declaration of Human Rights (UNDHR) also in 1948; from Article 15(b) of the 1966 International Covenant on Economic, Social and Cultural Rights (ICESCR), to the 2009 Venice Statement that resulted from the aforementioned meeting.

There are important changes in the language of these different legal documents. From 'intellectual progress, especially scientific discoveries' in the American Declaration of the Rights and Duties of Man to 'scientific advancement and its benefits' in UNDHR, to the 'right to enjoy the benefits of scientific progress and its applications' in ICESCR. It is to this last version – known as REBSP and making explicit reference to scientific progress – that I turn my attention to in this Chapter. The Venice meeting lamented the patchy implementation of the Article 15(b) almost half century after ICESCR and highlighted the need to clarify the core content of REBSP and associated notions of 'science and knowledge as "global public goods"' (Venice Statement, 2009, p. 7). In its final form, point 7, the Venice statement asserts the necessity to

> *clarify the nature of scientific knowledge*, progress or advancement and who decides on goals, policies, allocation of resources and possible conflicts between freedom of research and the protection of other human rights and human dignity. In addition, whereas the individual right to enjoy the benefits of scientific progress and its applications

DOI: 10.4324/9781003165859-24

must be respected, the rights of communities to share in these benefits must be recognized as equally important.

(p. 14, emphasis added)

In this chapter, I focus on 'science and knowledge as "global public goods"' and I take some preliminary steps towards a different way of asking philosophical questions about scientific progress. For there cannot be a meaningful discussion about scientific progress and its nature – or, needless to say, about a human right such as REBSP – until one elucidates the nature of scientific knowledge. Or better, the idea that one can have a philosophical discussion about scientific progress by insulating the very notion and associated ones (such as scientific knowledge) from the wider debate about its role in society engenders inward-looking discussions that risk missing the main point about why we care (and should care) about scientific progress.

My take here is that we should care about scientific progress mostly because without a better understanding of it, the very human right to enjoy the benefits of scientific progress remains an abstract concept, with a patchy implementation at the level of individual nation states and a nebulous core content. REBSP risks resembling more wishful thinking than actual legal ground for action whenever such human right is violated or neglected.

In this chapter I argue that legal obstacles to the implementation of REBSP, and in particular the difficulty mentioned by the Venice Statement in reconciling 'the individual right to enjoy the benefits of scientific progress and its applications' with the 'rights of communities to share in these benefits' originates from a pervasive, well-entrenched and ultimately impoverished picture of scientific knowledge.

I make a plea for replacing such an impoverished picture with a different one that in my view can deliver on 'science and knowledge as "global public goods"'. I highlight how this new picture of scientific knowledge is best equipped to deliver on the normative content of REBSP and has the potential to shed new light on some of the traditional controversies concerning the 'rights of communities to share in these benefits'.

Such rights should not be an afterthought in a legal framework that understands science and knowledge mainly in terms of individualistic discoveries and privately funded innovations. They should be embedded, and naturally follow from, the very fabric of scientific knowledge production as a social and collective human endeavour. Appreciating what makes scientific knowledge a progressive kind of enquiry implies understanding how *many situated epistemic communities contribute to reliable knowledge production over time*. Or so I shall argue, building here on my work on perspectival realism (Massimi, 2022b).

Where to start then? Let me start by comparing two images of scientific progress broadly construed – what I call the 'manifest image' and the 'philosophical image' – how they are related to one another and what I take to be missing in both. Scientific progress is one of those notions that has entered public discourse and regularly features in media releases as well as in science policy and political speeches. What is the manifest image of scientific progress?

In daily parlance, scientific progress is often associated with scientific discovery: devising a new scientific theory; designing a new piece of technology; creating a new drug, or innovation. Scientific fields are regarded as making progress (or not) depending on whether new particles are discovered, new cures for diseases are found, more complete theories are introduced. Lack of progress (real or perceived) is accompanied by absence of a full understanding of some mechanisms, or shortage of new discoveries or innovations. Milestones are set in research agendas and research funding distributed depending on their respective potential for making progress.

The idea that scientific progress is essential to the welfare of a nation and that research funding should be distributed in accordance to such potential for progress has a long tradition in science policy, back to *Science, the Endless Frontier* in 1945 by President Roosevelt's scientific advisor, Vannevar Bush (1945). The Summary of the Report opened with a section entitled 'Scientific Progress is Essential'. But nowhere in the report Bush explained what scientific progress was; or, why the pursuit of scientific progress – in and of itself – results in better economy, better welfare and better national security.

As Philip Kitcher (2001) in Chapter 11 has argued, *Science, the Endless Frontier* was operating with an elitist view of progress whereby an elite of scientists are entrusted with the task of pursuing fundamental research along the Baconian model of the *New Atlantis* with its envisaged House of Salomon. Bush's model has served American society well with its governance vision of the state (rather than individual philanthropists) funding universities and scientific research more in general. But the Bush model has also revealed the glaring shortcomings of what might be called the 'manifest image' of scientific progress.

Scientific progress – in and of itself – does not necessarily translate into societal progress. The entanglement of scientific and military purposes behind the Manhattan project and similar ones in the following Cold War years was a powerful reminder of it. Democratic societies, with better quality of life and better welfare for their citizens, require and demand scientific knowledge that serves their needs. But *whose* needs? The needs of whoever produces knowledge and owns patent rights to specific innovations? Or those of the general public, who despite its 'right to enjoy the benefits of scientific progress' (REBSP) is often at the receiving end of commercialised innovations? What makes scientific knowledge progressive? And how can scientific knowledge be expected to serve democratic societies and to advance them?

Here enters the 'philosophical image'. For long time, scientific progress has been identified with getting closer to the truth – the so-called verisimilitude account, see Niiniluoto (1984, 2014). The intuitively simple idea is that progress tracks truth. When compared to an earlier theory, a scientific theory is said to make progress if it preserves true claims and discards false ones (think of relativity theory vis-à-vis Newtonian mechanics).

This theory-centric approach to scientific progress seems however to leave out many other examples where progress seems to do more with technological advances and innovations than with putting forward a new theory. Thomas Kuhn's puzzle-solving account was meant to remedy for this pitfall. Scientific discovery has to do with the ability to solve puzzles and problems, where sometimes those problems originate from scarce data, or questionable assumptions – see for example (Kuhn, 1957) for a discussion related to the passage from Ptolemaic to Copernican astronomy.

Defenders of the verisimilitude account have been unmoved by Kuhn. They have worried that focussing on problems detracts attention away from what really matters – namely, truth. After all, how many scientific advances were made that solved problems but were based on false principles/false assumptions? Think, for example, of Carnot's ideal cycle, key to steam engines, and yet based on the false assumption that heat was an immaterial substance called caloric.

The more recent noetic account – see (Dellsén, 2018) – has gone some way towards addressing these concerns besetting both the verisimilitude and the puzzle-solving account. By focussing on understanding and relaxing the demand on truth as the epistemic norm, progress can be understood in terms of 'improved ability to use an unjustified (but correct) theory to explain or predict phenomena' (Dellsén, 2018, p. 7).

Without entering into the details of any of these specific accounts (and leaving here out others for reasons of space), one strikingly common feature is the attention paid to individual theories or individual problems to be solved, or the ability of an individual theory or research programme to improve on their predecessor. The 'philosophical image' of scientific progress bears the hallmark of individualism. Progress is by and large an individualistic achievement: that is, it is the achievement of either some individual person who puts forward a new theory or produces a scientific innovation; or of some (small or large) individual group of persons sharing for example a scientific paradigm able to solve more puzzles. This much is shared by the 'manifest image' of scientific progress. Bush's idea of investing in basic (i.e. fundamental) research done by individual scientists with the hope of a positive trickle-down effect on society as a whole was born out of the same individualistic approach that has fuelled the 'philosophical image'.

But what both the 'manifest image' and the 'philosophical image' have left untouched are the aforementioned fundamental questions: What makes scientific knowledge progressive? And how can scientific knowledge be expected to serve the needs of democratic societies and advance them? *Whose* needs? Those of the IP rights/patents owners? Or those of the general public at the receiving end of them? In posing these questions, I want to signal my broad alignment with the view that takes progress as linked somehow to scientific knowledge (rather than truth, or puzzle-solving or understanding). At the same time, I part way from the traditional epistemic account of progress, see for example (Bird, 2007), because I am not interested in defining progress as the accumulation of knowledge, but in progress as a distinctive feature of scientific knowledge that needs be explained (rather than taken for granted as a methodological fiat).

I see the individualistic approach to scientific progress as tacitly underpinning the REBSP and indeed the whole tradition behind the human rights approach to scientific advancements and progress. Already in the UNDHR, Article 27(1) celebrated the individual right to 'share in scientific advancement and its benefits' and coupled it with Article 27(2) which established the 'right to the protection of the moral and material interests resulting from any scientific, literary and artistic production of which he is the author'. Equally, the ICESCR of 1966 intertwined Article 15(b) concerning REBSP with Article 15(c) which reiterated the right of everyone to 'benefit from the protection of the moral and material interests resulting from any scientific, literary or artistic production of which he is the author'.

To be clear here: I am not saying that the problem with REBSP has necessarily to do with the patent system, current IP rights provision, TRIPS (Trade-Related Aspects of Intellectual Property Rights), or more broadly with the private/public funding divide in which inevitably scientific progress and innovations take place. Or better, in the measure in which the problem with REBSP may indeed have to do with some of these aspects, there are experts in law, economics, social studies of science and distributive justice who are better placed than me to address and tackle these specific aspects of the private vs. public divide – see for example (Sunder, 2007; Biagioli et al., 2011; Biagioli, 2019; Miller & Taylor, 2018; Patten, 2022).

All I am saying is that philosophers of science have a role to play in this debate. For what is fundamentally at stake in the Venice statement lamenting the patchy implementation of REBSP to date is the lack of clarity about the core content of REBSP and in particular about the '*nature of scientific knowledge*, progress or advancement and . . . possible conflicts between freedom of research and the protection of other human rights and human dignity' (Venice Statement, 2009, p. 14). Clarifying the *nature of scientific knowledge* – on which discussions about individualistic achievements vs collective rights to share in benefits depend – is one of the primary tasks for philosophers of science. I turn to it in the next section, where I give a diagnosis of three main problems I see at play in a well-entrenched view on scientific knowledge; and explain how these problems in turn feed into both the manifest and the philosophical image of scientific progress to date.

2. Three Problems About Scientific Knowledge Production and Its Progressive Nature

The idea that progress is an individualistic achievement (of either one scientist who brilliantly devises a new theory; or, a team of scientists that skilfully come up with the right technological innovation or discovery) camouflages a more complex and nuanced social dynamics always at play in science. The further idea that scientific progress for the public good is just the natural consequence of pursuing one's own research interests naively passes the camouflage off for a universal norm in science policy: 'Let us foster scientific progress and societal progress will follow naturally'.

As already mentioned, the history of science has often put the lie to the latter idea. Too often scientific programmes that were regarded as progressive at their time advanced hideous social agendas (e.g. think of eugenics as just one example). It is wishful thinking to believe that the relation between scientific progress and societal progress is a linear tricke-down one. Nor, a fortiori, should scientific progress be conceptualised as a mostly individualistic pursuit and achievement (of either an individual scientist or group of scientists at a particular time).

No one – no individual or particular scientific group or community – can ratify their own claims of knowledge as constituting scientific progress, on pain of particular groups passing off as progressive deleterious or harmful advancements. Rather than being worried about theories with false or speculative assumptions being passed off as progressive in the name of puzzle-solving, I think we should worry more about the risk of passing off as progressive pieces of scientific research that may feed into harmful and discriminatory research agendas. For this very simple reason alone, I think it is a mistake to take scientific progress as an individualistic achievement, pace the manifest and philosophical image.

Three main problems affect the received view of scientific knowledge on which philosophical discourse on progress (and its individualistic underpinning) has been built. They revolve around the following questions:

1. *Who produces scientific knowledge?*
2. *How does scientific knowledge grow?*
3. *What makes scientific knowledge progressive?*

Let us take a quick look at each of them, starting from 1.

2.1. *Who Produces Scientific Knowledge? The Injustice of Epistemic Trademarking*

Scientific knowledge production is typically associated with someone's name: for example Newton's laws, Lavoisier's oxygen, Maxwell's equations, Carnot's cycle, and Boltzmann's constant, just to mention a few examples from the history of physics. Attaching names to a scientific result (be it equations, laws, constants, models, particles, chemical elements, etc.) is a way of rightly recognising authorship and tracing back the original idea to its legitimate owner.

This is of course a common and uncontroversial practice, and key to copyright laws and patent rights (where applicable). Indeed, it is a recognised universal right as per ICESCR Article 15(c) to protect 'the moral and material interests resulting from any scientific, literary or artistic production of which he is the author'. Labelling scientific discoveries and innovations in this way is not only a recognised universal right but it is also epistemically innocuous.

However, a problem arises when an (epistemically innocuous) label or mark becomes what in Massimi (2022b, Chapter 11, Section 11.5 on which I draw here) I have called an 'epistemic trademark': namely, when there is an *epistemic overstretch* from a particular scientific output or achievement of someone, who is legitimately recognised as its intellectual owner, to wider swathes of knowledge claims that for various historically contingent reasons become associated with that specific label or mark.

For example, under the mark 'Newtonian mechanics' one would typically include not just Newton's laws codified in the *Principia* but also more broadly a certain understanding of the mechanical motion of bodies as distinct from Aristotelian physics. Such understanding of mechanical motion began to emerge and can be traced back via Galileo in seventeenth-century Pisa and Oresme in fourteenth-century France to medieval scholars like Abu Al-Barakāt in Baghdad, among others (without mentioning the contribution of authors like Émilie Du Châtelet to the field, for example). While these historical lineages are not lost to the historians of science, they can easily get lost outside historical circles as soon as the term 'Newtonian mechanics' is coined and gains traction in common parlance among scientists and the wider public. A mark has become an 'epistemic trademark', in my idiolect, overstretching and encompassing much wider swathes of knowledge claims – contributed through complex historical lineages and social dynamics – by various authors at various times and places.

I have elsewhere argued that epistemic trademarking is a way of avoiding 'consumer confusion' and 'dilution by blurring' – a way of ringfencing the uniqueness and originality of a particular body of knowledge claims over others. In this example, 'Newtonian mechanics' is a way of demarcating and ringfencing a body of knowledge from another one dominant in previous centuries, namely 'Aristotelian physics'. But the cost of this manoeuvre is that it tends to obfuscate complex and nuanced historical lineages and trademark them *as if* they were the *exclusive* achievement of a single individual:

> A label or mark in scientific discourse (e.g. "Maxwell's equations") becomes an *epistemic trademark* (e.g. "Maxwellian electromagnetic

theory") when it ends up *concealing* the complex historical lineages and *blurring* the epistemic contributions of various communities. To be clear, in these examples there is no culpability on the part of individual epistemic agents (be it Maxwell, or similar) in the process of transforming a mark into an epistemic trademark . . . *epistemic trademarking* is first and foremost a structural phenomenon of how scientific narratives (or a particular kind thereof) get off the ground and tacitly enter public discourse as a result of specific epistemic norms that codify scientific knowledge production in particular societies.

(Massimi, 2022b, pp. 356–357)

Some of these epistemic norms favour, for example, textual knowledge over oral knowledge, theoretical knowledge over artisanal knowledge, knowledge produced by a particular social group over others. Epistemic trademarking is a way of saying 'Newtonian mechanics' is really one and the same as Newton's laws as much as 'Maxwellian electromagnetic theory' is really one and the same as Maxwell's equations, and this is what one often finds in scientific narratives. The epistemic trademark shows its efficacy by branding large bodies of knowledge as epistemic goods that can be easily recognised and commodified for the use of a particular consumer audience. An example is in Richard Feynman's assessment of Maxwellian electromagnetic theory as reducible to Maxwell's equations:

It was not yet customary in Maxwell's time to think in terms of abstract fields. Maxwell discussed his ideas in terms of a model in which the vacuum was like an elastic solid. He also tried to explain the meaning of his new equation in terms of the mechanical model. There was much reluctance to accept his theory, first because of the model, and second because there was at first experimental justification. Today, we understand better that what counts are the equations themselves and not the model used to get them. We may only question whether the equations are true or false. This is answered by doing experiments, and untold numbers of experiments have confirmed Maxwell's equations. If we take away the scaffolding he used to build it, we find that Maxwell's beautiful edifice stands on its own. He brought together all of the laws of electricity and magnetism and made one complete and beautiful theory.

(Feynman's lectures, 18–1 The Maxwell equations, online from: www.feynmanlectures.caltech.edu/II_18.html)

The problem is that taking away 'the scaffolding', as Feynman put it, amounts to more than simply getting rid of an inadequate (and ultimately false) ether model in this particular case. It implies severing the very contribution of a number of epistemic communities – not just the ether modellers at the time, like FitzGerald and MacCullagh, but also experimentalists like Faraday and Ampère, without mentioning the role of artisanal communities like kelp-makers and glass blowers involved in producing lead-free glass tubes important for building cathode rays – whose contributions to the understanding of the laws of electricity and magnetism proved pivotal.

2.2. How Does Scientific Knowledge Grow? The Injustice of Epistemic Severing

Elsewhere (Massimi, 2022b, Ch 11, pp. 349–350) I have argued that epistemic trademarking presupposes another epistemic injustice, namely, epistemic severing:

> *Epistemic severing* affects *narratives* about scientific knowledge production that tend to surgically excise the contributions of particular epistemic communities. This might happen both across communities that might share the same 'scientific perspective' – see (Massimi, 2022b; Giere, 2006) – *and* across communities belonging to culturally diverse scientific perspectives. Severing is an act of informational injustice in how scientific knowledge production *gets narrated* in scientific textbooks and canons.

Epistemic severing differs from what might be called epistemic 'blinkering'. Inevitably, in any scientific narrative decisions are made about what to include and what to leave out, what piece of info to foreground and what to background depending on their relevance to the narrative and important ethical considerations too. If writing a popular science piece on Maxwell's electromagnetic theory, it might not be necessary to enlarge the scope of the narrative and give a fine-grain historical picture of the context that made it possible to flourish. However, in recounting the historical narrative in a history of science textbook, it is important not to omit the fine-grain contextual details; or worse, to give the historiographical impression that this particular achievement is simply the product of a lone genius.

I see epistemic severing as 'the act of cutting off specific historically and culturally situated communities to historically remove or blur their contributions to what I call 'historical lineages' in the scientific knowledge production' (Massimi, 2022b, p. 350). I understand the notion of historical lineages as the 'open-ended, ever-growing, and irreducibly entwined body of scientific knowledge claims grounded in well-defined scientific practices and in their experimental, modelling, and technological resources' (ibid., 341). Historical lineages and the situated knowledge they embed are a way of counteracting a pervasive picture of how scientific knowledge grows that can be found in philosophy of science: namely, the idea that scientific knowledge grows by replacing one old theory with a better one;

one degenerating scientific programme with a progressive one (to use Lakatos' terminology); or, with Thomas Kuhn, one old scientific paradigm beset by anomalies with a new one with higher puzzle-solving.

This ubiquitous picture reinforces the impression that scientific knowledge grows in silos and that progress is a matter of surgically excising faulty claims and replacing them with better ones (however one wants to read 'better' in this context – i.e. getting closer to the truth, more puzzle-solving or else). That scientific knowledge grows in silos has been one of the most widespread and too often unquestioned assumptions in this debate. But as an image of how scientific knowledge historically grows it remains highly questionable, as I explain in the next section.

2.3. What Makes Scientific Knowledge Progressive? Progress 'Sub Specie Aeternitatis' and 'From Here Now'

Zooming into the internal dynamics of how these processes of historical replacement and substitution operate, the question as to whether progress has indeed been made seems characterised by some kind of 'atemporal chauvinism', for lack of a better word. In asking whether the replacement of T_1 with T_2 amounts to scientific progress, one is not just assuming that T_1 and T_2 are sufficiently well insulated theories produced by different individuals (e.g. Ptolemy versus Copernicus, Aristotle versus Galileo, Newton versus Einstein). One is also typically assuming that the replacement in question amounts to scientific progress *sub specie aeternitatis*. Elsewhere (Massimi, 2016) I have described this as 'the view of success from nowhere (or, the view of success from God's eye)' whereby 'We might never be in the position of achieving such complete true knowledge of nature, but it acts as a regulative idea of scientific inquiry to assess success and failure at any given historical time' (p. 760).

A more modest variant of it is what I have called 'success from here now', whereby progress is not assessed vis-à-vis some regulative ideal *sub specie aeternitatis* but from the standpoint of 'here now' as a privileged epistemic standpoint. On this other view, although neither Copernican astronomy nor Galilean physics was a perfectly accurate description of the relevant phenomena, they constituted nonetheless progress when assessed from here and now. One is here reminded of the need to weed out the false claims of past theories from the true posits that remained intact as knowledge grew. Varieties of semi-realism and selective realism have flourished against this backdrop – see for example (Chakaravartty, 1998; Lyons, 2006).

The problem with both these takes is that they tacitly assume that epistemic agents do occupy such privileged standpoints, either *sub specie aeternitatis* or from 'here and now'. And while the former is hard to reclaim for any historically and culturally situated community of epistemic agents, the latter risks smuggling in contextual and perspectival standards 'here and now' *as if* they were *sub specie aeternitatis*.

3.   The Deontic Framework: Where Perspectival Realism Meets Cosmopolitan Rights

In reply to these three problems, I suggest a number of moves that I have already hinted at in the previous sections. In response to *Who produces scientific knowledge?* – and with an eye to avoiding epistemic trademarking – I suggest taking situated communities as the starting point of these conversations. Scientific knowledge production is never the product of the lone genius or isolated individual (no matter how brilliant) or (small or large) group of people. It starts instead from historically and culturally situated scientific practices of many epistemic communities at any one time.

As an answer to 2. *How does scientific knowledge grow?* – against the injustice of epistemic severing – in my book *Perspectival Realism*, I have defended a different picture of science which starts from a plurality of historically and culturally situated scientific perspectives and analyses the growth of scientific knowledge in terms of how these perspectives have *methodologically intersected* with one another and *historically interlaced* in historical lineages.

By *interlacing* with one another in highly complex and non-linear historical lineages, scientific perspectives allow us to track the evolution of knowledge concerning some phenomena by looking at how some of the tools, artefacts, and experimental techniques changed use and function as a result of this interlacing. In my book I give the example of the early Chinese wet and dry compasses which originally situated in geomantic practices became eventually navigation tools and allowed knowledge of phenomena concerning the Earth magnetic field. Going back to the history of electromagnetic theory, one could similarly make the point that the exhausted glass tubes, used among others for Crookes's speculative radiometer science – see (Gay, 1996) – became in the following decades the key scientific tool (cathode rays) for the study of electrical phenomena.

Paying attention to how a plurality of situated perspectives have historically interlaced has a twofold function. First, against the picture of insulated siloes, it celebrates the social and deeply collective nature of scientific knowledge. Second, it reinstates epistemic communities that too often have been severed in scientific narratives, especially under-represented communities that 'on various grounds (e.g. class, ethnicity, gender) are not the dominant, ruling, scientific-canon-writing ones' (Massimi, 2022b, p. 351).

Epistemic severing and epistemic trademarking are the very same reasons why today it continues to be hard to acknowledge the contribution of underrepresented communities to the production of scientific advancements and their benefits. The patchy implementation of REBSP lamented by the Venice Statement and the tension between the individual rights to enjoy the benefits of scientific progress and 'the rights of communities to share in these benefits' are, in my view, nothing but the natural consequence of a ubiquitous image of scientific knowledge production (and narratives thereof) blighted by epistemic severing and epistemic trademarking.

Coming to 3. *What makes scientific knowledge progressive?*, in Massimi (2016, pp. 764–765) I made some tentative steps to answer the question by appealing to a notion of 'success from within' as 'the ability of a theory to perform adequately with respect to standards that are appropriate to the scientific perspective of the time, when assessed from the point of view of other scientific perspectives'. The key idea is that a scientific claim meets the criterion of 'success from within' iff the propositional content of the claim is true (i.e. it corresponds to the way things are) and it satisfies the standards of performance adequacy in the original scientific perspective when assessed from the viewpoint of other (subsequent) scientific perspectives. This perspectival take is meant to counteract both the presumption of a God's eye view on scientific progress ('success from nowhere') and the presumption of 'success from here now' whereby we by our current lights would stand better than our predecessors in assessing what counts as scientific progress.

In Massimi (2022b), I have further clarified how scientific knowledge production is ultimately the ability to *reliably* identify modally robust phenomena, that is stable (qua lawlike) events in nature that can be identified and re-identified by a number of situated epistemic communities from different datasets via many perspectival data-to-phenomena inferences (see Massimi, 2022b, Ch 6). Ultimately, the propositional content of a knowledge claim is true if it corresponds to modally robust phenomena in my phenomena-first ontology. And progress is assessed perspectivally by considering how claims of knowledge continue to be retained (or withdrawn) when assessed from the point of view of other scientific perspectives.

I made also the point that different communities may have the epistemic upper hand in the identification of particular phenomena: for example, local beekeepers know best how to identify the phenomenon of pollination peak for a particular plant in their region – see (Massimi, 2022a). And I have urged to see natural kinds as open-ended groupings of historically identified phenomena so as to give communities their due without the risk of identifying one particular phenomenon as more foundational than any other in a truly Neurahian spirit (Massimi, 2022b, Ch 8).

At this point one might ask: is not there a suspicion of passing off as scientific knowledge some claims that although may be reliably produced were not justified by the particular historically situated scientific perspectives in question? In other words, is not there a risk of passing off as scientific knowledge some claims about ether, caloric, or maybe even Crookes' radiometer speculations in the name of interlacing scientific perspectives?

I think this way of posing the question forces a false dichotomy on us: either go multicultural, embrace the social and perspectival nature of scientific knowledge *at the cost of* jeopardising, say, Popper's demarcation criterion; *or*, uphold the latter and continue to assume that scientific progress is a matter of replacing one by one false theories with better new ones, pace situated epistemic communities and their interlaced scientific perspectives.

Here enters the human rights approach – or deontic framework as I'd like to call it – to scientific progress, at a really important cross-junction between perspectival realism and existing landmark legislative documents that have long established REBSP. This is the cross-junction where a new philosophical image of scientific knowledge has the potential to shed light on the core content of REBSP and some of the thorny questions about its patchy implementation to date. Conversely, the human rights approach to progress encapsulated by REBSP can in turn help re-align epistemic narratives about scientific progress beyond the well-trodden path of individualistic achievements and siloed systems of knowledge that get replaced over time with better ones.

Most importantly, this cross-junction between a new image of scientific knowledge and core content of REBSP can help address the aforementioned epistemic injustices – epistemic severing and epistemic trademarking – and put centre stage the multicultural roots of scientific knowledge and the reasons why communities *have the right* to enjoy and share in scientific progress and its applications. What is to be said then about this fertile cross-junction between the epistemology of science and human rights legislation? In what follows, for reasons of space, I will only make two main remarks:

(I)   I suggest that if one understands the core content of REBSP along the lines of an image of scientific knowledge as that proposed by my perspectival realism, REBSP is best understood as a *ius cosmopoliticum* or 'cosmopolitan right', namely, a right that pertains to everyone as a world citizen.

(II) Understood as a 'cosmopolitan right', REBSP is no longer trapped into the tension between the individual right to enjoy in scientific progress and the 'rights of the communities' to share in progress and its benefits as the Venice Statement still lamented in 2009.

Let me briefly sketch the contours of (I) and (II), starting with (I). I started this chapter by mentioning a lacuna flagged by the Venice Statement and concerning the need to '*clarify the nature of scientific knowledge,* progress or advancement' with an eye to addressing why more than half century after the landmark UNDHR and ICESCR, the REBSP remains at the mercy of nation states. The rise of 'vaccine nationalism' that the COVID-19 pandemic has glaringly brought to light – and the remaining gap in vaccine supplies for the Global South – is just one manifestation of how the implementation of REBSP continues to face roadblocks.

Bioprospecting and biopiracy, namely, the unauthorised use of knowledge about flora, fauna (and more broadly the harvesting of genetic resources) from indigenous people and local communities (IPLC) for commercial purposes, is yet another painful manifestation. Can a different take on the core content of REBSP help address these structural inequities in access to and benefit sharing when it comes to scientific progress?

A deontic framework is an important step in this direction. By 'deontic' I mean a philosophical framework that takes scientific progress explicitly as a *matter of right and duty* at the same time. For there can only be duties (or obligations) when there are rights. The emergence of the idea of science as a "global public good" in recent times (see Boulton 2021; for a philosophical discussion see Massimi 2022c) has proceeded in parallel with a renewed interest in the legal and institutional foundations of REBSP (see Shaver 2015). A recent literature in international human rights law has increasingly focused on REBSP under the name of 'Right to Science' (RtS) – see (Besson, 2015a and essays in this special issue; Besson 2015b; Porsdam & Porsdam Mann, 2021; Donders & Tararas, 2021; Porsdam Mann et al., 2018). Some of this literature has gone back to other legislative tools such as, for example (Committee on Economic, Social and Cultural Rights, 2020) and has elucidated both the international obligations related to scientific progress and its applications, and the challenges in translating such international human rights law into domestic law. For example, there are still countries like the United States that have signed but not ratified to this day (April 2022) the ICESCR – see for a discussion (Porsdam et al., 2021). And even among the 170 countries that have signed and ratified the ICESCR, the threefold state obligations to '*respect* the right, to *protect* it, and, finally, to *fulfil* it' (ibid. p. 232) may not necessarily be implemented.

The problem originates from the difficulty of translating an international human rights law into domestic law of various state parties. The onus of the legal implementation is left effectively to individual nations where often socio-economic poverty, private interests, political lobbies stand on the way to the implementation of REBSP and the wider RtS. Even in the more recent (Committee on Economic, Social and Cultural Rights, 2020), the emphasis lies entirely on state parties to

> direct their own resources and coordinate actions of others to ensure that scientific progress happens and that its applications and benefits are distributed and are available, especially to vulnerable and marginalized groups. This requires, inter alia, instruments for the diffusion of science (libraries, museums, Internet networks, etc.), a strong research infrastructure with adequate resources, and adequate financing of scientific education.
>
> (ibid. Article 16)

Moreover, states are reminded of the 'duty of international cooperation' which is

> essential because of the existence of deep international disparities among countries in science and technology. If it is necessary, owing to financial or technological constraints, developing States should resort to international assistance and cooperation, with a view to complying with their obligations under the Covenant. Developed States should contribute to the development of science and technology in developing countries, adopting measures to achieve this purpose, such as allocating development aid and funding towards building and improving scientific education, research and training in developing countries, promoting collaboration between scientific communities of developed and developing countries to meet the needs of all countries and facilitating their progress while respecting national regulations.
>
> (ibid. Article 79)

Framed thus and so, however, there is a risk that the duty to implement REBSP reduces at best to the goodwill of individual states to ratify the ICESCR and, more precisely, to the goodwill of 'developed States' to contribute – through 'development aid and funding' – 'to the development of science and technology in developing countries'. The onus of the legal enforcement lies squarely within individual states which 'should establish effective mechanisms and institutions, where they do not already exist, to prevent violations of the right . . . As this right can be threatened or violated not only by actions of the State but also through omission, remedies must be effective in both cases' (ibid., Article 89).

In my view, the only way of remedying these shortcomings in the implementation of REBSP to date is to relocate its governance from domestic law (in those states which have signed and ratified ICESCR), or even international human rights law, to cosmopolitan law. I contend that if one endorses an image of scientific knowledge as that proposed by my perspectival realism, REBSP is best understood as a *ius cosmopoliticum* or 'cosmopolitan right' rather than international human rights law. What is the difference, one may ask?

The idea of cosmopolitan rights has a long-standing tradition in political philosophy, and in recent times has been brought back to the general attention thanks to the work of Pauline Kleingeld and Seyla Benhabib (Benhabib, 2006), among many others. The idea of *ius cosmopoliticum* can be found in Kant's *Toward Perpetual Peace* and *Metaphysics of Morals* – see (Kant, 1795; Kant, 1797). In *Toward Perpetual Peace*, Kant saw each person as amenable to three different kinds of rights: (1) rights that pertain to '*citizens of a state* governing the individuals of a people (*ius civitatis*)'; (2) international rights 'governing the relations of states among one another (*ius gentium*)'; (3) '*cosmopolitan right*, to the extent that individuals and states, who are related externally by the mutual exertion of influence on each other, are to be regarded as citizens in a universal state of humankind (*ius cosmopoliticum*)' AAVIII: 349 (Kant, 1795) (English translation, p. 73).

Most of the discussions surrounding cosmopolitan rights in Kant's original context focused on the right to hospitality for foreigners based on the idea of a '*peaceful* . . . universal community of all people on earth who can come into active relations with one another' not as a 'philanthropic (ethical)' idea but rather as 'a principle of *right*' AAVI:352 (Kant, 1797), English translation, p. 146. The far-reaching legal implications of this notion for holding accountable injustices and crimes that go beyond both national and international law are well known and have been discussed in a long-standing tradition back to Hannah Arendt, Karl Jaspers, Jürgen Habermas, Jeremy Waldron, among others.

But so far the debate on cosmopolitan rights has surprisingly not been applied to science and scientific progress. If the diagnosis I have offered in this chapter is on the right track, one can begin to see why the debate on cosmopolitan rights has left REBSP untouched. For one would need to first embrace a different philosophical image of science and scientific knowledge as genuinely 'global public goods' compared to the image that has been prevalent in both the manifest and the philosophical image to date. Perspectival realism provides such an alternative philosophical image where scientific knowledge, its progress, and benefits can legitimately be regarded as 'global public goods' to the extent that they are the product of myriad historically and culturally situated epistemic communities whose scientific perspectives have historically interlaced.

That scientific progress and its applications should be regarded as 'public goods' openly accessible to all might seem uncontroversial. But that they should be regarded as a 'global public good' is less so. Consider typical examples of public goods: for example, public parks, motorways, military defence systems – the sort of goods that nation states provide to their citizens via taxation systems. Each public good might benefit some portion of the population more than others. City dwellers might benefit more from public park than country dwellers; and car drivers might benefit from good motorways more than non-car drivers. Considerations of this nature typically drive debates about what a fair system of national taxation might look like in terms of benefits-costs to the citizen of a state – see for example (Miller & Taylor, 2018) for a discussion.

Now one may think of scientific progress and its applications along the same lines of public parks and motorways as 'public goods': e.g. vaccines, medicines, fMRI scans, particle accelerators, telescopes, digital innovations, and myriad other examples resulting from scientific research paid through taxpayers' money via research councils funded projects and initiatives. Taken as 'public goods' within the confines of national boundaries, it would seem that nation states have duties towards their own citizen to make sure they enjoy the benefits of scientific progress and its applications.

However, understood thus and so, it is not very clear why nation states should have equally binding duties towards foreign nationals to make sure they too enjoy the benefits of scientific progress and its applications. For example, at the time of a pandemic, when access to vaccines becomes a priority, should individual nation states give priority to their own citizen before shipping any extra supply to foreign countries? And should this duty towards one's own citizen trump any other consideration, including stark disparities in vaccine supplies and vaccination rates across countries and especially in the Global South?

My point is the following: what makes scientific progress and its applications different from say public parks, or motorways, or similar is that the former are 'global public goods' whose *right to enjoy* extends well beyond national boundaries and nation states' jurisdictions (including international relations among them). These are rights that ought to be enjoyed by *everyone everywhere* all else being equal (i.e. assuming equal access to them and equal ability to enjoy their benefits in the absence of, say, structural socio-economic stark inequalities).

Thus, treating scientific progress and its applications as a 'global public good' means relocating its normative core content from the realm of domestic law (where national systems of taxations determine funds allocated to different public goods and the extent to which individual states might or might not be willing to 'contribute to the development of science and technology in developing countries') to the realm of cosmopolitan rights. Hence, my suggestion that REBSP ought to be treated as a cosmopolitan right, namely, a right that pertains to everyone not in virtue of either being citizen of a nation state or in virtue of international law, but in virtue of a *ius cosmopoliticum*.

This is not at all a moot shift. Nor is it one that can be justified on the back of traditional images of science and scientific progress 'from nowhere' so to speak. 'Success from nowhere' as well as the presumption of 'success from here now' would hardly square with the idea of scientific advancements as 'global public goods' and hence of REBSP as a cosmopolitan right. For they are both the disguised philosophical expression of what is historically a Global North take on scientific progress and its applications. They tend to treat implicitly scientific advancements as 'public goods' like public parks and motorways, but not necessarily as 'global public goods'.

I think it is indicative that the operating definition of science that still lies at the core of the REBSP and the more recent (Committee on Economic, Social and Cultural Rights, 2020) 'General Comment' tacitly and unwittingly buys into a view of 'success from nowhere':

> the word "science" signifies the enterprise whereby *humankind, acting individually or in small or large groups*, makes an organized attempt, by means of the objective study of observed phenomena and its validation through sharing of findings and data and through peer review, to discover and master the chain of causalities, relations or interactions; brings together in a coordinated form subsystems of knowledge by means of systematic reflection and conceptualization;

and thereby furnishes itself with the opportunity of using, to its own advantage, understanding of the processes and phenomena occurring in nature and society.

(ibid., Article 4. Emphases added)

There is mention of 'objective study of observed phenomena', of 'sharing of findings and data and through peer review', of 'subsystems of knowledge' and 'systematic reflection and conceptualization' as if all this epistemic feat were the product of a culturally deracinated and historically decontextualised 'humankind, acting individually or in small or large groups' rather than the outcome of myriad historically and culturally situated epistemic communities, whose scientific perspectives have historically interlaced.

Unsurprisingly, as a consequence of this take, partaking in REBSP becomes then mostly a matter of duties of nation states to 'disseminate' scientific progress – from the centre of production to the periphery – for the enjoyment of everyone, as stressed by the aforementioned Articles 16 and 79 of (Committee on Economic, Social and Cultural Rights, 2020).

But if one understands the normative core content of REBSP by taking as a starting point a different image of scientific knowledge as the one proposed by my perspectival realism, partaking in REBSP ceases to be a matter of 'disseminating' knowledge from developed countries to developing countries where local duties give way to international duties. Partaking in REBSP – I maintain – is a cosmopolitan right, a right that *everyone everywhere* can reclaim as their own in virtue of belonging to a seamless multicultural web of interlaced scientific perspectives through which only scientific knowledge becomes possible.

Re-aligning REBSP with cosmopolitan rights is therefore only possible by embracing a richer and more substantive philosophical image of scientific knowledge: one that places epistemic communities and their situated knowledge centre stage; that does not treat knowledge production as a siloed exercise of surgically excising theories or parts thereof; and that does not assess progress *sub specie aeternitatis* but as something that every community at any time can reclaim as its own contribution.

Coming to point (II), re-interpreted as a cosmopolitan right on the back of perspectival realism, REBSP becomes a powerful tool to fight the injustices of epistemic severing and epistemic trademarking. Because where there are rights, there are obligations. And REBSP qua *ius cosmopoliticum* brings along with it the cosmopolitan obligation to ensure that scientific progress and its benefits are indeed accessible to a world citizenship rather than to a few privileged developed countries. To be counteracted, vaccine nationalism and biopiracy – see on the latter for example (Mgbeoji, 2006) – require more than 'access and benefit sharing' (ABS) mechanisms with developing countries. They require changing some deeply seated presuppositions behind 'community sharing', which should not be understood as an act of philanthropy or as a 'dissemination' from the centre to the periphery, from developed countries to developing ones; but instead as a cosmopolitan obligation towards fellow citizens, no matter in which nation states or under which international law order they happen to be living in.

This shift requires the creation of institutions and governance systems that can secure the implementation of such cosmopolitan obligations analogous to The Hague International Court of Justice, which over decades has secured the implementation of cosmopolitan obligations in fighting crimes against humanity. If REBPS continues to be delegated to nation states for its implementation, the intended promise of being a 'universal human right' risks remaining wishful thinking. In our after-COVID world, the time is ripe for making sure that vaccine nationalism does not repeat itself and REBSP is not paid lip service. Philosophers of science have a voice to contribute to these important on-going debates. A deontic framework for scientific progress, which starts from a perspectival and multicultural image of scientific knowledge and reinterprets REBSP as a cosmopolitan right, is a good starting point for continuing these conversations.

Acknowledgements

I am grateful to the editor Yafeng Shan for kindly inviting me to contribute to this volume. The material here presented builds and expands upon some of the key ideas in my monograph Massimi (2022b) and presented in an abridged form in Massimi (2022d). In particular, the material in Sections 2.1 and 2.2 is reproduced from Section 11.5 of Chapter 11 in Massimi (2022b).

References

Benhabib, S. (2006). *Another Cosmopolitanism: With Commentaries by Jeremy Waldron, Bonnie Honig, Will Kymlicka*, R. Post (ed.). Oxford University Press.

Besson, S. (2015a). Mapping the Issues. Introduction. *European Journal of Human Rights* 2015/4. 403–410.

Besson, S. (2015b). Science without borders and the boundaries of human rights: who owes the human right to science? *European Journal of Human Rights* 2015/4. 462–485.

Biagioli, M. (2019). Weighing intellectual property: Can we balance the costs and benefits of patenting? *History of Science*, 57, 140–163.

Biagioli, M., Woodmansee, M., & Jaszi, P. (2011). *Making and Unmaking Intellectual Property: Creative Production in Legal and Cultural Perspective*. University of Chicago Press.

Bird, A. (2007). What is scientific progress? *Noûs*, 41, 64–89.

Boulton, G.S. (2021). Science as a global public good. A position paper of the International Science Council. 21pp. https://council.science/wp-content/uploads/2020/06/Science-as-a-global-public-good_v041021.pdf

Bush, V. (1945). *Science, The Endless Frontier*. MIT Press.

Chakaravartty, A. (1998). Semirealism. *Studies in History and Philosophy of Science Part A, 29*, 391–408.

Committee on Economic, Social and Cultural Rights, E. 12/GC/25. (2020). *General Comment No. 25 (2020) on Science and Economic, Social and Cultural Rights (Article 15 (1) (b), (2), (3) and (4) of the International Covenant on Economic, Social and Cultural Rights)*. United Nations. https://digitallibrary.un.org/record/3899847

Dellsén, F. (2018). Scientific progress: Four accounts. *Philosophy Compass, 13*, e12525.

Donders, Y., & Tararas, K. (2021). Mainstreaming science and human rights in UNESCO. In *The Right to Science. Then and Now*, edited by H. Porsdam and S. Porsdam Mann (pp. 124–139). Cambridge University Press.

Gay, H. (1996). Invisible resource: William Crookes and his circle of support, 1871–81. *The British Journal for the History of Science, 29*, 311–336.

Giere, R. (2006). *Scientific Perspectivism*. Chicago and London: The University of Chicago Press.

Kant, I. (1795). *Toward Perpetual Peace and Other Writings on Politics, Peace, and History*, with essays by J. Waldron, M.W. Doyle, A. Wood, & P. Kleingeld (ed.), D.L. Colclasure (trans.). Yale University Press.

Kant, I. (1797). *Metaphysics of Morals, in Toward Perpetual Peace and Other Writings on Politics, Peace, and History*, with essays by J. Waldron, M.W. Doyle, A. Wood, & P. Kleingeld (ed.), D.L. Colclasure (trans.). Yale University Press.

Kitcher, P. (2001). *Science, Truth, and Democracy*. Oxford University Press.

Kuhn, T. S. (1957). *The Copernican Revolution*. Harvard University Press.

Lyons, T. (2006). Scientific realism and the stratagema de divide et impera. *British Journal for the Philosophy of Science, 57*, 537–560.

Massimi, M. (2016). Three tales of scientific success. *Philosophy of Science, 83*, 757–767.

Massimi, M. (2022a). Perspectival ontology: Between situated knowledge and multiculturalism. *The Monist, 105*, 214–228.

Massimi, M. (2022b). *Perspectival Realism*. Oxford University Press.

Massimi, M. (2022c). Big Science, Scientific Cosmopolitanism and the Duty of Justice. In *Debating the Societal and Economic Impact of Big Science in the 21st century*, edited by P. Charitos and T. Arabatzis, Institute of Physics.

Massimi, M. (2022d). Perspectives on scientific progress. *Nature Physics 18*, 604–606.

Mgbeoji, I. (2006). *Global Biopiracy: Patents, Plants and Indigenous Knowledge*. UBC Press.

Miller, D., & Taylor, I. (2018). Public goods. In *The Oxford Handbook of Distributive Justice*, edited by S. Olsaretti. Oxford University Press.

Niiniluoto, I. (1984). *Is Science Progressive?* D. Reidel.

Niiniluoto, I. (2014). Scientific progress as increasing verisimilitude. *Studies in History and Philosophy of Science (Part A), 75*, 73–77.

Patten, A. (2022). Public good fairness. In *Political Philosophy Here and Now: Essays in Honour of David Miller*, edited by Daniel Butt, Sarah Fine, & Zofia Stemplowska. Oxford University Press.

Porsdam, H., & Porsdam Mann, S. (2021). *The Right to Science. Then and Now*. Cambridge University Press.

Porsdam Mann, S., Donders, Y., Mitchell, C., et al. (2018). Advocating for science progress as a human right. *Proceedings of the National Academy of Sciences*, *115*(43), 10820–10823.

Porsdam Mann, S., Donders, Y., & Porsdam, H. (2021). The right to science in practice. A proposed test in four stages. In *The Right to Science. Then and Now* (pp. 231–245). Cambridge University Press.

Shaver, L. (2015). The right to science: ensuring that everyone benefits from scientific and technological progress. *European Journal of Human Rights*, 2015/4. 411–430.

Sunder, M. (2007). The invention of traditional knowledge. *Law and Contemporary Problems*, *70*, 97–124.

"The Right to Enjoy the Benefits of Scientific Progress and Its Application." (2009). Paris. https://unesdoc.unesco.org/ark:/48223/pf0000185558.

Index

Printed in the United States
by Baker & Taylor Publisher Services